TRIGONOMETRY

TRIGONOMETRY

Third Edition

John D. Baley
Cerritos College

Gary Sarell
Cerritos College

The McGraw-Hill Companies, Inc.
New York   St. Louis   San Francisco   Auckland   Bogotá   Caracas
Lisbon   London   Madrid   Mexico City   Milan   Montreal   New Delhi
San Juan   Singapore   Sydney   Tokyo   Toronto

TRIGONOMETRY

This book is printed on acid-free paper.

2 3 4 5 6 7 8 9 0 SEM SEM 9 0 9 8 7 6

ISBN 0-07-005188-7

The editor was Karen M. Minette;
the production supervisor was Richard A. Ausburn.
The photo editor was Kathy Bendo.
Quebecor-Semline was printer and binder.

Library of Congress Catalog Card Number: 95-78947

About the Authors

Gary L. Sarell is a Professor of Mathematics at Cerritos College. He received his Master's Degree in Mathematics from California State University, Long Beach. He has been a full time instructor at Cerritos College since 1986. Gary has also taught at the high school level and at California State University. He has been involved in the production of multimedia educational packages including video tapes, video disks, and computer software.

While in college Gary was nationally ranked in track and field, and continues to compete today in the Master's division. He also coaches the cross country, and track and field teams at Cerritos College

John D. Baley graduated from John Carroll University in Cleveland, Ohio, with a degree in Physics, in 1961. After moving to California, he taught calculus for two years at a high school for gifted students and then spent the next five years teaching at Jordan High School in Watts. After earning a M.S. and Ed.D. at the University of Southern California in part-time study he was employed at Cerritos College where, along with Martin Holstege he established the Mathematics Learning Center in 1970. Dr. Baley was chairman of the Aviation Department at Cerritos College and has taught computer science courses. He continues to supervise the Mathematics Learning Center while writing textbooks and developing computer-assisted instruction.

In his spare time John is a Certified Flight Instructor and an active skier.

To The Instructor

Our research indicates that four-color artwork is not universally accepted as necessary or even pedagogically useful. Actually, a number of instructors asserted that four-color production distracts from the mathematics and does not warrant the additional expense. In response to these reactions, we have put together a special verson of *TRIGONOMETRY*, third edition. This text has been designed in a one-color format. McGraw-Hill is passing the savings in production costs on to the student by offering this proven, well-established textbook at a reduced price.

Contents

Preface

This text is an in-depth course in trigonometry designed to be understood and used by students. Although the development of trigonometry begins on page one, we, the authors, realize that many students may have completed algebra and geometry some time ago. Therefore, we have included algebra and geometry reminders throughout the text where we know from teaching experience that many students need help in recalling ideas that are necessary to develop trigonometry.

While it assumes no previous knowledge of trigonometry, this book shows how trigonometry can be used in many fields. It also develops algebra skills so that students will be thoroughly prepared to continue their study of mathematics and science.

 The use of graphing calculators has been incorporated throughout the text to reduce the labor of calculations and to expand the students' understanding of concepts and give students the opportunity to explore relationships.

Features: Organization and Pedagogy

This book is divided into twelve chapters to allow instructors greater flexibility in arranging course outlines to conform to academic calendars and student needs.

Each Chapter Includes:

Preview: This gives the students an overview of how the chapter relates to previous chapters and how the coming chapter will be developed.

Sections: Each chapter is divided into sections of material that approximate one hour of classroom lecture time.

Numerous Examples: Over 200 examples illustrated with graphics demonstrate the concepts presented in the text. Students are given one or more examples of each task they are expected to perform.

Ample Exercises: Over 1800 problems, usually in matched odd-even sets, give the students the opportunity to apply the concepts and practice the skills taught in the text. Problem sets are a critical part of the learning process because they not only give the student needed practice but they also show the student the results of subtle variations. By doing problems a student gets an opportunity to internalize mathematics and see the effects of changes in a parameter.

Many graphing calculator exercises are integrated into the problem sets and examples to help the students develop these abilities.

Applications: Trigonometry provides more opportunities for students to see actual applications of the mathematics they are studying than any lower-level course they have encountered. This text uses applications in examples and problems to show students that trigonometry is a very useful branch of mathematics in engineering, electronics, geology, optics, aviation, surveying, construction, forestry, navigation, and physics.

Highlighted Definitions, Properties, Theorems, and Rules: Those features, which are essential to remember and understand, are highlighted in boxes throughout the text both to draw the students' attention to important concepts and to make it easy for students to reference key ideas.

Key Ideas: Each chapter concludes with a summary of key ideas to aid students in organizing their knowledge and help them prepare for exams.

Review Tests: To further insure students that they have mastered the concepts of each chapter, there is a review test that closely approximates the questions that are likely to be asked on an exam.

Special Features of This Text

Cumulative Reviews: There are cumulative reviews after Chapters 6 and 12. These reviews are an excellent opportunity for students to get an overview of the course as they study for midterm and final exams.

Conversational Bubbles: These bubbles are spread throughout the book but generally appear in the context of a worked example. Teachers will quickly recognize that these bubbles anticipate and verbalize the questions the students are likely to ask as they are learning the material. Response bubbles provide the answers that experienced teachers are likely to give.

Geometry and Algebra Review: It would be nice if every student beginning this course had a complete mastery of algebra and geometry, but for many students it has been several years since they have taken these courses. The geometry and algebra reviews are placed in the text to help students refresh key ideas just before these ideas are needed in trigonometry.

 Using Your Calculator: This is a series of special features that show students how to efficiently use their choice of a graphing or a scientific calculator to solve problems. Each calculator feature tells students what they need to know and when they need to know it.

Trigonometric Identities: Identities are one of the most critical topics of a complete trigonometry course. Trigonometric identities are essential for many calculus and physical applications. They not only provide an excellent review of algebraic skills but can be used to teach advanced algebraic manipulation techniques that students will need in later courses. Unfortunately, identities can also be one of the most frustrating topics for both teacher and students. Some texts expect students to fail at identities and they avoid challenging students with serious identities. This text solves the problem of mastering the identities by teaching students the entire hierarchy of skills that are needed to prove trigonometric identities. These skills, which include problem analysis and algebraic manipulation, are carefully developed in Chapters 5, 6, and 7. This text does not avoid identities; it teaches students how to prove them at a very high level.

Graphing Trigonometric Functions: Trigonometric functions are graphed using the concepts of a *generic box*, which simplifies the task while providing unifying ideas about period and the effect of parameters on a function. These ideas are applicable throughout mathematics.

New Features of the Third Edition

We have made changes based on comments of users of the second edition to incorporate new technology and improve the readability and teachability of the text.

The major changes are:

 Throughout the Book: Calculator hints are now in two columns where appropriate to help the students with scientific and graphing calculator use. Some sections have been retitled to more accurately reflect their content. Additional problems requiring the use of a graphing calculator have been added. Group writing exercises have been incorporated into many problem sets and throughout the book to help students develop writing skills in a cooperative learning context.

Chapters With Significant Changes

Chapter 2: Explanations have been clarified to help students understand how angles are measured in aviation and nautical navigation.

 Chapter 3: Examples and problem sets have been expanded to include the use of graphing calculators. The problem sets have been altered in such a way so that students using graphing calculators not only draw graphs but also investigate the effect of parameters on those graphs. An understanding of the concepts of trigonometric functions is required to solve these problems.

A new section on qualitative analysis of trigonometric functions has been added. It is designed to help students develop their analytical skills and enhance their understanding of the effects of each parameter on the graphs of the trigonometric functions.

 Chapters 5, 6, 7: Graphing calculators are used to compare the graphs of possible identities. If the graphs match, students are encouraged to use algebraic methods to verify the identity.

 Chapter 8: The trace function of the graphing calculator is used to approximate solutions to trigonometric equations. The motion of a piston in a cylinder is also analyzed using the graphing calculator.

Chapter 10: The explanation of how angles are measured in aviation has been clarified. A section on algebraic manipulation of vectors has been added and leads very nicely into a discussion of the dot product of two vectors. The dot product is then used to calculate work in physical applications. It is a much stronger vector approach than has been used in past editions. Calculator hints about the storage of variables that help students resolve vectors using components have been added.

Chapter 12: This chapter has been substantially rewritten to improve the students' visualization of functions in polar coordinates. A section on parametric equations has been added. Graphing calculators are used to graph polar functions using parametric equations or as polar functions in calculators with that ability.

Class Testing

This edition has been class tested through four revisions by over 600 students in both lecture and semi-independent mathematics classes. The authors are grateful to the many students and instructional aides who provided useful feedback, which helped to improve this text.

Ancillaries

Student's Solution Manual: A student solution manual with selected problems worked step by step is available. This manual also provides additional hints and explanations about how the problems were solved.

Instructor's Manual: An instructor's manual provides four forms of each chapter test, cumulative tests, and final tests.

Acknowledgments

The authors wish to thank Elizabeth Hamman for her consistant efforts to make this book a quality production. She learned the computer typesetting system TEX and used it along with her significant artistic talents to typeset the text you see. Invaluable assistance was also provided by Hai Vo, who mastered the skill of using PjCTEX to produce the graphs in the book. Hai Vo, Karen Nguyen, Nikki Nguyen, and Phuong Nguyen all worked hard to type the book.

The accuracy of the answers and the tests in the instructor's manual are largely the work of Steven Malabicky who enhanced the item bank and produced the tests. Late in the production of the text we were joined by a student from the Czech Republic, Karel Vondra, whose considerable computer skills helped us scan images and utilize encapsulated PostScript files to incorporate the more complex figures into the book.

We would like to extend particular thanks to the following people at McGraw Hill: Karen M. Minette, Associate Editor of Mathematics, Jack Maisel, Editing Supervisor, and Rich Ausburn, Production Supervisor.

The authors are also especially grateful to the reviewers from colleges throughout the country who carefully read, critiqued, and provided many excellent suggestions that have been incorporated into the book you are holding. The quality of this text has been improved by the efforts of:

James Arnold	University of Wisconsin, Milwaukee
Jerald Ball	Chabot College
Paul W. Britt	Louisianna State University
Elizabeth Cauley	Pensacola Junior College
Bettyann Daley	University of Delaware
Marjorie Freeman	University of Houston, Downtown
Virginia Hamilton	Ball State University
Louis Hoelzle	Bucks County Community College
Michael Karelius	American River College
Anna Marie Lallement	Pennsylvania State University
Peter Lindstrom	North Lake College
Thomas McCabe	Harper College
Beverly Rich	Illinois State University
Vicki J. Schell	Pensacola Junior College
Ken Seydel	Skyline College
Cynthia Siegel	University of Missouri, St. Louis
Ann Thorne	College of DuPage
Sharon Walker	Arizona State University
Wilton Clarke	La Sierra University
Marc Franco	South Seattle Community College
Jack Goebel	Montana Technical College
Donna Hoops	Johnson County Community College
Vijay Joshi	Virginia Intermont College
Ruben Leon	Cerritos College
Willian M. Mays	Gloucester County College
Elizabeth M. Morrison	Valencia College—West Campus
Sunny Norfleet	St. Petersburg Junior College
Elaine Parks	Laramie County College
Raymond Uribe	San Francisco State University
Kelly Wyatt	Umpqua Community College

Chapter 1

Measurement of Angles, Arcs, and Sectors

Contents

Preview

Trigonometry is a branch of mathematics with many applications. Early trigonometric applications included land surveys, building construction, astronomy, and navigation. Later developments in optics, electronics, mechanics, engineering, and communications using radio waves or fiber optics increased trigonometric applications.

There are several valid ways to begin the study of trigonometry. We will start by looking at the coordinates of a point rotated through an angle because most applications of trigonometry involve analysis of this type of motion. As a point rotates through an angle, it traces an arc. This first chapter will show you how to determine the length of this arc or any arc that is part of a circle, for example, the curve of a freeway bridge. It will also show you how to find the area of a sector of a circle. In the last section, we will study the velocity of an object as it travels in a circle. The object could be your foot on a bicycle pedal, a satellite in orbit, the pendulum of a grandfather's clock, or a horse on a carousel.

1

1.1 Using Radians, Degrees, or Grads to Measure Angles

Many points travel in circular motion. Some practical examples of circular motion or rotation are the moving armature of an electric motor or generator, a piston on a crankshaft, an automobile traveling around a curve, and a hand on the end of an arm. To study rotation, we need to define a few terms.

Definition 1.1A Ray $\overrightarrow{AB}$

Ray $\overrightarrow{AB}$ is the set of points consisting of line segment $\overline{AB}$ and all other points, X, such that point B lies between point X and point A.

□ Figure 1.1

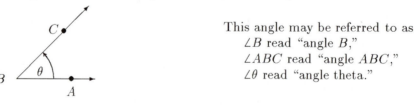

Point C is not on $\overrightarrow{AB}$ because B is not between C and A.

B is between X and A, therefore point X is on ray AB.

A ray is a half-line with one end point. It continues in one direction forever.

Definition 1.1B Angle

An angle consists of two rays with a common end point called the *vertex*.

□ Figure 1.2

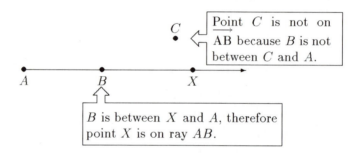

This angle may be referred to as
$\angle B$ read "angle B,"
$\angle ABC$ read "angle ABC,"
$\angle \theta$ read "angle theta."

Most letters in the Roman alphabet have standard meanings, such as t for time. Therefore, to identify angles we use Greek letters. Common letters used in this text are:

α Alpha, β Beta, γ Gamma, θ Theta.

We can think of an angle as a ray rotated about its end point. The angle consists of the starting or initial position and the ending or terminal position of the ray. The size of the angle is determined by the amount of rotation.

Definition 1.1C Standard Position for an Angle

In a Cartesian or rectangular coordinate system an angle is in standard position if its vertex is at the origin and its initial side lies on the positive x-axis.

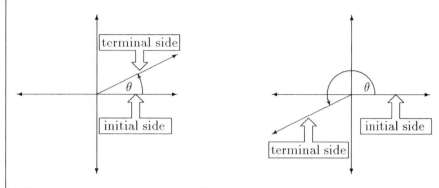

Do the lengths of the sides of an angle determine how big an angle is?

No, the sides of all angles are rays which extend indefinitely. The size of the angle is a measure of how much the terminal side was rotated from the initial side, not the length of its sides.

Geometry Reminder

About Angles

Because angles are defined as two rays with a common end point, an angle is a set of points.

A B

Strictly speaking, we cannot say that $\angle A$ above is equal to $\angle B$ because each angle is a different set of points.

We can say that if you pick up $\angle A$, you could move it so that it fits exactly over $\angle B$. To express this idea we say $\angle A \cong \angle B$. An equal sign with a wavy line above it is read "congruent," so this is read "angle A is congruent to angle B." In trigonometry, we are interested in comparing the amount of rotation or the size of two angles. To compare, we use a number called the measure of an angle for each angle.

Then we can say: $m\angle A = m\angle B$

In trigonometry, however, we make this comparison so frequently that we usually refer to "angle θ " rather than "the measure of angle θ." Throughout this book, unless we specifically say otherwise, we will use θ to mean the measure or "size" of $\angle \theta$ rather than the set of points that makes up $\angle \theta$.

Geometry Reminder

About Circles

A circle is defined as the set of points in a plane all the same distance from a fixed point. The fixed point is called the center of the circle. The given distance is called the radius.

□ Figure 1.5

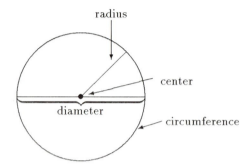

Neither the center nor the radius is part of the circle; they are used only to define the circle.

The circle is a set of points.

Each of the following are distances associated with a circle.

Radius—distance from the center of the circle to a point on the circle.

Diameter—distance across the circle passing through the center.

$$d = 2r$$

Circumference—distance around the circle.

$$C = 2\pi r \qquad \text{or} \qquad C = \pi d$$

The Measure of Angles

The size or measure of an angle is a real number associated with the amount of rotation. There are four basic ways to measure rotation. All are based on a complete circle.

1. REVOLUTIONS: 1 revolution = 1 circle rotation.   A point rotated through a complete circle is called 1 revolution.

□ Figure 1.6

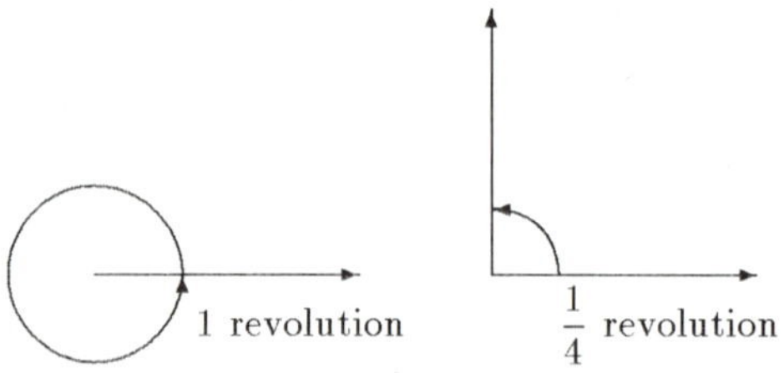

1 revolution $\frac{1}{4}$ revolution

2. DEGREES: Earliest astronomers thought a year was 360 days long, so 360 degrees seemed a logical number of divisions for a circle.

Chapter 1 Measurement of Angles, Arcs, and Sectors

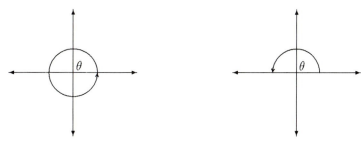

$360° = $ rotation in
a complete circle

$180° = $ rotation
through half a circle

Counterclockwise rotations are considered positive. Clockwise rotations are considered negative.

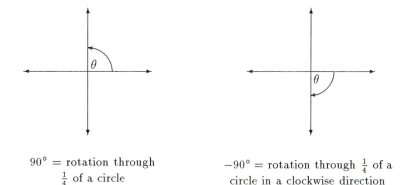

$90° = $ rotation through
$\frac{1}{4}$ of a circle

$-90° = $ rotation through $\frac{1}{4}$ of a
circle in a clockwise direction

Traditionally each degree can be divided into sixty minutes and each minute further divided into sixty seconds. However, calculators use decimal fractions to divide degrees.

3. GRADS: Your calculator may have a "grads" key. The grad system is used by the military for artillery calculations. It divides a circle into 400 grads. This text will not use grads.

4. RADIANS: In higher mathematics, the most convenient measure of an angle is the ratio of the intercepted arc to the radius of a circle. The measure of a central angle that intercepts an arc on a circle equal to the length of a radius is one radian. To draw an angle of one radian follow these steps:

   Step 1.

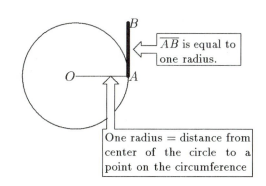

$\overline{AB}$ is equal to one radius.

One radius = distance from center of the circle to a point on the circumference

1.1 Using Radians, Degrees, or Grads to Measure Angles

Step 2.

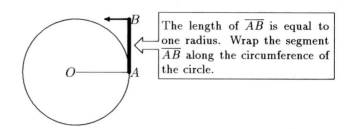

The length of $\overline{AB}$ is equal to one radius. Wrap the segment $\overline{AB}$ along the circumference of the circle.

Step 3.

☐ Figure 1.11

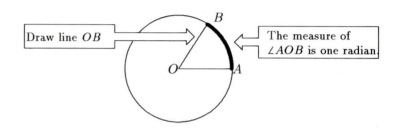

Draw line OB

The measure of $\angle AOB$ is one radian.

In general

$$\text{Angle in radians} = \frac{\text{length of intercepted arc}}{\text{length of radius}}$$

☐ Figure 1.12

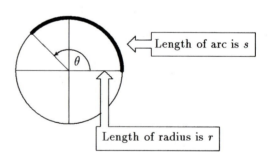

Length of arc is s

Length of radius is r

$$\theta = \frac{s}{r}$$

Where θ = measure of angle in radians

s = length of intercepted arc

r = radius of circle

Chapter 1 Measurement of Angles, Arcs, and Sectors

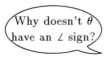

Why doesn't θ have an $\angle$ sign?

$\angle\theta$ is the set of points that makes the figure. In this book θ refers to the number used to measure the angle.

Next we will determine the number of radians in one revolution. The circumference of a circle is $2\pi r$; therefore, an angle in radians equal to one revolution is as follows:

$$\text{One revolution} = \theta = \frac{\text{length of arc intercepted in one revolution}}{\text{length of radius}}$$
$$\theta = \frac{2\pi r}{r}$$
$$\theta = 2\pi$$
$$\theta \approx 6.2832 \qquad \text{Because } \pi \approx 3.1416$$

2π what?

Just 2π. Radians don't have dimensions. Notice any units used to measure the arc length in the numerator cancel with the units used to measure the length of the radius in the denominator.

Some common angles and their radian measure are drawn in Figure 1.13 to help you visualize radian measure.

□ Figure 1.13

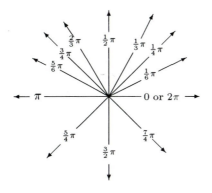

How do you know where to draw a line to make an angle such as $\frac{1}{3}\pi$?

Right now, we only estimate where $\frac{1}{3}\pi$ falls. In Chapter 2 we will be more specific.

Group Writing Activity

Given a circle, a straight edge, and some string, how would you locate an angle of 2 radians? Locate an angle of 5 radians.

Geometry Reminder

About Angles

Angles are frequently classified by their measure or size.

Acute Angle—An angle with a measure greater than zero and less than 90° or $\frac{1}{2}\pi$ is called an acute angle.

Right Angle—An angle with measure of 90° or $\frac{1}{2}\pi$ is called a right angle.

Obtuse Angle—An angle with a measure greater than 90° or $\frac{1}{2}\pi$ and less than 180° or π is called an obtuse angle.

Straight Angle—An angle with a measure of 180° or π is called a straight angle.

Complementary Angles—Two positive angles, the sum of whose measures is 90° or $\frac{1}{2}\pi$, are called complementary angles.

Supplementary Angles—Two positive angles, the sum of whose measures is 180° or π, are called supplementary angles.

Problem Set 1.1A

By mentally dividing a straight angle into equal parts, draw a freehand sketch of each of the following angles.

1.	90°	**2.**	−45°	**3.**	45°	**4.**	−30°	**5.**	30°	**6.**	120°
7.	−150°	**8.**	315°	**9.**	π	**10.**	$-\frac{1}{2}\pi$	**11.**	$\frac{1}{4}\pi$	**12.**	$\frac{1}{3}\pi$
13.	$-\frac{7}{4}\pi$	**14.**	$-\frac{1}{6}\pi$	**15.**	$\frac{4}{3}\pi$	**16.**	$-\frac{7}{6}\pi$	**17.**	$-\frac{5}{3}\pi$	**18.**	$-\frac{11}{6}\pi$

19. To the nearest multiple of 30° or 45° give the positive angle measure of angles A through G in the figure.

20. To the nearest multiple of 30° or 45° give the measures of angles A through G as negative angles.

21. To the nearest multiple of $\frac{1}{6}\pi$ or $\frac{1}{4}\pi$ give the measures of angles A through G as positive angles.

22. To the nearest multiple of $\frac{1}{6}\pi$ or $\frac{1}{4}\pi$ give the measures of angles A through G as negative angles.

23. In the same units of measure, give the complementary angle to

    A. $\frac{1}{4}\pi$ B.   15°

24. In the same units of measure, give the supplementary angle to

    A. $\frac{1}{6}\pi$ B. $\frac{1}{4}\pi$ C.   45°   D.   20°

Coterminal Angles

Two angles in standard position with the same terminal side are called coterminal angles. By adding or subtracting an angle equal to one rotation, you get a coterminal angle.

□ Figure 1.14

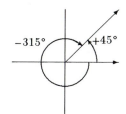

45° and −315° are coterminal angles.
+405° is also coterminal with 45°

Example 1 □ Name the smallest non-negative angle that is coterminal with 560°.
Because 560° is larger than one revolution, subtract one revolution (360°).

□ Figure 1.15

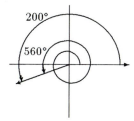

$$560° − 360° = 200°$$

200° is the smallest non-negative angle coterminal with 560° because they both have the same terminal side and $0° \leq 200° < 360°$. □

Example 2 □ What is the smallest non-negative angle that coincides with −300°?
In absolute value −300° is less than one revolution, but it is negative. To find the smallest positive coterminal angle, add one revolution (360°).

□ Figure 1.16

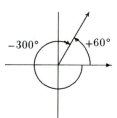

$$−300° + 360° = 60°$$

60° is coterminal with −300°. □

Example 3 □ What is the smallest non-negative angle coterminal with an angle of $\frac{14}{3}\pi$?

Because $\frac{14}{3}\pi$ is larger than two revolutions and smaller than three revolutions, subtract two revolutions ($2 \cdot 2\pi$ or 4π radians).

□ Figure 1.17

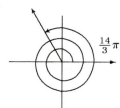

$$\frac{14}{3}\pi − 4\pi =$$
$$\frac{14}{3}\pi − \frac{12}{3}\pi = \frac{2}{3}\pi$$

$\frac{2}{3}\pi$ is coterminal with $\frac{14}{3}\pi$. □

1.1 Using Radians, Degrees, or Grads to Measure Angles

Example 4 □ Name the smallest non-negative angle that is coterminal with $-\frac{7}{4}\pi$ in radians.

In absolute value $-\frac{7}{4}\pi$ is less than one revolution, so add one revolution (2π radians).

□ Figure 1.18

$$-\frac{7}{4}\pi + 2\pi =$$
$$-\frac{7}{4}\pi + \frac{8}{4}\pi = \frac{1}{4}\pi$$

$\frac{1}{4}\pi$ is coterminal with $-\frac{7}{4}\pi$ because they have a common terminal side. □

Example 5 □ What is the smallest non-negative angle that coincides with $-\frac{23}{6}\pi$ radians?

In absolute value $-\frac{23}{6}\pi$ is more than one revolution but less than two revolutions. Therefore add two revolutions to find a coterminal angle less than one revolution.

$$-\frac{23}{6}\pi + 4\pi =$$
$$-\frac{23}{6}\pi + \frac{24}{6}\pi = \frac{1}{6}\pi$$

$$-\frac{11}{6}\pi \text{ is coterminal with } \frac{1}{6}\pi$$

□

In general, for negative angles, add a multiple of 2π that will produce a sum between 0 and 2π. For positive angles, subtract a multiple of 2π that yields a result between 0 and 2π.

 ## Selecting Your Calculator

At a minimum, you will need a scientific calculator to complete this course. To take full advantage of the text you will need a graphing calculator. If you can not afford a graphing calculator at this time or if your instructor chooses to teach the course without using a graphing calculator, we recommend you purchase a basic scientific calculator as soon as possible.

Conversion from One System of Angle Measurement to Another

Some operations such as measurement of the length of an arc or integrals in calculus lend themselves naturally to radian measure. Other operations like navigation and construction are commonly done using degree measure. To facilitate switching between the two systems of measurement we need conversion formulas. The measure of one revolution can be a starting point for conversion formulas because the basic quantity being measured is revolutions.

Radians to Degrees

Both 360° and 2π radians equal one revolution. Therefore:

$$
\begin{aligned}
2\pi \text{ radians} &= 360° \\
\pi \text{ radians} &= 180° \\
1 \text{ radian} &= \frac{180°}{\pi} \\
1 \text{ radian} &\approx 57.2958°
\end{aligned}
$$

Chapter 1 Measurement of Angles, Arcs, and Sectors

Example 6 □ Find the number of degrees in 2.7 radians.

$$1 \text{ radian} = \frac{180°}{\pi}$$

$$2.7 \text{ radians} = 2.7 \cdot \frac{180°}{\pi}$$

$$\approx 154.7°$$

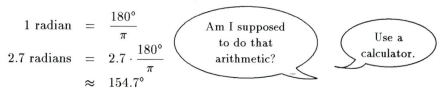

□

Example 7 □ Find the number of degrees in $\frac{1}{6}\pi$ radians.

$$\text{Because } \pi \text{ radians} = 180°$$

$$\frac{1}{6}\pi \text{ radians} = 30°$$

□

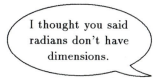

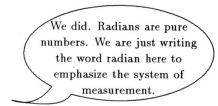

Degrees to Radians

A surveyor laying out an expressway on a ramp might want to know the length of an arc that spans 120°. As we will see in this section, the most convenient way to find an arc length is using radian measure.

To convert from degrees to radians, use the measure of one revolution.

$$360° = 2\pi \text{ radians}$$

$$\frac{360°}{360} = \frac{2\pi \text{ radians}}{360}$$

$$1° = \frac{\pi}{180} \text{ radians}$$

Example 8 □ Find the number of radians in 90°.

$$\text{Because } 1° = \frac{\pi}{180} \text{ radians}$$

multiply both sides by 90 to get

$$90 \cdot 1° = 90 \cdot \frac{\pi}{180} \text{ radians}$$

$$90° = \frac{1}{2}\pi \text{ radians}$$

□

Problem Set 1.1B

Label each of the following angles in radians. All angles are some multiple of $\frac{1}{6}\pi$, $\frac{1}{4}\pi$, $\frac{1}{3}\pi$, or $\frac{1}{2}\pi$.

1.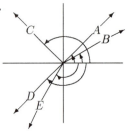

2.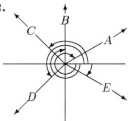

Name in degrees the smallest non-negative angle that is coterminal with each of the following angles.

3. $600°$ **4.** $-120°$ **5.** $-580°$ **6.** $900°$ **7.** $-430°$

8. $-160°$ **9.** $875°$ **10.** $-640°$ **11.** $515°$ **12.** $-310°$

Find in radians the smallest non-negative angle that is coterminal with each of the following angles.

13. $-\frac{2}{3}\pi$ **14.** $\frac{13}{6}\pi$ **15.** $\frac{17}{6}\pi$ **16.** $-\frac{7}{6}\pi$ **17.** $\frac{11}{4}\pi$

18. $\frac{14}{3}\pi$ **19.** $-\frac{5}{2}\pi$ **20.** $-\frac{11}{4}\pi$ **21.** 2π **22.** -4π

Find the degree measure for each of these angles given in radian measure.

23. 2π **24.** $-\frac{1}{3}\pi$ **25.** π **26.** $\frac{2}{3}\pi$ **27.** $\frac{3}{4}\pi$

28. $\frac{1}{6}\pi$ **29.** $-\frac{11}{6}\pi$ **30.** $\frac{1}{2}\pi$ **31.** $-\frac{1}{4}\pi$ **32.** $\frac{25}{2}\pi$

Find the radian measure as a fraction of π for each of these angles given in degree measure.

33. $150°$ **34.** $60°$ **35.** $-270°$ **36.** $330°$ **37.** $450°$

38. $-335°$ **39.** $690°$ **40.** $900°$ **41.** $-495°$ **42.** $1740°$

1.2   Length of an Arc and Area of a Sector of a Circle

Length of an Arc

The first section of this chapter defined the measure of an angle in radians. This definition related radius, angle, and arc length. Using it we can find the length of an arc if it is part of a circle.

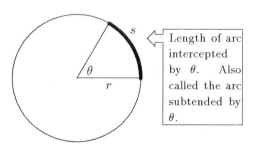

$$\theta = \frac{s}{r} = \frac{\text{length of intercepted arc}}{\text{length of radius}}$$

Solving for s, the length of the arc,

$$s = r\theta$$

Arc Length

Example 9 □ What is the length of the arc intercepted by an angle of $\frac{1}{3}\pi$ radians on a circle with a 9-inch radius?

$$
\begin{aligned}
s &= r\theta \\
s &= (9 \text{ inches})\left(\frac{1}{3}\pi\right) \qquad \text{Substituting} \\
s &= 3\pi \text{ inches} \\
s &= 9.42 \text{ inches}
\end{aligned}
$$

For this formula to work, θ must be measured in radians.

□

Example 10 □ A car wheel has a radius of 11 inches from the center of the wheel to the road. How many revolutions must it make to travel 200 feet?

This time we know the arc length is 200 feet, but we must express this distance in inches because the radius is given in inches.

$$
\begin{aligned}
s &= r\theta \\
200 \text{ ft} \cdot \frac{12 \text{ inches}}{1 \text{ ft}} &= (11 \text{ inches})\theta \\
\frac{2400 \text{ inches}}{11 \text{ inches}} &= \theta \qquad \text{Dividing both sides by 11 inches} \\
218.18 &\approx \theta
\end{aligned}
$$

This is angle θ in radians. Find the number of revolutions by converting 218.18 radians to revolutions.

$$
\begin{aligned}
\text{number of revolutions} &= 218.18 \text{ radians} \cdot \frac{1 \text{ revolution}}{2\pi \text{ radians}} \\
&\approx 34.7 \text{ revolutions}
\end{aligned}
$$

□

Example 11 □ A pump for an oil well raises and lowers a cable in a shaft by rotating a long arm. The amount of travel of the cable can be adjusted by changing the angle of rotation or by changing the length of the rotating arm. If the arm rotates through a 30° angle, how long should the rotating arm be if the cable is to travel 8 feet on each stroke?

The tip of the arm is designed to be a part of a circle so the distance the cable moves is equal to the length of arc on a circle.

In the formula $s = r\theta$, θ must be expressed in radians. $30° = \dfrac{\pi}{6}$ radians.

□ Figure 1.20

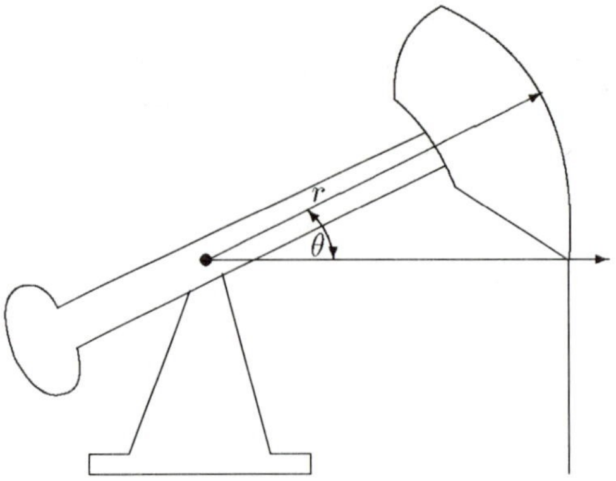

$$s \;=\; r\theta \qquad \text{becomes}$$
$$8 \text{ ft} \;=\; r\left(\frac{\pi}{6}\right)$$
$$\left(\frac{6}{\pi}\right) 8 \text{ ft} \;=\; r$$
$$15.28 \text{ ft} \;\approx\; r$$

An arm 15.28 feet long rotating through an angle of $30°$ will move the cable 8 feet up and down in the well.

□

Reminder about Conversion Factors

Conversion factors can be thought of as ratios that are equal to 1. Twelve inches and one foot both measure the same distance.

$$\text{Therefore } \frac{12 \text{ inches}}{1 \text{ foot}} = 1$$

Because conversion factors are ratios equal to one, we can multiply by a conversion factor without changing the value of an expression.

Example 12 □ Change $88 \dfrac{\text{ft}}{\text{sec}}$ to $\dfrac{\text{miles}}{\text{hour}}$.

$$\frac{88 \text{ ft}}{1 \text{ sec}} = \frac{88 \text{ ft}}{1 \text{ sec}}$$

This is a true equation because it is an identity.

$$\frac{88 \text{ ft}}{1 \text{ sec}} = \frac{88 \text{ ft}}{1 \text{ sec}} \cdot \frac{60 \text{ sec}}{1 \text{ min}} \cdot \frac{60 \text{ min}}{1 \text{ hour}} \cdot \frac{1 \text{ mile}}{5280 \text{ ft}}$$

⇧ multiplication by one ⇧ multiplication by one ⇧ multiplication by one

$$= 60 \frac{\text{miles}}{\text{hour}}$$

□

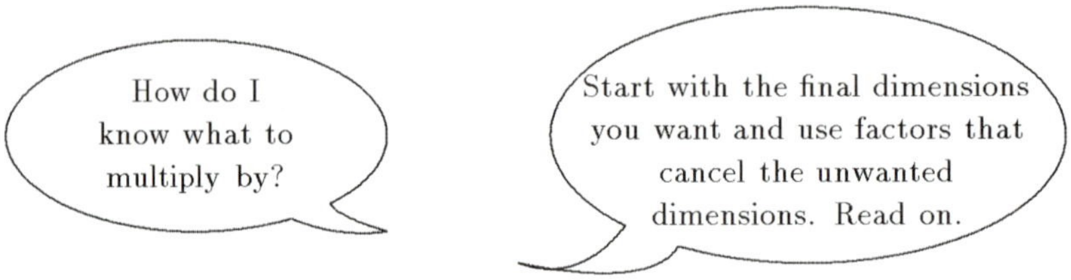

How do I know what to multiply by?

Start with the final dimensions you want and use factors that cancel the unwanted dimensions. Read on.

Chapter 1 Measurement of Angles, Arcs, and Sectors

Example 13 □ If you have $80\frac{\text{cents}}{\text{lb}}$ and you want $\frac{\text{dollars}}{\text{ton}}$ think this way

$$\frac{80 \text{ cents}}{1 \text{ lb}} \cdot \frac{1 \text{ dollar}}{100 \text{ cents}} \cdot \frac{2000 \text{ lbs}}{1 \text{ ton}} = \frac{\$1600}{\text{ton}}$$

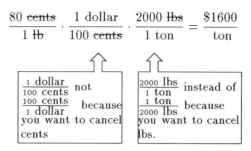

$\frac{1 \text{ dollar}}{100 \text{ cents}}$ not $\frac{\frac{100 \text{ cents}}{1 \text{ dollar}}}{}$ because you want to cancel cents

$\frac{2000 \text{ lbs}}{1 \text{ ton}}$ instead of $\frac{1 \text{ ton}}{2000 \text{ lbs}}$ because you want to cancel lbs.

□

Example 14 □ The unit of measurement used for navigation is the nautical mile, which is 6080 feet. This measure was chosen because it is approximately 1′ (minute) of latitude on the earth's surface. From this information, the radius of the earth can be found.

Because there are 60′ in one degree, there are 600 minutes or 600 nautical miles in 10° of latitude on the earth's surface.

To apply $s = r\theta$, we must convert 10° to radians.

□ Figure 1.21

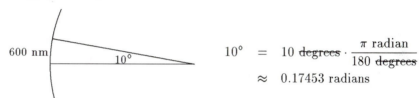

$$10° = 10 \text{ degrees} \cdot \frac{\pi \text{ radian}}{180 \text{ degrees}}$$

$$\approx 0.17453 \text{ radians}$$

Because

$$
\begin{aligned}
s &= r\theta \\
600\text{nm} &\approx r(0.17453) \qquad \text{Substituting for } r \text{ and } \theta \\
\frac{600 \text{ nm}}{0.17453} &\approx r \\
3437.8 \text{ nm} &\approx \text{ radius of the earth}
\end{aligned}
$$

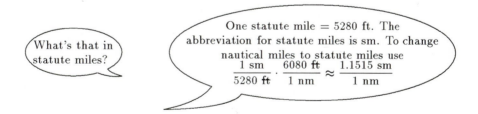

What's that in statute miles?

One statute mile = 5280 ft. The abbreviation for statute miles is sm. To change nautical miles to statute miles use
$$\frac{1 \text{ sm}}{5280 \text{ ft}} \cdot \frac{6080 \text{ ft}}{1 \text{ nm}} \approx \frac{1.1515 \text{ sm}}{1 \text{ nm}}$$

$$
\begin{aligned}
\text{Radius of the earth} &\approx 3437.8 \text{ nm} \cdot \frac{1.1515 \text{ sm}}{1 \text{ nm}} \\
&\approx 3958.6 \text{ statute miles} \\
\text{The diameter of the earth} &\approx 7917 \text{ statute miles.}
\end{aligned}
$$

□

About 230 BC Eratosthenes used a fraction of the earth's circumference to measure the radius of the earth. He noticed that in Syene at noon on the summer solstice, a vertical stick did not cast a shadow. At a later summer solstice he was in Alexandria and observed that the sun's rays were inclined one fiftieth of a complete circle to the vertical. Believing that Syene and Alexandria were on the same line of longitude and knowing the distance between these two cities, he estimated the circumference of the earth to be 50 times the distance from Alexandria to Syene.

Area of a Sector

How is the energy of a radar beam distributed as it sweeps back and forth over the approach course to an airport? How many square inches of viewing surface are cleaned as a windshield wiper moves in an arc in front of a driver? We can answer these and other questions by extending our knowledge of arc lengths to find the area of a sector of a circle.

A sector of a circle is the area enclosed by a central angle and the arc it intersects. Because the area of a sector of a circle is a fraction of the area of the entire circle, we can set up a proportion.

□ Figure 1.22

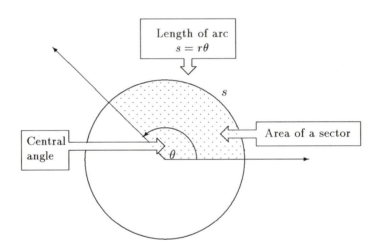

$$\frac{\text{Area of sector}}{\text{Area of circle}} = \frac{\text{arc length of sector}}{\text{circumference of circle}}$$

Substituting:

$$\frac{\text{Area of sector}}{\pi r^2} = \frac{r\theta}{2\pi r}$$

Multiplying both sides by πr^2 yields:

$$\text{Area of a sector} = \frac{\pi r^2 \cdot r\theta}{2\pi r}$$
$$= \frac{1}{2}r^2\theta$$

□

Definition 1.2A Area of a Sector of a Circle

The area of a sector of a circle of radius r, bounded by a central angle θ and an intercepted arc s, is:

$$A = \frac{1}{2}r^2\theta$$

Where r = radius of circle

θ = central angle in radians

Example 15 □ A 10-inch-diameter pizza pie is cut into six equal slices. Find the area of each slice.

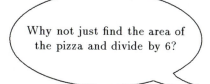

The central angle of each sector is $\frac{1}{6}$ of a circle, therefore

$$\theta = \frac{1}{6} \cdot 2\pi$$
$$= \frac{1}{3}\pi$$

The radius is $\frac{1}{2}$ the diameter or $r = 5$ inches.

The area of one slice is:
$$A = \frac{1}{2}r^2\theta$$
$$= \frac{1}{2}(5 \text{ inches})^2\frac{1}{3}\pi$$
$$A \approx 13 \text{ square inches}$$

□

Example 16 □ A windshield wiper of a car has a blade that is 16 inches long on the end of an 18-inch arm. If the arm rotates through an arc of 105°, how many square inches of windshield are cleaned by the blade on each pass of the wiper?

□ Figure 1.23

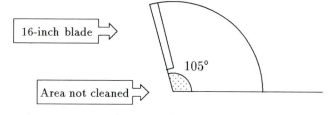

The area of usable windshield will be the area of the sector swept by the arm minus the area below the part of the arm without a blade.

First convert 105° to radians:

$$\text{central angle in radians} = 105 \text{ degrees} \cdot \frac{\pi \text{ radians}}{180 \text{ degrees}}$$
$$= \frac{7}{12}\pi \text{ radians}$$

If r_1 equals the length of the arm and r_2 equals the length of the arm without a blade, we can find the usable area.

$$A = \frac{1}{2}r_1{}^2\theta - \frac{1}{2}r_2{}^2\theta$$
$$= \frac{1}{2}(18)^2\left(\frac{7}{12}\pi\right) - \frac{1}{2}(2)^2\left(\frac{7}{12}\pi\right)$$

$A \approx 293$ square inches to the nearest square inch $\qquad \square$

Problem Set 1.2

Solve the following problems.

1. On a circle with a radius of 2.10 cm, find the length of the arc intercepted by an angle of $\frac{1}{4}\pi$ radians. Give your answer correct to two decimal places.

2. Find the length of the arc on a circle with radius 5.6 inches if the angle θ is 2.16 radians. Give your answer correct to one decimal place.

3. The part of the circumference of a circle cut by an angle with its vertex at the center of a circle is called the arc subtended by the central angle. On a circle of radius 3.4 inches, find the length of an arc, within 0.1 inches, subtended by a central angle of $120°$.

4. During the Los Angeles Olympics, a bicycle wheel had a diameter of approximately 0.66 meters. How many revolutions did the wheel make in a 30 km race (1 km = 1000 m)? Express your answer correct to the nearest 10 revolutions.

5. The pendulum on a grandfather clock is 120 cm long. It swings side to side through an arc of 0.35 radians. How far does the tip of the pendulum travel in one swing? Express your answer correct to the nearest cm.

6. A space shuttle 240 miles above the earth orbits the earth every 5 hours. How far (to the nearest 10 miles) does the shuttle travel in 1 hour? (The diameter of the earth is approximately 8000 miles.)

7. The minute hand of a clock is 3.4 inches long. How far does its tip move in 20 minutes? Give your answer correct to one decimal place.

8. On a circle of diameter 9.4 feet, find the length of the arc subtended by a central angle of $240°$.

9. A highway curve is laid out as an arc of a circle of radius 1840 feet. How long is the curve if it subtends a central angle of 100°? Give an answer correct to the nearest 10 feet.

10. A 40–foot pendulum hanging in the science building at Western University swings through an angle of 20°. Find the length of arc through which it swings. Give your answer correct to the nearest tenth of a foot.

11. In 2 hours, the tip of the hour hand in the Santa Royale train station travels 30 cm. To the nearest cm, how long is the hour hand?

12. If the moon subtends an angle of approximately 0.52° at the earth's surface, and the distance of the moon from the earth is approximately 240,000 miles, find the length (to the nearest 100 miles) of arc from one edge of the moon to the other edge. (Note: Because of the great distance of the moon from the earth, this arc length will be approximately the same as the diameter of the moon.)

13. Carpet is to be laid in a room that is in the form of a circular sector with a central angle of 105°. If the length (radius) of the room is 80 feet, what is the area of the carpet to be laid?

14. A rotating sprinkler is used to water a lawn. If the sprinkler shoots the water out 36 feet and the sprinkler rotates through an angle of 75°, what is the area of lawn covered by the sprinkler? Give your answer to the nearest square foot.

15. The pendulum on a grandfather's clock is 90 cm long. If it swings side to side through an arc of 20°, what is the area of the sector covered by the pendulum arm? Give your answer to the nearest square cm.

16. The on-ramp from a service street to a freeway is laid out as an arc of a circle of radius 660 feet. If the curve subtends an angle of 140°, what is the area of the sector formed by the service street and the freeway? Give your answer to the nearest square foot.

17. A rotating sprinkler is placed in the vertex of a pie-shaped plot of ground. If the sprinkler shoots the water out 46 feet and the length of arc of the watered area is 66 feet, to the nearest square foot, how much area does the sprinkler water?

18. Two popular restaurants famous for the quality of their pies use 8-inch pie tins for their pies. One restaurant cuts their pies into six equal slices, whereas the other one cuts them into five equal slices. To the nearest one-tenth of a square inch, what is the difference in the area of the slices of pie sold by the two restaurants?

19. A pie-shaped concert hall is constructed so that the back wall is an arc of a circle and the side walls meet at a point behind the stage. The length of arc along the back wall is 148 feet. The front of the stage is 75 feet from the back wall and is 30 feet from the point where the side walls meet behind the stage. How much area is available for seating in the concert hall? Give your answer to the nearest square foot.

20. The bases on a baseball diamond form a square. The fence along the outer edge of the outfield of a baseball field is circular in shape and is 290 feet from home plate. The outer edge of the infield is also circular in shape and is 135 feet from home plate. How much area is covered by the outfield from the foul-ball line along the first base line to the foul-ball line along the third base line? Give your answer to the nearest square foot.

Group Writing Activity

A pizza shop wants to feature "double size" slices. Does it make a difference if they double the central angle of the slice or if they double the length of the radius of the slice? Why?

1.3   Circular Motion

Angular Velocity

In section 1.2 we learned how to determine the length of an arc. Something moving along an arc, say a car on an on-ramp or a satellite in orbit, has both a linear and an angular velocity. Before we distinguish between these, we need to look at velocity in general.

Velocity is defined as directed distance traveled divided by time elapsed. If a car traveling on an interstate highway at a constant speed covers 130 miles in 2 hours, its velocity is:

$$\text{velocity} = \frac{130 \text{ miles}}{2 \text{ hours}}$$
$$= 65 \frac{\text{miles}}{\text{hour}}$$

Properly speaking, velocity is a vector. A vector is a quantity with both amount and direction. Direction as well as amount is important. 130 miles north is very different from 130 miles south. In Chapter 10 we will talk more about vectors.

Linear velocity is velocity in a straight line. Angular velocity is a measure of the amount of rotation in a period of time. We usually use the Greek letter ω (omega) to represent angular velocity.

Definition 1.3B Angular Velocity

Angular velocity is defined as the amount of rotation per unit of time. If rotation is measured in radians:

$$\omega = \frac{\text{angle of rotation in radians}}{\text{time elapsed}}$$
$$= \frac{\theta}{t}$$

Example 17 □ Find the angular velocity in radians per minute of a $33\frac{1}{3}$ rpm record.

It will be necessary to convert revolutions per minute to radians to answer this question.

$$\text{one revolution} = 2\pi \text{ radians}$$
$$33\frac{1}{3} \text{ revolutions} = 33\frac{1}{3}\,\text{revolutions} \cdot 2\pi \frac{\text{radians}}{\text{revolution}}$$
$$= 66\frac{2}{3}\pi \text{ radians}$$

The angular velocity is:

$$\omega = \frac{\text{amount of rotation}}{\text{time elapsed}}$$

$$\omega = \frac{66\frac{2}{3}\pi \text{ radians}}{\text{minute}}$$

□

Linear Velocity of a Point in a Circular Motion

In the case of straight line motion the direction remains the same. For a point traveling along the arc of a circle, the direction of motion is constantly changing.

□ Figure 1.24

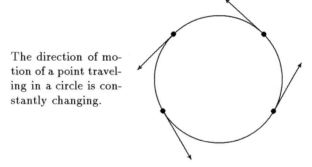

The direction of motion of a point traveling in a circle is constantly changing.

Because the direction of a point traveling in a circle is constantly changing, we define the instantaneous linear velocity of a point in circular motion. To completely describe instantaneous velocity requires calculus. In this course we will ignore the direction of travel and talk about speed, which is the magnitude of the instantaneous velocity. In many cases we are interested in speed rather than velocity. To determine the damage to a car we only need to know it hit the tree at 50 miles per hour. Whether the car was traveling east or west at the time is unimportant.

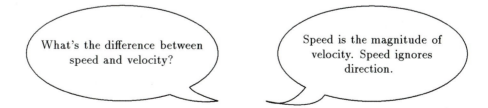

What's the difference between speed and velocity?

Speed is the magnitude of velocity. Speed ignores direction.

If you have ever been on the end of a whip on an ice skating rink, you will recognize that the speed of a point rotating in a circle is a function of the length of the radius of the circle.

Just as velocity is distance divided by time, the linear speed of a point on a circle is the arc length traversed divided by the time elapsed.

$$v = \frac{\text{length of arc}}{\text{time elapsed}}$$

The length of arc traversed is given by the formula $s = r\theta$. Substituting this expression for length of arc and t for time elapsed yields:

$$v = \frac{s}{t}$$
$$v = \frac{r\theta}{t}$$

Because $\frac{\theta}{t} = \omega$, we can substitute ω for $\frac{\theta}{t}$ in this equation to get:

$$v = r\omega$$

Definition 1.3C Linear Speed

The linear speed of a point in constant circular motion is:

$$v = \frac{\text{length of arc traversed}}{\text{time elapsed}} = \frac{s}{t}$$
$$v = \frac{r\theta}{t}$$
$$v = r\omega$$

Where s = length of arc

r = radius

θ = central angle

t = time

ω = angular velocity

Example 18 □ Find the linear speed of the needle relative to a record at the time it is 10 inches from the center of the record playing at $33\frac{1}{3}$ rpm.

In Example 17 we calculated the angular velocity of a $33\frac{1}{3}$ rpm record to be:

$$\omega = 66\frac{2}{3}\pi \frac{\text{radians}}{\text{minute}}$$

From the definition of linear speed we know: $v = r\omega$. Substituting 10 inches for r and $66\frac{2}{3}\pi$ for ω

$$v = r\omega \quad \text{becomes:}$$
$$v = 10 \text{ inches} \cdot \frac{66\frac{2}{3}\pi}{\text{minute}}$$
$$v \approx 2094 \frac{\text{inches}}{\text{minute}}$$

□

Example 19 □ Find the angular velocity of a stone imbedded in the tread of a tire with a radius of 13 inches on a car going 55 mph.

Because the car is going 55 mph the linear speed of a point on the outside of a tire must also be 55 mph or the tire would be skidding along the road.

We can use the definition of linear speed: $v = r\omega$.

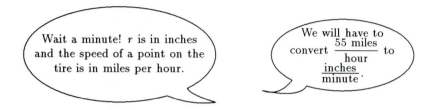

$$v = \frac{55 \text{ miles}}{1 \text{ hour}} \cdot \frac{1 \text{ hour}}{60 \text{ minutes}} \cdot \frac{5280 \text{ feet}}{1 \text{ mile}} \cdot \frac{12 \text{ inches}}{1 \text{ foot}}$$

$$v = 58080 \frac{\text{inches}}{\text{minutes}}$$

Note this is equivalent to multiplication by one.

Solving $v = r\omega$ for ω yields:

$$\omega = \frac{v}{r}$$

Substituting for v and r:

$$\omega = \frac{58080 \frac{\text{inches}}{\text{minute}}}{13 \text{inches}}$$

$$\approx 4468/\text{minute}$$

We can convert the angular velocity of this tire to revolutions per minute. Each revolution is 2π radians.

$$\omega = 4468 \frac{\text{radians}}{\text{minute}}$$

$$\approx 4468 \frac{\text{radians}}{\text{minute}} \cdot \frac{1 \text{ revolution}}{2\pi \text{ radians}}$$

$$\approx 711 \frac{\text{revolution}}{\text{minute}}$$

□

Problem Set 1.3

Express your answers to the degree of accuracy that you can reasonably expect from the information given.

1. Find ω if $\theta = 5$ radians and $t = 2$ seconds.

2. Find ω if $\theta = 5$ revolutions and $t = 4$ seconds.

3. Find t if $\omega = 90$ radians per second and $\theta = 270$ radians.

4. Find t if $\omega = 120$ radians per second and $\theta = 80$ radians.

5. Find θ if $\omega = 100$ radians per second and $t = 4$ seconds.

6. Find θ if $\omega = 100$ revolutions per second and $t = 2$ seconds.

7. Find v if $r = 3$ inches and $\omega = 120$ radians per minute.

8. Find v if $r = 10$ centimeters and $\omega = 129$ revolutions per second.

9. Find r if $\omega = 60$ revolution per minute and $v = 100$ feet per minute.

10. Find r if $\omega = 60$ radians per second and $v = 1200$ centimeters per minute.

11. Find the angular velocity of the hour hand of a clock.

12. Find the angular velocity in radians per minute of a ferris wheel that takes one and a half minutes to make one turn.

13. To the nearest mile per hour, find the linear speed of a seat at a distance of 25 feet from the center of a ferris wheel that takes two minutes to make a revolution.

14. To the nearest foot per minute, find the linear speed of a point on the minute hand of the town hall clock that is 8 feet from the center.

15. Assuming the earth to be a perfect sphere with a diameter of 8000 miles, find the angular velocity and linear speed of a point at the equator.

Chapter 1 Measurement of Angles, Arcs, and Sectors

16. The city of Los Angeles is at 34° north latitude. Using trigonometry from Chapter 2 of this book, it is possible to determine that Los Angeles is about 3315 miles from the earth's axis of rotation. Determine the angular velocity and linear velocity of the city of Los Angeles due to the earth's rotation.

17. What is the angular velocity in radians per second of a 28-inch bicycle wheel if the bicycle is traveling 20 miles per hour?

18. What is the linear speed in miles per hour of a bicycle with 28-inch wheels turning at 180 revolutions per minute?

19. One reason why airplane propellers are noisy is that the tip of the propeller is traveling at close to the speed of sound, which is approximately 660 feet/second. Calculate the linear speed of the tip of a propeller that is 30 inches from the center of the hub at 2500 rpm.

20. If the linear speed of the tip of an airplane propeller is to be no more than 600 feet/second, what is the maximum angular velocity in rpm of a propeller with a diameter of 6 feet?

21. A conveyor belt is to travel at 10 feet per second. How big should the drive pulleys be if they are to turn at 200 revolutions per minute?

22. How fast will a conveyor belt passing over rollers with a diameter of 20 inches travel if the rollers are turning at a rate of 50 radians per minute?

For problems 23–24: Suppose a bicycle chain is connected to the rear sprocket as shown.

□ Figure 1.25

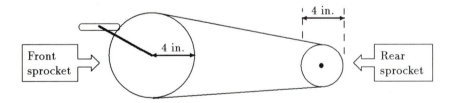

23. A good pedaling speed for a touring bicycle is one revolution per second. At this speed, what will be the angular velocity of the rear wheel in the picture above? How fast will a bicycle with 28-inch wheels be traveling at this angular velocity?

24. What pedaling speed in rotations per minute must a racer maintain to make the rear wheel turn at 150 revolutions per minute?

1.1 **1.** An angle in standard position has its vertex at the origin and its initial side on the x-axis.

2. Angles are measured in revolutions, degrees, grads, and radians.

3. The length of an arc is given by the formula $s = r\theta$, where r is the radius, θ is the angle in radians, and s is the length of the arc.

4. Coterminal angles are angles in standard position with the same terminal side.

5. Conversion formulas:

$$1 \text{ radian} = \frac{180°}{\pi} \qquad\qquad 1° = \frac{\pi}{180}\text{radians}$$

1.2 **1.** To change from one unit of measure to another multiply by one, expressed as the conversion factor.

$$\frac{88 \text{ ft}}{\text{sec}} \cdot \frac{60 \text{ sec}}{\text{min}} \cdot \frac{60 \text{ min}}{\text{hr}} \cdot \frac{1 \text{ mile}}{5280 \text{ ft}} = 60\frac{\text{miles}}{\text{hr}}.$$

2. The formula $s = r\theta$ gives the length of an arc on a circle, where the angle θ is measured in radians.

3. The area of a sector of a circle is given by the formula $A = \frac{1}{2}r^2\theta$, where θ is the central angle measured in radians and r is the radius of the circular sector.

1.3 **1.** Angular velocity is given by the formula $\omega = \frac{\theta}{t}$, where ω is the angular velocity, θ is the angle of rotation in radians, and t is the time elapsed.

2. Linear velocity is given by the formulas $v = \frac{s}{t}$, $v = \frac{r\theta}{t}$ or $v = r\omega$, where s is the arc length, r is the radius, θ is the central angle, t is the time, and ω is the angular velocity.

Chapter 1 Review Test

Name the smallest non-negative angle that is coterminal with each of the following angles. **(1.1)**

1. $540°$ **2.** $-240°$ **3.** $-\dfrac{13}{6}\pi$ **4.** $\dfrac{25}{3}\pi$

Find the degree measure of each of these angles given in radian measure. **(1.1)**

5. $\dfrac{13}{6}\pi$ **6.** $\dfrac{33}{4}\pi$

Find the exact radian measure of each of these angles given in degree measure. **(1.1)**

7. $330°$ **8.** $630°$

9. A ramp on a freeway is circular in shape. If the radius is 42 meters and the central angle is $225°$, what is the length of the ramp? Give an answer correct to 2 decimals. **(1.2)**

10. When a clock hand measuring 15.6 cm in length moves through 40 minutes, how far does the tip move? Give your answer correct to 2 decimal places. **(1.2)**

11. The minute hand of a clock is 5.5 centimeters long. To the nearest tenth, find the area of the sector covered by the minute hand of the clock in 25 minutes. **(1.2)**

12. A highway curve is laid out as an arc of a circle of radius 2140 feet. What is the area of the sector of the circle formed by the curve if it subtends a central angle of $115°$? Give your answer correct to the nearest 10 square feet. **(1.2)**

13. On Martin's car the total length of the windshield wiper arm, plus the blade, is 31 inches and it turns through a $110°$ angle. If the blade alone is 18 inches long, to the nearest square inch how much area of the windshield is cleaned by the wiper blade? **(1.2)**

14. Find ω if $\theta = 12$ radians and $t = 5$ seconds. **(1.3)**

15. Find t if $\omega = 120$ radians per second and $\theta = 480$ radians. **(1.3)**

16. Find θ if $\omega = 160$ radians per second and $t = 4$ seconds. **(1.3)**

17. Find θ if $\omega = 80$ revolutions per second and $t = 5$ seconds. **(1.3)**

18. Find ω in revolutions per day if $v = 600$ miles per hour and $r = 4240$ miles. **(1.3)**

19. Find v if $r = 12$ meters and $\omega = 90$ revolutions per minute. **(1.3)**

20. Find the linear speed of the tip of the second hand in a watch that is 1.5 centimeters in length. Give your answer to one decimal place. **(1.3)**

21. To the nearest revolution per minute, find the angular velocity a 28-inch bicycle wheel that has a linear speed of 25 miles per hour. **(1.3)**

22. A conveyor belt driven by a 16-inch pulley is to travel at 12 feet per second. To the nearest revolution per minute how fast must the drive pulley turn? **(1.3)**

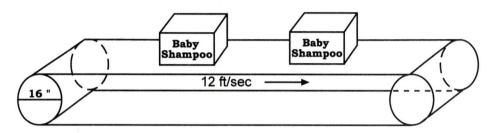

Chapter 2

The Trigonometric Functions

Contents

Preview

In this chapter, we will use a point on the terminal side of an angle to define the six trigonometric functions used throughout this course. To avoid dependence on calculators and provide reference points to easily visualize examples, we will examine 0°, 30°, 45°, 60° and 90° angles. We will then use the location of a point on the terminal side of each of these frequently used angles to find values for their trigonometric functions. In the third section of this chapter, we will apply the definitions of the trigonometric functions to right triangles. Here you will see how trigonometry is useful in surveying, optics, mechanics, and architecture. Then, we will show you how to find the three sides and three angles of any right triangle as long as you know at least one side and an acute angle. We conclude the chapter with a section on circular functions which shows how the trigonometric functions can be derived as functions of a real number.

2.1   Definition of the Six Trigonometric Functions

In the last chapter, we studied central angles of circles and the arcs they intercepted. In this chapter we will look at the point (x, y) on the terminal side of an angle θ. We will examine how the coordinates of this point change as a function of θ, the central angle.

Consider the coordinates of a point at a distance r from the origin on the terminal side of an angle θ in standard position as it rotates about its end point.

☐ Figure 2.1

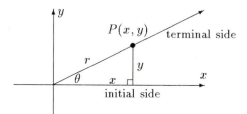

As θ changes, the x and y coordinates of point P also change. The value of r, however, remains constant. From the three numbers $x, y,$ and r it is possible to form six ratios. Note that in the figure above a right triangle is formed with hypotenuse of length r and with sides of length $|x|, |y|$. By the Pythagorean Theorem, $x^2 + y^2 = r^2$.

This is true regardless of the quadrant in which the terminal side of θ lies.

2.1 Definition of the Trigonometric Functions

$$\text{sine } \theta \;=\; \frac{y}{r} \qquad \text{read "sign theta"—abbreviated } \sin\theta$$

$$\text{cosine } \theta \;=\; \frac{x}{r} \qquad \text{read "co-sign theta"—abbreviated } \cos\theta$$

$$\text{tangent } \theta \;=\; \frac{y}{x} \qquad \text{read "tan-jent theta"—abbreviated } \tan\theta$$

$$\text{cosecant } \theta \;=\; \frac{r}{y} \qquad \text{read "co-see-kant theta"—abbreviated } \csc\theta$$

$$\text{secant } \theta \;=\; \frac{r}{x} \qquad \text{read "see-kant theta"—abbreviated } \sec\theta$$

$$\text{cotangent } \theta \;=\; \frac{x}{y} \qquad \text{read "co-tan-jent theta"—abbreviated } \cot\theta$$

Where θ is an angle and (x, y) are the coordinates of a point on the terminal side of θ at a distance r from the vertex.

These ratios are the six trigonometric functions.

Chapter 2 The Trigonometric Functions

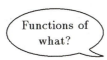

Functions of what?

They are functions of the measure of the angle θ. Change the value of θ and you change the value of each of these trigonometric functions.

What if I move point P along the terminal side of the angle?

The ratios remain the same. Their value depends only on the size of the angle. Read on.

□ Figure 2.2

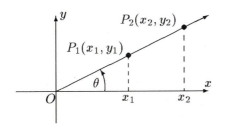

The ratios remain the same because $\triangle P_1 O X_1$ is similar to $\triangle P_2 O X_2$.

□ Figure 2.3

Geometry Reminder

Similar Triangles

Two triangles are similar if they have two equal corresponding angles. In two triangles that are similar, the ratios of any two corresponding sides are equal.

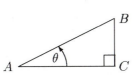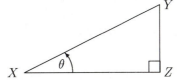

The triangles above have been drawn so that $\angle A \cong \angle X$ and $\angle C$ and $\angle Z$ are both right angles. Therefore $\triangle ABC$ is similar to $\triangle XYZ$.

Because the two triangles are similar, the ratios of any two corresponding sides are equal as well.

For example,

$$\frac{BC}{AC} = \frac{YZ}{XZ} \qquad \text{and} \qquad \frac{AB}{AC} = \frac{XY}{XZ}$$

2.1 Definition of the Six Trigonometric Functions 31

Algebra Reminder

Quadrants

The Cartesian coordinate system divides a plane into four parts. These parts are called quadrants as indicated below.

□ Figure 2.4

	y	
Quadrant II		Quadrant I
		x
Quadrant III		Quadrant IV

We usually use Roman numerals to label the quadrants, but we sometimes call them first, second, third, and fourth quadrants. Notice that the quadrants are numbered in order as the terminal side of a positive angle in standard position rotates. The axes do not lie in any quadrant.

Example 1 □ What are the quadrants where $\cos\theta$ is negative?

Cos θ is defined as $\dfrac{x}{r}$. The radius is always considered to be positive.

Because x is negative in the II and III quadrants, $\cos\theta$ is negative in these quadrants.

□ Figure 2.5

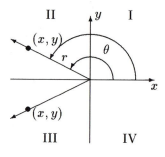

□

Example 2 □ What is the range of values of θ where both $\sin\theta$ and $\cos\theta$ are negative?

The y coordinate of any point below the x-axis is negative.

□ Figure 2.6

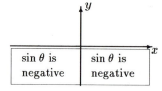

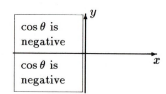

Sin θ is negative in the III and IV quadrants.

Cos θ is negative in the II and III quadrants.

Chapter 2 The Trigonometric Functions

Therefore both $\sin\theta$ and $\cos\theta$ are negative only in quadrant III. The range of values for θ in the third quadrant is $180° < \theta < 270°$.

☐ Figure 2.7

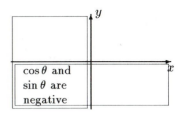

cos θ and sin θ are negative

☐

Calculating Values of Trigonometric Functions

The six trigonometric functions can be calculated for any angle θ where the function is defined if a point on the terminal side of the angle is known.

Example   3 ☐ Find the values of the six trigonometric functions of an angle θ if one of the points on its terminal sides is $(-3, -4)$.

First we sketch the angle.

☐ Figure 2.8

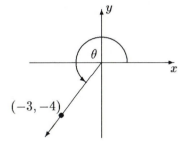

We know (x, y). What is r?

The Pythagorean Theorem tells us that for right triangles the square of the hypotenuse is equal to the sum of the squares of the other two sides. We use it below.

To find r, examine the right triangle in the third quadrant and apply the Pythagorean Theorem.

☐ Figure 2.9

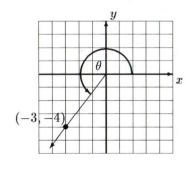

$$r^2 = |x|^2 + |y|^2$$
$$r^2 = |-3|^2 + |-4|^2$$
$$r^2 = 9 + 16$$
$$r = \sqrt{25}$$
$$r = 5$$

Looks like the equation of a circle.

It is. The equation of a circle is an application of the Pythagorean Theorem.

Now that we know x, y, and r, we can write the trigonometric functions directly from their definitions:

$$\sin \theta = \frac{y}{r} = \frac{-4}{5} \qquad\qquad \csc \theta = \frac{r}{y} = \frac{5}{-4} = -\frac{5}{4}$$

$$\cos \theta = \frac{x}{r} = \frac{-3}{5} \qquad\qquad \sec \theta = \frac{r}{x} = \frac{5}{-3} = -\frac{5}{3}$$

$$\tan \theta = \frac{y}{x} = \frac{-4}{-3} = \frac{4}{3} \qquad\qquad \cot \theta = \frac{x}{y} = \frac{-3}{-4} = \frac{3}{4}$$

□

Example 4 □ Find the values of the remaining trigonometric functions of θ if $\cos \theta = \frac{12}{13}$ and $\tan \theta = -\frac{5}{12}$.

Because $\cos \theta = \frac{x}{r}$, it can be positive in either the first or fourth quadrant. Because $\tan \theta = \frac{y}{x}$, it can be negative in either the second or fourth quadrant. The problem states $\cos \theta$ is positive and $\tan \theta$ is negative.

To satisfy both conditions θ must be in the fourth quadrant.

□ Figure 2.10

One point on the terminal side of θ would look like the figure on the right.

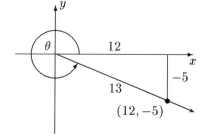

Now we can write the remaining trigonometric functions directly from the figure.

$$\sin \theta = \frac{y}{r} = \frac{-5}{13} \qquad\qquad \sec \theta = \frac{r}{x} = \frac{13}{12}$$

$$\csc \theta = \frac{r}{y} = \frac{13}{-5} = -\frac{13}{5} \qquad\qquad \cot \theta = \frac{x}{y} = \frac{12}{-5} = -\frac{12}{5}$$

□

Problem Set 2.1

Use the Pythagorean Theorem to find the distance r from the origin for each of the following points.

  1. $(3, 4)$ **2.** $(-4, 3)$ **3.** $(5, -12)$

Find the values of the six trigonometric functions of an angle θ in standard position for the following coordinates of points on the terminal side of the angle.

  4. $(2, -2)$ **5.** $(-3, 2)$

  6. $(12, -5)$ **7.** $(-24, 7)$

                                              Chapter 2 The Trigonometric Functions

Indicate the quadrant(s) in which the terminal side of angle θ must lie in order that:

8. $\cos \theta$ is negative

9. $\tan \theta$ is positive

10. $\csc \theta$ is positive

11. $\cot \theta$ is negative

12. $\tan \theta$ and $\cos \theta$ are positive

13. $\sec \theta$ is negative and $\sin \theta$ is positive

For the following give the range of θ in which both conditions are fulfilled:

14. $\sin \theta$ is positive and $\tan \theta$ is negative

15. $\sec \theta$ is negative and $\csc \theta$ is positive

16. $\cot \theta$ is negative and $\sin \theta$ is positive

17. $\tan \theta$ is negative and $\csc \theta$ is negative

18. $\csc \theta$ is negative and $\cot \theta$ is negative

Find the values of the remaining trigonometric functions of θ if:

19. $\cos \theta = \dfrac{3}{5}$ and $\tan \theta = -\dfrac{4}{3}$

20. $\sin \theta = -\dfrac{5}{13}$ and $\cot \theta = -\dfrac{12}{5}$

21. $\csc \theta = -\dfrac{25}{7}$ and $\tan \theta = -\dfrac{7}{24}$

22. $\cos \theta = -\dfrac{3}{5}$, θ in quadrant II

23. $\tan \theta = -\dfrac{5}{12}$, θ in quadrant IV

24. $\sin \theta = \dfrac{4}{5}$, θ in quadrant II

25. $\sin \theta = -\dfrac{\sqrt{8}}{3}$, θ in quadrant III

26. $\cos \theta = \dfrac{1}{2}$, θ in quadrant IV

27. $\csc \theta = 5$, θ in quadrant II

28. $\sec \theta = -3$, θ in quadrant II

2.2 Values of the Trigonometric Functions for 0°, 30°, 45°, 60°, 90°, 180° Angles

The previous section evaluated trigonometric functions when the coordinates of only one point on the terminal side of an angle were known. The measure of the angle was not mentioned. Usually we know an angle and need to find the value of the trigonometric functions of that angle. In practice we will find the values of the trigonometric functions using a scientific calculator. But we can find the trigonometric functions of 30°, 45°, 60° angles by examining a few triangles you studied in geometry. We will use these angles in examples and problems both because these values occur frequently and because we can easily memorize them for future reference. Let's start with a review of some geometry.

Geometry Reminder

About Triangles

1. Standard labels for a right triangle

Lower-case letters indicate the sides. The side opposite the right angle is called the hypotenuse; it is usually labeled side c.

Capital letters indicate vertices of the triangle. We also call this Angle B.

Square angle symbol indicates a right angle with a measure of $90°$.

The right angle is usually labeled Angle C.

2. The Pythagorean Theorem for any right triangle is: $a^2 + b^2 = c^2$. In words, the square of the hypotenuse is equal to the sum of the squares of the other two sides.

3. An isosceles triangle has two sides equal; the angles opposite these sides are also equal.

If side AC = side BC then $\angle A = \angle B$

4. The sum of the angles in any triangle is $180°$.

5. Consider right $\triangle ABC$ with two sides of length 1: $\angle C$ is given as a $90°$ angle; therefore $\angle A + \angle B = 90°$ because the sum of the angles in a triangle is $180°$. $\angle A = \angle B$, therefore $\triangle ABC$ is isosceles and $\angle A = \angle B = 45°$. Side c is the hypotenuse of $\triangle ABC$.

The length of side c in a $45°$ isosceles triangle can be computed as follows:

$$
\begin{aligned}
c^2 &= a^2 + b^2 \\
&= 1^2 + 1^2 \\
&= 2 \\
c &= \sqrt{2}
\end{aligned}
$$

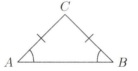

$$c = \sqrt{2}$$

$$
\begin{aligned}
\sin 45° &= \frac{a}{c} = \frac{\sqrt{2}}{2} \\
\cos 45° &= \frac{b}{c} = \frac{\sqrt{2}}{2} \\
\tan 45° &= \frac{a}{b} = \frac{1}{1}
\end{aligned}
$$

Chapter 2 The Trigonometric Functions

6. An altitude or height of a triangle is a line from a vertex perpendicular to the opposite side. A median is a line from a vertex to the midpoint of the opposite side.

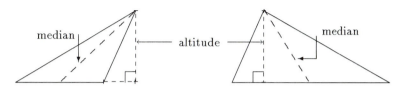

In an isosceles triangle, the altitude is the same as the median.

7. Consider an equilateral triangle with each side 2 units long. The altitude, y, divides the base into two equal segments each of length 1 unit. $\angle A = \angle C = 60°$ because all three angles of an equilateral triangle are equal.

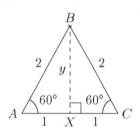

$\angle A = 60°$ and $\angle BXA = 90°$, $\angle XBA = 30°$.

Now examine $\triangle AXB$ in the figure above.

By the Pythagorean Theorem

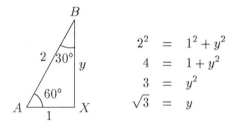

$$2^2 = 1^2 + y^2$$
$$4 = 1 + y^2$$
$$3 = y^2$$
$$\sqrt{3} = y$$

In any $30°-60°-90°$ triangle.

a) The length of the hypotenuse is twice the length of the side opposite the 30° angle.

b) The length of the side opposite the 60° angle is $\sqrt{3}$ times the length of the side opposite the 30° angle.

We can use our knowledge of $30°- 60°- 90°$ triangles to find the trigonometric functions of $60°$. For example, the ratios of the sides of a $30°- 60°- 90°$ triangle tell us that if $\theta = 60°$, the coordinates of a point at $r = 2$ will be $\left(1, \sqrt{3}\right)$.

☐ Figure 2.17

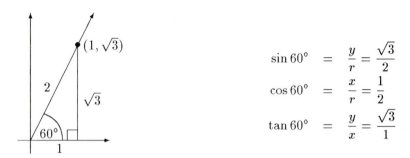

$$\sin 60° = \frac{y}{r} = \frac{\sqrt{3}}{2}$$

$$\cos 60° = \frac{x}{r} = \frac{1}{2}$$

$$\tan 60° = \frac{y}{x} = \frac{\sqrt{3}}{1}$$

Redrawing the $30°- 60°- 90°$ triangle with the $30°$ angle at the origin allows us to find the trigonometric functions of a $30°$ angle:

☐ Figure 2.18

$\dfrac{\sqrt{3}}{2}$

$\dfrac{1}{\sqrt{3}}$

$$\sin 30° = \frac{1}{2}$$

$$\cos 30° = \underline{\hspace{2cm}}$$

$$\tan 30° = \underline{\hspace{2cm}}$$

Algebra Reminder

Rationalizing the Denominator

Tan $30° = \dfrac{1}{\sqrt{3}}$. We usually try to avoid writing expressions with a radical in the denominator. Rewriting a fraction so that its denominator is a rational number is called rationalizing the denominator. To rationalize a denominator multiply by 1.

$$\frac{1}{\sqrt{3}} = \frac{1}{\sqrt{3}}\left(\frac{\sqrt{3}}{\sqrt{3}}\right) \quad \Longleftarrow \boxed{\frac{\sqrt{3}}{\sqrt{3}} \text{ is equal to one}}$$

$$= \frac{\sqrt{3}}{3}$$

Chapter 2 The Trigonometric Functions

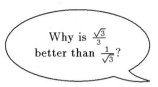

Historically denominators that are rational numbers were preferred because before calculators it was more efficient to calculate

$$\frac{\sqrt{3}}{3} \approx \frac{1.732}{3} \approx 0.577 \quad \text{than} \quad \frac{1}{\sqrt{3}} \approx \frac{1}{1.732} \approx 0.577.$$

The next three examples show how to use the symmetry of a circle to find the trigonometric functions for multiples of $30°$.

Example 5 □ Find the values of all six trigonometric functions of $150°$.

□ Figure 2.19

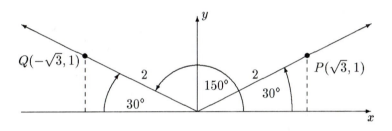

Point Q is on the terminal side of a $150°$ angle. To find the coordinates of Q, look at point P on the terminal side of a $30°$ angle with coordinates $(\sqrt{3}, 1)$. Point Q is symmetrical with respect to the y-axis to point P. Therefore Q has coordinates $(-\sqrt{3}, 1)$.

We can use the Pythagorean Theorem to find $r = 2$ because r forms the hypotenuse of a right triangle.

For point Q, $x = -\sqrt{3}$, $y = 1$, $r = 2$. The trigonometric functions of $150°$ are written:

$$\sin 150° = \frac{1}{2} \qquad \cos 150° = \frac{-\sqrt{3}}{2} \qquad \tan 150° = \frac{1}{-\sqrt{3}} \cdot \left(\frac{-\sqrt{3}}{-\sqrt{3}}\right)$$

$$\csc 150° = \frac{2}{1} \qquad \sec 150° = \frac{2}{-\sqrt{3}} \qquad = -\frac{\sqrt{3}}{3}$$

$$= -\frac{2\sqrt{3}}{3} \qquad \cot 150° = -\frac{\sqrt{3}}{1}$$

□

We found the values for the trigonometric functions of 150° by using the coordinates of a point in a 30° angle. The 30° angle is called a reference angle for the 150° angle.

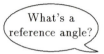

It's the angle used to find the values of trigonometric functions for angles outside the range 0° to 90°.

Definition 2.2 Reference Angle

The reference angle for any angle placed in standard position is the angle θ' (pronounced "theta prime") which is the positive acute angle formed by the terminal side of θ and the x-axis.

□ Figure 2.20

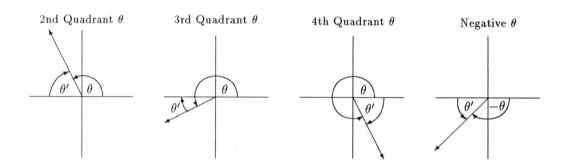

To find the values of the trigonometric functions of any angle we find its reference angle θ'. From the coordinates of the reference angle and our knowledge of symmetry of a circle, we determine the trigonometric functions of θ.

Example 6 □ Give the reference angle for each of the following angles.

30°, 45°, 60°

a. 150° _____   **b.** 225° _____   **c.** 300° _____

60°, 30°, $\frac{1}{6}\pi$

d. −120° _____   **e.** −330° _____   **f.** $\frac{5}{6}\pi$ _____

$\frac{1}{3}\pi$, $\frac{1}{6}\pi$, $\frac{1}{4}\pi$

g. $-\frac{4}{3}\pi$ _____   **h.** $-\frac{11}{6}\pi$ _____   **i.** $-\frac{9}{4}\pi$ _____

□

Chapter 2 The Trigonometric Functions

Example 7 □ Find the values of all six trigonometric functions for an angle with a measure of $\frac{5}{3}\pi$ radians.

There is a point Q on the terminal side of a $\frac{5}{3}\pi$ angle that is symmetrical with point P on a $\frac{1}{3}\pi$ angle. Because $\frac{1}{3}\pi$ measures the same angle as $60°$, we know there is a point on the terminal side of a $\frac{1}{3}\pi$ angle where $r = 2$, $x = 1$, and $y = \sqrt{3}$.

□ Figure 2.21

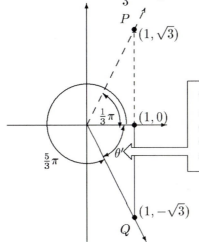

The angle labeled θ' is the reference angle. It is symmetrical to $\frac{1}{3}\pi$. We obtain the coordinates of point Q on the terminal side of $\frac{5}{3}\pi$ by using symmetry.

For point Q

$$x = 1, \qquad y = -\sqrt{3}, \qquad r = 2$$

Therefore the values of the trigonometric functions of $\frac{5}{3}\pi$ are:

$$\sin \frac{5}{3}\pi = \frac{-\sqrt{3}}{2} \qquad\qquad \cos \frac{5}{3}\pi = \frac{1}{2} \qquad \tan \frac{5}{3}\pi = \frac{-\sqrt{3}}{1}$$

$$\csc \frac{5}{3}\pi = \frac{2}{-\sqrt{3}} \cdot \frac{-\sqrt{3}}{-\sqrt{3}} \qquad \sec \frac{5}{3}\pi = \frac{2}{1} \qquad \cot \frac{5}{3}\pi = \frac{1}{-\sqrt{3}} \cdot \frac{-\sqrt{3}}{-\sqrt{3}}$$

$$= -\frac{2\sqrt{3}}{3} \qquad\qquad\qquad\qquad\qquad\qquad\qquad = -\frac{\sqrt{3}}{3}$$

□

Example 8 □ Find the values of the six trigonometric functions of $\frac{1}{2}\pi$.

□ Figure 2.22

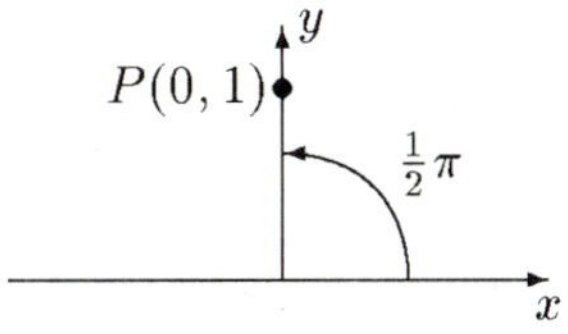

$P(0,1)$ y

$\frac{1}{2}\pi$

x

A $\frac{1}{2}\pi$ angle has its terminal side on the y-axis.

Consider the point $P(0, 1)$ which lies on the terminal side of a $\frac{1}{2}\pi$ angle. For P,

$$x = 0, \qquad y = 1, \qquad r = 1$$

Therefore the values of the trigonometric functions of $\frac{1}{2}\pi$ are:

$$\sin \frac{1}{2}\pi = \frac{1}{1} = 1 \qquad \tan \frac{1}{2}\pi = \frac{y}{x} = \frac{1}{0} \text{ (undefined)} \qquad \csc \frac{1}{2}\pi = \frac{1}{1} = 1$$

$$\cos \frac{1}{2}\pi = \frac{0}{1} = 0 \qquad \cot \frac{1}{2}\pi = \frac{0}{1} = 0 \qquad\qquad\qquad \sec \frac{1}{2}\pi = \frac{r}{x} = \frac{1}{0} \text{ (undefined)}$$

2.2 Values of the Trigonometric Functions for 0°, 30°, 45°, 60°, 90°, 180° Angles   41

Why are $\tan \frac{1}{2}\pi$ and $\sec \frac{1}{2}\pi$ undefined?

The x coordinate of any point on the terminal side of the angle $\frac{1}{2}\pi$ is zero. Because $\tan \theta$ and $\sec \theta$ require division by x, these functions are undefined along the y-axis.

□

Example 9 □ Find the values of the six trigonometric functions of π.

π radians lie along the negative x-axis. One point on the negative x-axis is $(-1, 0)$. At $(-1, 0)$, $x = -1$, $y = 0$, $r = 1$.

□ Figure 2.23

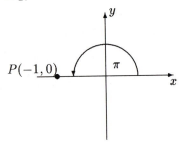

$$\sin \pi = \frac{0}{1} \qquad \cos \pi = \frac{-1}{1} \qquad \tan \pi = \frac{0}{-1}$$

$$\csc \pi = \text{undefined} \qquad \sec \pi = \frac{1}{-1} \qquad \cot \pi = \text{undefined}$$

□

Problem Set 2.2

Give the reference angle for each of the following angles:

1. $120°$ **2.** $210°$ **3.** $-330°$ **4.** $-315°$ **5.** $-225°$ **6.** $-405°$

7. $\frac{5}{3}\pi$ **8.** $\frac{11}{6}\pi$ **9.** $\frac{1}{3}\pi$ **10.** $\frac{7}{4}\pi$ **11.** $-\frac{5}{6}\pi$ **12.** $-\frac{3}{4}\pi$

Without using your calculator, find the values of the six trigonometric functions of the following angles given in degrees:

13. $120°$ **14.** $210°$ **15.** $-180°$ **16.** $-300°$ **17.** $225°$ **18.** $135°$

Find the values of the six trigonometric functions of the following angles given in radians:

19. $\frac{3}{2}\pi$ **20.** $\frac{4}{3}\pi$ **21.** $-\frac{5}{6}\pi$ **22.** $\frac{7}{4}\pi$ **23.** $-\frac{7}{6}\pi$ **24.** $\frac{11}{6}\pi$

Group Writing Activity

Which functions are not defined along the negative x-axis? Why? Why is $\tan \theta$ undefined for angles whose terminal side lies along the negative y-axis? Why do undefined values of the trigonometric functions only occur along the coordinate axes?

Chapter 2 The Trigonometric Functions

2.3 Trigonometric Functions for Right Triangles

The earliest applications of trigonometry dealt with applications of trigonometric functions where θ is between $0°$ and $90°$. In these cases, $x, y,$ and r form a right triangle in the first quadrant. Hence, the ideas in this section are sometimes called right triangle trigonometry.

Another Definition of the Trigonometric Functions

Picture point P on the terminal side of any angle in the first quadrant. By drawing a line from P to the x-axis, we can form a right triangle.

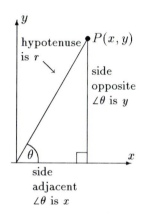

Definition 2.3A Trigonometric Functions for Right Triangles

For angles between $0°$ and $90°$ the trigonometric functions are:

$$\sin \theta = \frac{y}{r} = \frac{\text{opposite side}}{\text{hypotenuse}} \qquad \csc \theta = \frac{r}{y} = \frac{\text{hypotenuse}}{\text{opposite side}}$$

$$\cos \theta = \frac{x}{r} = \frac{\text{adjacent side}}{\text{hypotenuse}} \qquad \sec \theta = \frac{r}{x} = \frac{\text{hypotenuse}}{\text{adjacent side}}$$

$$\tan \theta = \frac{y}{x} = \frac{\text{opposite side}}{\text{adjacent side}} \qquad \cot \theta = \frac{x}{y} = \frac{\text{adjacent side}}{\text{opposite side}}$$

The definitions above are also convenient if the angle θ is not in standard position.

Before starting the first example, find the value of tangent θ on your calculator using the following method:

Using Your Scientific Calculator

To approximate the value of the tangent of $45°$ on your calculator, follow these steps.

1. Press CLR or AC when you first turn on your calculator so that the calculator can clear itself.

2. Your calculator probably has these three modes for trigonometry:

   DEG for Degrees

   RAD for Radians

   GRAD for Grads

   To find the tangent of an angle measured in degrees, your calculator must be in DEG mode. Refer to your manual if necessary.

3. To find the value of $\tan 45°$, press TAN, type 45, then press ENTER or EXE. The number 1 should appear. If you get 1.6197752, check the MODE to be sure the calculator is in DEG mode.

Using Your Graphing Calculator

To approximate the value of the tangent of $45°$ on your graphing calculator, follow these steps.

1. Press CLEAR when you first turn on your calculator so that the calculator can clear itself.

2. Your calculator probably has these two modes for trigonometry:

   Radians and Degrees

   To set the proper mode, first press the MODE key. This should display a menu. Use the arrow key to select DEGREE and press ENTER or EXE. Next press CLEAR to get a blank screen. Then press TAN 45 followed by ENTER or EXE.

3. The number 1 should appear. If you get 1.6197752, check to see that you are in degree mode.

Finding the Missing Side in a Triangle

Example 10 □ How tall is a flagpole that casts a 12-foot shadow when the sun is at an angle of 75° above the horizon?

We know the side adjacent to the 75° angle, and we are looking for the opposite side, so use the definition of tan 75°.

$$\tan 75° = \frac{\text{opposite side}}{\text{adjacent side}}$$

To find the value of tan 75°, make sure your calculator is in degree mode (you may need to refer to the calculator manual). Then press tan 75 and press ENTER or EXE.

□ Figure 2.24

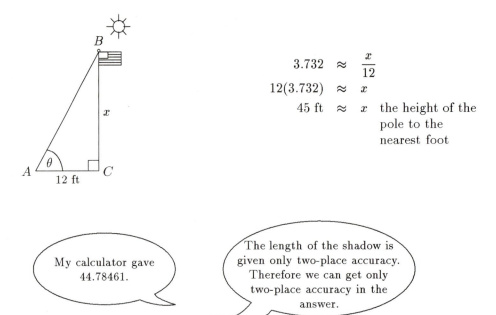

$$3.732 \approx \frac{x}{12}$$
$$12(3.732) \approx x$$
$$45 \text{ ft} \approx x \quad \text{the height of the pole to the nearest foot}$$

My calculator gave 44.78461.

The length of the shadow is given only two-place accuracy. Therefore we can get only two-place accuracy in the answer.

□

Comments About Calculators and Accuracy

In this text we want you to focus on the ideas of trigonometry and algebra. Therefore, we will make some assumptions to help keep page clutter down while we try to be realistic about accuracy. To do this we will consider two kinds of figures, theoretical and physical.

In theoretical figures, all measurements will be considered exact. A right triangle with sides of 3, 4, and 5 units is an example of a theoretical triangle. In theoretical figures we will give exact answers, or we will specify the degree of accuracy desired.

In physical figures and applied problems we are dealing with measured values. Measured values are always approximations; hence, in these problems we will follow the appendix on accuracy to determine the correct number of significant digits. Briefly:

  a. No answer can have more significant digits than the least accurate of the beginning data. **Computations cannot create accuracy.**

     Chapter 2 The Trigonometric Functions

b. Zeros will be considered significant unless indicated otherwise. That is, 3000 feet indicates four significant digits, while 3000 feet to the nearest hundred feet indicates two-digit accuracy.

c. Angles will be treated as follows:

"Nearest degree" implies two significant digits of accuracy.

"Nearest tenth of a degree" implies three significant digits of accuracy.

That still leaves the problem of how to treat all those digits on your calculator display. Consider the example above where we determined the height of a flagpole from its shadow. The shadow was measured to two-digit accuracy. The 75° angle is also expressed to two-digit accuracy. However, your calculator will probably provide 3.7320508 as the tangent of 75°. You might reason that because the data in the problem is expressed to two-digit accuracy, you should use 3.7 or 3.73 as the tangent of 75°. If you were doing the calculation by hand, this would be a reasonable approach. Using a calculator is a different story:

a. There is no savings of effort by doing arithmetic to fewer places.

b. To get a two-digit value for tangent 75°, you would have to find tangent 75° to the full accuracy of the calculator and then rekey the value to the two-digit accuracy. This is more work than using the full accuracy.

c. While 75° is a measurement, tangent 75° is, theoretically at least, an exact number. Therefore, we should use all the accuracy available from the calculator when expressing tangent 75°.

In this text we will do the following:

a. Use the full accuracy of the calculator for all computations.

b. We will normally show only the first four places from the calculator display. This should be enough digits so you know you are pressing the correct keys.

c. We will round our final answer to the accuracy of the least accurate piece of measured data.

SEE APPENDIX FOR MORE DETAIL ABOUT ACCURACY.

Example 11 □ A surveyor wants to know the distance x across the pond pictured below.

To create a convenient right triangle, the surveyor measures a distance AC of 100 ft perpendicular to the desired BC along the ground. Then the surveyor measures angle CAB. With this information, the distance across the pond can be found.

□ Figure 2.25

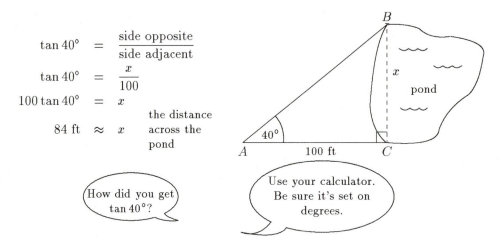

$$\tan 40° = \frac{\text{side opposite}}{\text{side adjacent}}$$

$$\tan 40° = \frac{x}{100}$$

$$100 \tan 40° = x$$

$$84 \text{ ft} \approx x \quad \text{the distance across the pond}$$

How did you get tan 40°?

Use your calculator. Be sure it's set on degrees.

□

Example 12 □ A kite on the end of a 250-foot string is at an angle of 35° above horizontal. How high is the kite above the ground? Assume the string is a straight line.

□ Figure 2.26

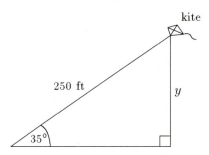

This time we know the measurement of one angle and the hypotenuse. We want to find the side opposite the angle. So we can use the definition of $\sin \theta$.

$$\sin \theta = \frac{\text{opposite side}}{\text{hypotenuse}}$$

$$\sin 35° = \frac{y}{250}$$

$$250 \sin 35° = y$$

$$140 \text{ ft} \approx y$$

□

Chapter 2 The Trigonometric Functions

Finding Values for Inverse Trigonometric Functions

In Example 12 we know one angle and one side of a right triangle. With that information we calculated the length of another side of the triangle. We would also like to be able to find the measure of an angle of a right triangle when we know only the lengths of two sides.

To find the measures of angles from a trigonometric function of the angle, we must define inverse trigonometric functions.

Definition 2.3B   Arcsin x, Arccos x, Arctan x

The Arcsin x or $\sin^{-1} x$ is the angle θ with a sine of x. Similarly

$$\cos^{-1} x = \text{Arccos } x \quad = \quad \text{angle whose cosine is } x.$$
$$\tan^{-1} x = \text{Arctan } x \quad = \quad \text{angle whose tangent is } x.$$

(This definition is always correct for angles between $0°$ and $90°$. In a later section, we will discuss the range and domain of the inverse trigonometric functions.)

$$\text{Because the } \sin 30° \quad = \quad 0.5000,$$
$$\text{Arcsin } (0.5000) \quad = \quad 30°$$

Using Your Scientific Calculator

To use your calculator to find arcsin(0.5000) follow these steps:
1. Be sure your calculator is in degree mode.
2. Enter .5 on the keys.
3a. If your calculator has a SIN^{-1} key press it.
3b. Otherwise press the key marked 2ND or INV then press the SIN key.

Using Your Graphing Calculator

Most calculators use each key for more than one function, just as a typewriter uses a shift key for capital letters. Arcsin x is usually a SHIFT key or 2ND function key.

To use your calculator to find arcsin (0.5000) follow these steps:
1. Be sure your calculator is in DEG mode.
2. SIN^{-1} is probably above the SIN key. To access this function,
   a. Press 2ND or SHIFT.
   b. Press SIN^{-1}.
3. Type .5.
4. Press ENTER or EXE.
The display should show 30.

Example 13 □ Find Arctan (2.050) in degrees.

$$\text{Arctan } (2.050) = \theta \text{ is the same as } \tan\theta = 2.050$$
$$\theta \approx 64°$$

□

Example 14 □ For safety and noise abatement, landing aircraft are required to be 1500 feet above the ground when they are 25000 feet from the end of the runway. What is the angle of the glide slope that makes this possible?

□ Figure 2.27

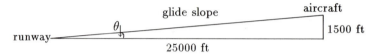

This time we know the opposite and adjacent sides and wish to find an angle.

$$\tan\theta = \frac{\text{side opposite}}{\text{side adjacent}}$$
$$\tan\theta = \frac{1500 \text{ ft}}{25000 \text{ ft}}$$
$$\tan\theta = 0.0600$$

If you know $\tan\theta$ and want to find θ, use the INV TAN or TAN^{-1} key on your calculator.

$$\tan\theta = 0.0600$$
$$\theta \approx 3.43°$$

□

Help with Problem Solving

What Do I Know?

A good starting point in solving any problem is to make explicit the answer to the question, "What do I know?"

Define what quantities are known. What are their units of measurement? What relationships are known?

These are several ways to do this:

1. Pretend you are explaining the problem to a friend over the phone. What can you tell him or her without actually reading the problem word by word?

2. Underline the quantities, units of measurement, and relationships in the problem.

3. Translate as much English grammar into algebraic grammar as possible. At this point, don't worry that you have more knowns than equations.

4. Draw a sketch of the problem. Be as accurate as possible. A line that is 10 units long should be twice as long as a line 5 units long. Approximate angles as closely as possible.

Chapter 2 The Trigonometric Functions

Example 15 □ An airplane's distance measuring equipment (DME) measures the slant range from the aircraft to the navigation aid. If an aircraft is 17.56 miles slant range distance from a navigation aid and is flying two miles above the ground, what is the distance along the ground from the airplane to the navigation aid?

□ Figure 2.28

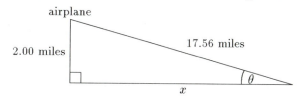

airplane

2.00 miles

17.56 miles

θ

x

We could use the Pythagorean Theorem to solve this problem. However, we'll use the definition of the $\sin\theta$ to illustrate an idea.

$$\sin\theta = \frac{\text{opposite side}}{\text{hypotenuse}}$$

$$= \frac{2.00 \text{ miles}}{17.56 \text{ miles}}$$

$$\sin\theta \approx 0.1139$$

We know the $\sin\theta$ but we need the value of θ with a sine of 0.1139. This is called Arcsin x. To find Arcsin (0.1139) use the INV SIN key or the SIN^{-1} key on your calculator.

$$\theta \approx 6.5°$$

Now we can use $\cos 6.5°$ to find the ground distance.

$$\cos 6.5° = \frac{\text{adjacent side}}{\text{hypotenuse}}$$

$$\cos 6.5° \approx \frac{x}{17.56}$$

$$17.45 \text{ miles} \approx x$$

□

Example 16 □ A circle has a radius of 12 inches. A 40° angle is drawn with its vertex at the center of the circle. What is the length of the chord $\overline{AB}$ connecting the points where the angle intercepts the circle?

□ Figure 2.29

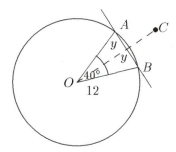

We are attempting to find the length of line segment AB. Notice that triangle AOB is not a right triangle. However, we can construct two right triangles if we bisect the 40° central angle with line segment OC.

Line segment OC will divide $\overline{AB}$ into two equal line segments. It will also form a right angle with $\overline{AB}$.

Now we can use trigonometry to find half the length of $\overline{AB}$, which we labeled y.

$$\sin 20° = \frac{\text{opposite side}}{\text{hypotenuse}}$$
$$\sin 20° = \frac{y \text{ inches}}{12 \text{ inches}}$$
$$12 \text{ inches}(\sin 20°) = y$$
$$4.1 \text{ in} \approx y$$

The distance between points A and B is twice the length of y.

$$\text{length of } AB \approx 2(4.1 \text{ in})$$
$$\approx 8.2 \text{ inches}$$

□

Problem Set 2.3

Use your calculator to find the following to the nearest tenth of a degree.

1. Arcsin (0.6018)   **2.** Arccos (0.6018)   **3.** Arctan (1.0)

Use your calculator to find the following to the nearest hundredth of a radian.

4. Arcsin (0.7810)   **5.** Arctan (1.6825)   **6.** Arccos (– 0.6012)

Solve each of the following problems. Make a diagram for each and label the given information.

7. A 12.0-foot ladder leaning against a wall has its base 4.0 feet from the wall. What angle, to the nearest one tenth of a degree, does the ladder make with the ground and how far up the wall is the top of the ladder, to one decimal place?

8. A building is equipped with a ramp to make it wheelchair accessible. The ramp is 30 feet long and is inclined at an angle of 4° with the ground. To the nearest tenth of a foot, how high does it rise above the ground?

9. A rectangular swimming pool is 15.0 feet by 40.0 feet. A rope is drawn diagonally across the top of the pool. To the nearest tenth of a degree, what angle does the rope make with the longer side and how long is the rope to one decimal place?

Chapter 2 The Trigonometric Functions

10. An airplane is approaching an airport on a 6° glide slope. At the time that the aircraft's distance measuring equipment shows it is a slant range distance of 6.5 miles from the end of the runway, how far above the elevation of the runway is the airplane to the nearest hundredth of a mile?

11. A surveyor wishes to determine the width of a straight river. Starting at a point directly across the river from a large oak tree, the surveyor measured a distance of 120 meters upstream. If the angle between the line of sight from the upstream point to the tree and the river is 42°, how wide to the nearest meter is the river?

12. A telephone pole is supported with a guy wire 25 feet from its base. If the wire makes an angle of 62° with the ground, how tall to the nearest foot is the telephone pole and how long is the guy wire?

13. A fisher used an 8.0-foot rope to tie a rowboat to the dock. If the dock is 3.0 feet above the top of the boat where the rope is attached, to the nearest tenth of a foot, how far can the boat drift from the dock? (Do not use the Pythagorean Theorem.)

14. An airplane climbs with an airspeed of 120 mph at an angle of 5.5° with the horizontal. What is the altitude of the airplane after 5 minutes? (Give the answer to the nearest 10 feet.)

15. The Starship Enterprise launched two probes at an angle of 85° to each other. After the probes have traveled three light years, how far apart are the probes to two decimal places?

16. The angle of a roof above horizontal is 20°. If the distance from one side of the house to the other is 35 feet, how long must the boards that connect the side of the house to the peak of the roof be?

17. A 32.0-inch door is obstructed so that when it is open it makes an angle of 25° with its closed position. To the nearest tenth of an inch, what is the largest object that can be squeezed through the opening?

18. A 19-foot ladder is leaning against a wall. The base of the ladder is 6 feet from the wall. To the nearest foot find the height of the wall. (Use trigonometric functions.)

19. A straight highway in a mountainous area rises 16.0 feet for each 150 feet along the highway. Find the angle the highway makes with the horizontal to the nearest tenth of a degree.

2.4 Solving Right Triangles

In the previous section we showed many applications of right triangle trigonometry, usually looking for the value of one side or one angle. In the next section, we will demonstrate how to find all the angles of a right triangle if we are given one side and either an acute angle or another side. We will use several facts from geometry to do this. Therefore, we start with a geometry review.

Geometry Reminder

Parallel Lines

Two lines in the same plane are parallel if they do not intersect. A line crossing two parallel lines is called a transversal.

□ Figure 2.30

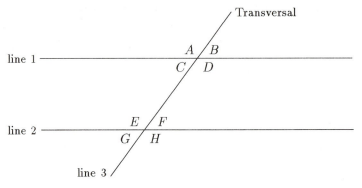

In the figure above: Line 1 is parallel to line 2

Line 3 is a tranversal

Any two angles in the same position relative to the parallel lines are called corresponding angles.

Corresponding Angles

□ Figure 2.31

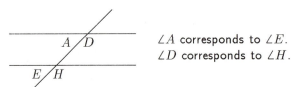

∠A corresponds to ∠E.
∠D corresponds to ∠H.

Theorem from Geometry

If two parallel lines are cut by a transversal, corresponding angles are equal.

Two non-adjacent angles on opposite sides of the transversal and between the lines are called alternate interior angles.

Alternate Interior Angles

□ Figure 2.32

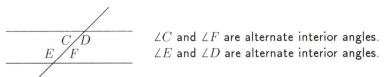

∠C and ∠F are alternate interior angles.
∠E and ∠D are alternate interior angles.

Theorem from Geometry

If two parallel lines are cut by a transversal, alternate interior angles are equal.

Two non-adjacent angles on opposite sides of the transversal and outside
the parallel lines are called alternate exterior angles.

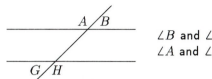

∠B and ∠G are alternate exterior angles.
∠A and ∠H are alternate exterior angles.

Theorem from Geometry
If two parallel lines are cut by a transversal, alternate exterior angles are equal.

Interior Angles of a Triangle

Angles may be designated either by the letter associated with their vertices, A,
B, C or by the Greek letters α (Alpha), β (Beta), γ (Gamma).
The angles of a triangle are always positive.

Theorem from Geometry

The sum of the interior angles of a triangle is 180°.

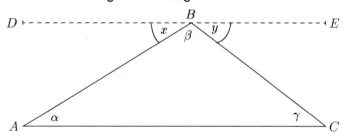

To prove this statement: Draw line DE parallel to side AC of the triangle
above.

$$\angle x = \angle \alpha \qquad \text{Alternate interior angles are equal.}$$
$$\angle y = \angle \gamma \qquad \text{Alternate interior angles are equal.}$$

Then $\angle x + \angle \beta + \angle y = 180°$ The sum of the angles in a straight angle is 180°.
$\angle \alpha + \angle \beta + \angle \gamma = 180°$ Substitution.

In a right triangle one angle is 90°. This means that the sum of the other
two angles must be 90°.

Example 17 □ One angle of a right triangle is 53°. Find the other angle.

Because this is a right triangle, the sum of the two acute angles must be 90°. Therefore, the remaining angle is $90° - 53° = 37°$. □

The Solution of a Triangle

Only six measurements completely describe a triangle. These six measurements are the three sides and the three angles. In a right triangle, if we know one side and either an acute angle or another side, we can find the other three pieces of information.

When all six pieces of information are known the triangle is "solved."

Example 18 □ Solve the right triangle ABC if $a = 10$ and $\beta = 30°$.

First draw a sketch:

□ Figure 2.35

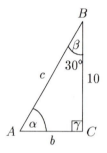

How do you know which way the triangle points?

I don't. I do know C is the right angle and the side opposite the right angle is the hypotenuse c. I also know that side a is opposite $\angle \alpha$ and side b is opposite $\angle \beta$.

First find α:

$$\begin{aligned} \alpha + \beta &= 90° \\ \alpha + 30° &= 90° \\ \alpha &= 60° \end{aligned}$$

To find side c

$$\begin{aligned} \sin \alpha &= \frac{\text{opposite}}{\text{hypotenuse}} \\ \sin 60° &= \frac{10}{c} \\ c &= \frac{10}{\sin 60°} \\ c &\approx 11.55 \end{aligned}$$

To find side b

$$\begin{aligned} \tan \alpha &= \frac{\text{opposite}}{\text{adjacent}} \\ \tan 60° &= \frac{10}{b} \\ b &= \frac{10}{\tan 60°} \\ b &\approx 5.77 \end{aligned}$$

□

Why not use side c and $\cos \alpha$ to find side b?

You could, but the value of side c includes some computation error. If we use it to find side b our calculation error is compounded.

Chapter 2 The Trigonometric Functions

Example 19 □ Solve the right triangle where $a = 10$ and $b = 21$.

Draw a figure.

□ Figure 2.36

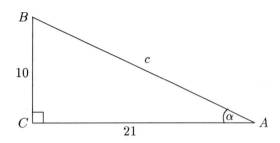

We could use the Pythagorean Theorem to find side c; however we will find α first.

$$\tan \alpha = \frac{\text{opposite side}}{\text{adjacent side}}$$
$$\tan \alpha = \frac{10}{21}$$
$$\alpha \approx 25.46°$$

Note: In this example, we will work all steps to the full accuracy of the calculator using the memory of the calculator. However, we will show only the first four places so you will know you are pressing the right keys.

Now, because the sum of the measures of the two acute angles of a right triangle is $90°$, we can find β.

$$\beta = 90° - \alpha$$
$$\approx 90° - 25.46°$$
$$\beta \approx 64.53°$$

Side c is part of the definition of both $\sin \alpha$ and $\cos \alpha$. Either can be used. We'll use $\sin \alpha$.

$$\sin \alpha = \frac{\text{opposite side}}{\text{hypotenuse}}$$
$$\sin 25.46° = \frac{10}{c}$$
$$c = \frac{10}{\sin 25.46°}$$
$$c \approx 23.25$$

We used the full accuracy of the calculator to avoid introducing rounding errors but, since the sides were given to two significant figures, we will round all our final answers to two significant figures.

We now know all six measurements which describe the triangle.

Sides	$a = 10$	$b = 21$	$c = 23$
Angles	$\alpha = 25°$	$\beta = 65°$	$\gamma = 90°$

□

Problem Set 2.4

Solve the following right triangles. In each case $\gamma = 90°$.

1. $a = 3.5$, $b = 2.6$ 2. $\alpha = 26°$, $b = 25$

3. $\beta = 12.4°$, $c = 72.4$ 4. $a = 57$, $c = 75$

5. $\alpha = 38°$, $a = 19$ 6. $\beta = 48.4°$, $a = 1.56$

7. $b = 9.34$, $c = 12.6$ 8. $\beta = 42.63°$, $b = 72.68$

9. $\alpha = 69.4°$, $c = 258$ 10. $a = 4.05$, $b = 6.13$

11. $a = 9.06$, $c = 15.76$ 12. $\alpha = 72.24°$, $a = 22.76$

13. If a leg of a right triangle is 56.76 m and the angle adjacent to that side is 56.76°, find the other parts of the triangle.

14. If an angle of a right triangle is 49.1° and the opposite side is 17.9 inches, find the other parts of the triangle.

2.5 Applications of Right Triangle Trigonometry

Directions Measured From the North-South Line

To measure an angle you need a reference line. On the ocean or in the middle of a forest there are no lines on the ground. To solve this problem, surveyors and sailors measure direction as an acute angle from the north-south line which they can establish with a magnetic compass.

N15°W means from north go 15° to the west. By general agreement, maps and charts are always drawn with north at the top of the page.

□ Figure 2.37

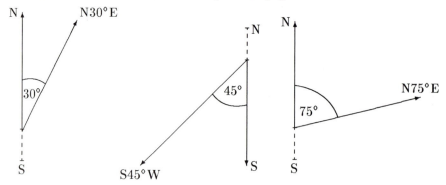

Legal descriptions of irregularly shaped property are given using a system called metes and bounds. This system describes a property starting at a landmark in terms of distances and directions measured from the north-south line until a closed line returning to the starting point is defined.

                   Chapter 2 The Trigonometric Functions

Example 20 □ A triangular property might be described this way:

> Beginning at the NE corner of 4th Street and Maple Avenue, proceed 80 feet along a line due east. Thence north 35° west to Maple Ave...

If 4th Street and Maple Avenue intersect at 90°, what are the outside measurements of this piece of property?

□ Figure 2.38

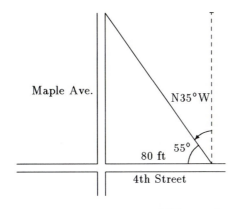

Because $90° - 35° = 55°$, N35°W means the interior angle of the triangle is 55°. To find the other sides of the triangle, we use the corresponding trigonometric functions of 55°.

$$\tan 55° = \frac{\left(\begin{array}{c}\text{distance along}\\\text{Maple Ave.}\end{array}\right)}{80 \text{ ft}}$$

$$\begin{array}{c}\text{distance along}\\\text{Maple Ave.}\end{array} = (80 \text{ ft})(\tan 55°)$$

$$\approx 114.3 \text{ ft}$$

$$\cos 55° = \frac{80 \text{ ft}}{\text{diagonal distance}}$$

$$\text{diagonal distance} = \frac{80 \text{ft}}{\cos 55°}$$

$$\approx 139.5 \text{ feet}$$

□

Angles of Elevation or Depression

Navigators use an instrument called a sextant. Surveyors use a transit. Both instruments are capable of measuring an angle above or below horizontal.

Definition 2.5 Angle of Elevation or Depression

The angle of elevation or depression of an object is the angle between a horizontal line and the observer's line of sight to the object.

□ Figure 2.39

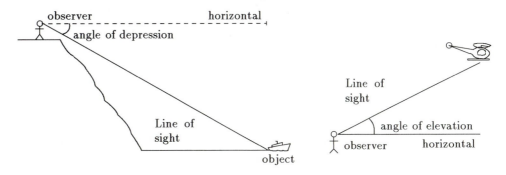

Example 21 ☐ An airplane is flying above the ocean with an altimeter correct to the nearest 100 feet. It indicates an altitude of 10000 feet above the ocean. The pilot observes that the angle of depression to a ship is 20.0°. What is the horizontal distance from the ship to a spot directly below the airplane?

☐ Figure 2.40

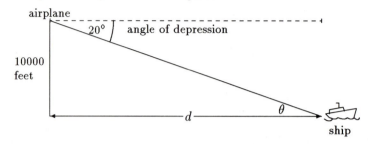

The distance d is parallel to the horizon; therefore, angle $\theta = 20°$ (if 2 parallel lines are cut by a transversal, alternate interior angles are equal).

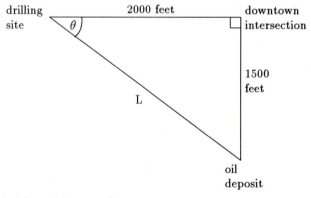

$$\tan \theta = \frac{10000}{d}$$

$$\tan 20° = \frac{10000}{d}$$

$$d = \frac{10000}{\tan 20°}$$

$$d \approx 27500 \text{ ft}$$

Why not use cot θ?

You could, but my calculator has only sin θ, cos θ, and tan θ.

☐

Example 22 ☐ Frequently oil deposits are located where it is impractical to place a drilling rig. For example, a deposit might be 1500 feet under a busy downtown intersection. To get to the oil a slant well is drilled from some accessible location. If the accessible location is 2000 feet away, at what angle of depression should the well be drilled to hit the oil deposit? How much pipe will be needed to reach the deposit?

☐ Figure 2.41

To find the angle of depression

$$\tan \theta = \frac{1500}{2000}$$

$$= 0.75$$

$$\theta \approx 36.9°$$

The length of the pipe

$$\cos \theta = \frac{2000}{L}$$

$$\cos 36.9° \approx \frac{2000}{L}$$

$$\frac{2000}{\cos 36.9°} \approx L$$

$$2500 \text{ feet} \approx \text{Length of slant well}$$

☐

Chapter 2 The Trigonometric Functions

Example 23 □ A surveyor on a mountain road 524 feet above a lake was asked to calculate the distance across the lake. The surveyor measured the angle of depression to the near and far edges of the lake to be 66° and 27° respectively. Determine the distance across the lake.

□ Figure 2.42

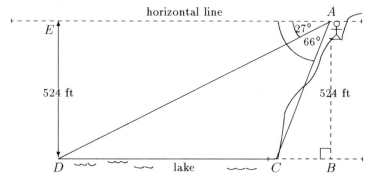

Distance AB was given as 524 feet. Since $\angle EAB$ is a right angle,

$$
\begin{aligned}
\angle CAB &= 90° - 66° \\
&= 24° \\
\tan \angle CAB &= \frac{\text{opposite side}}{\text{adjacent side}} \\
\tan 24° &= \frac{\text{length of } \overline{CB}}{524 \text{ feet}} \\
524 \text{ ft}(\tan 24°) &= \text{length of } \overline{CB} \\
233 \text{ feet} &\approx \text{length of } \overline{CB}
\end{aligned}
$$

$ABDE$ is a rectangle. Since opposite sides of a rectangle are equal, if the length of $\overline{AE}$ can be found, the length of $\overline{DB}$ also will be known.

Triangle AED can be used to find the length of $\overline{AE}$.

$$
\begin{aligned}
\tan \angle EAD &= \frac{\text{opposite side}}{\text{adjacent side}} \\
\tan 27° &= \frac{524 \text{ feet}}{\text{Length of } \overline{AE}} \\
\text{length of } \overline{AE} &= \frac{524 \text{ feet}}{\tan 27°} \\
\text{length of } \overline{AE} &\approx 1028 \text{ feet}
\end{aligned}
$$

Length of $\overline{DB}$ is equal to length of $\overline{AE}$. The distance across the lake is the difference between length of $\overline{DB}$ and length of $\overline{CB}$.

$$
\begin{aligned}
\text{Distance across the lake} &= 1028 \text{ ft} - 233 \text{ ft} \\
&= 795 \text{ ft}
\end{aligned}
$$

□

Example 24 □ To determine the height of a mountain, a surveyor measured the angle of elevation to the top of the mountain from two points 500 feet apart. The two angles of elevation were 35.0° and 40.0°. To the nearest 10 feet, how high is the mountain?

□ Figure 2.43

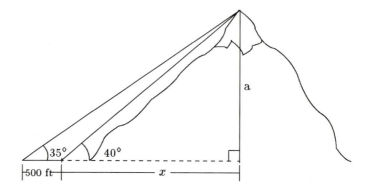

To solve this problem using right triangles we need to work two equations in two unknowns.

<table>
<tr><td align="center">Equation 1</td><td align="center">Equation 2</td></tr>
</table>

$$\tan 35° = \frac{a}{500 + x}$$
$$\tan 35°(500 + x) = a$$

$$\tan 40° = \frac{a}{x}$$
$$(\tan 40°)\,x = a$$

$$\tan 35°(500 + x) = (\tan 40°)\,x \longleftarrow$$
$$350.1 \approx 0.8391x - 0.7002x$$
$$350.1 \approx 0.1389x$$
$$2521 \approx x$$

To find a, substitute for x in equation 2.

$$0.8391x \approx a$$
$$0.8391(2521) \approx a$$
$$2120 \text{ ft} \approx a$$

□

Example 25 □ Two observers on the ground are located one mile apart. They observe a blimp directly above the road connecting them. One measures the angle of elevation to the blimp to be 53.00°; the other measures the angle of elevation to the blimp as 22.00°. Find the height of the blimp above the ground.

□ Figure 2.44

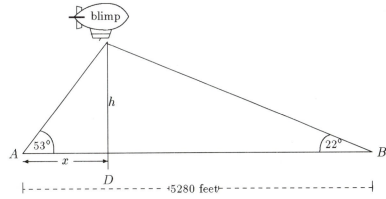

We have to find h, which is a line from the blimp perpendicular to the ground. If length $AD = x$, then length DB = _____ .

5280 − x

Equation 1 | **Equation 2**

$$\tan 53° = \frac{h}{x} \qquad\qquad \tan 22° = \frac{h}{5280 - x}$$

Substituting for h in equation 2.

$$\tan 22° \approx \frac{1.3720x}{5280 - x}$$

$$x \tan 53° = h \qquad\qquad 0.4040 \approx \frac{1.3270x}{5280 - x}$$

$$x(1.3720) \approx h$$

$$0.4040(5280 - x) \approx 1.3270x$$

$$0.4040(5280) - 0.4040x \approx 1.3270x$$

$$2133 \approx 1.3270x + 0.4040x$$

$$2133 \approx 1.731x$$

$$1232 \approx x$$

Now that x is known, we can substitute in equation 1 to find h.

$$x \tan 53° = h$$

$$1232(1.3270) \approx h$$

$$1635 \text{ ft} \approx h$$

□

Group Writing Activity

If the blimp is not directly over a straight line between the two observers, what is the minimum information you need to know to solve the problem and how would you do it?

Help with Problem Solving

What Would I Like to Know?

After you have identified what you know about a problem, another aid to problem solving is to identify what you would like to know. This question is usually explicit in the problem.

> For example: How old is Mary?
> How far did he travel?
> What is the angle of elevation?

But frequently there are other intermediate questions that lead to the final answer:
How is Mary's age related to John's age?
What determines how far he traveled?
Is there a formula for distance traveled?
How fast is he traveling?
How long did he travel?
Can I make the desired angle part of a triangle?
Did I know a relationship between the desired angle and a known angle or side?

Often you can find the answer to a question by answering easier questions that tell you what you need to know.

Problem Set 2.5

Solve the following problems. Make a diagram and label the given information. (If you make your diagram reasonably to scale, it will help you visualize the problem.)

1. A flagpole casts a 42.0-foot shadow. If the angle of elevation of the sun is 57.4°, what is the height of the flagpole to the nearest tenth of a foot?

2. A 37.0-foot tree casts a 27.0-foot shadow. What is the angle of elevation of the sun?

3. An airplane is flying at 4000 feet above the ground. If the angle of depression from the airplane to the beginning of the runway is 5.4°, what is the horizontal distance to the nearest tenth of a mile of the airplane to the beginning of the runway?

4. A balloon flying at 3000 feet above the ocean measures the angles of depression of each end of an island directly in front of the balloon to be 75.8° and 15.6°. What is the length of the island to the nearest tenth of a mile?

5. To measure the height of a tree, a 35-year-old surveyor walked a short distance from the tree and found the angle of elevation to be 43.9°, then walked 20 feet farther and found the angle of elevation to be 37.6°. Find the height of the tree to the nearest foot.

6. A 110-lb biologist standing on a cliff 320 feet above the ocean observed a whale surfacing at an angle of depression of 16.5°. The next time the whale surfaced, the angle of depression was 10.5°. How far, to the nearest foot, did the whale swim between sightings?

7. Two observers on the same straight road, 7220 feet apart, observe a balloonist over a line directly between them. Their measure of the angles of elevation at 3:30 PM are 35.6° and 58.2° respectively. Within 10 feet, what is the height of the balloonist above the ground?

8. An observer stands 120 feet away from a church and measures the angles of elevation of the top and bottom of the steeple to be 24.4° and 18.2° respectively. What is the height of the steeple to the nearest foot?

9. A 40-foot flagpole stands on the top of a hill. To measure the height of the hill, a surveyor standing at the base of the hill finds the measure of the angle of elevation of the top and bottom of the flagpole to be 49.7° and 39.3° respectively. Find the height of the hill to the nearest foot.

10. A blimp is flying parallel to a road and is 2460 feet directly above it. The angles of depression of two parked cars on the road are 34.9° and 26.5°. To the nearest foot, how far apart are the parked cars? (NOTE: The blimp is above a line between the parked cars.)

11. A fly, buzzing about the room, calculates that the angle of depression of the base of an 8-foot wall is 33.7° and the angle of elevation of the top of the wall is 12.5°. Find to the nearest foot the horizontal distance the fly is from the wall.

12. A group of Boy Scouts on a straight trail headed N30.6°E found the trail led through a briar patch. They chose to walk 156 meters due east along the south edge of the briar patch, then due north along the east edge of the patch back to the trail. How much farther did they walk to avoid walking through the briar patch? (Calculate to the nearest meter.)

13. In the San Gabriel Mountains in California, lookout station Running Deer is 18.0 kilometers due west of station Lazy Bear. The bearing from Running Deer to a fire directly south of Lazy Bear is S38.4°E. To the nearest tenth of a kilometer, how far is the fire from Running Deer? From Lazy Bear?

14. A ship is 15 miles due west of lighthouse A on the island of Hawaii. If lighthouse B is 6.2 miles due south of lighthouse A, what is the bearing to the nearest tenth of a degree of the ship from lighthouse B? To the nearest tenth of a mile, how far is the ship from lighthouse B?

 Group Writing Activity

Use a piece of graph paper to draw a map of the property described below using the method of metes and bounds.

Starting from the monument (a brass stake placed in the ground by a surveyor) at the NE corner of Washington and Grant, proceed due east 150 ft, then N20°E for 100 ft. At that point, proceed along an arc of radius 206.8 ft centered at the starting point through 35° counter clockwise. Thence 206.8 feet S28°W to the starting point.

How many square feet of land are bounded by this plot of property?

2.6   Circular Functions

We have defined the trigonometric functions in terms of a point (x, y) on the terminal side of an angle at a distance r from its vertex. This definition of the trigonometric functions inputs an angle measured in degrees or radians and outputs a real number. We can extend this concept to produce a definition of the trigonometric functions that accepts a real number as its input, or domain, and outputs a real number.

One way to define trigonometric functions with a domain of real numbers is as follows.

On a circle with radius r and central angle θ, if we let the length of r equal 1 and let θ assume all values, then the point (x, y) traces a unit circle.

□ Figure 2.45

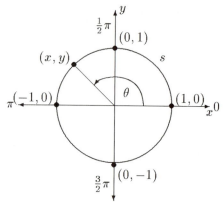

Point (x, y) on the terminal
side of an angle θ.

A unit circle is a circle with a radius
equal to one unit centered at the origin.

It is possible to locate point (x, y) on a unit circle using either the central angle θ or the arc length s measured clockwise from the x-axis. If we use the arc length s to locate a point (x, y) on the unit circle, bearing in mind that $r = 1$, we can make the following definitions of the circular functions.

Definition of the Circular Functions

On a unit circle with $r = 1$, each length of arc starting at the x-axis corresponds to a point (x, y) on the circle. Using an arc of length s to locate a point (x, y), we define the following functions:

$$\sin s = y \qquad \cos s = x \qquad \tan s = \frac{y}{x}$$

$$\csc s = \frac{1}{y} \qquad \sec s = \frac{1}{x} \qquad \cot s = \frac{x}{y}$$

The number s is the length of an arc on a unit circle from the x-axis to (x, y). Because $r = 1$ and $s = r\theta$, this length is exactly equal to the measure of the central angle it intercepts, in radians. Therefore, these definitions of the circular functions correspond exactly to the definitions of the trigonometric functions as angles measured in radians.

If the central angle θ intersects the unit circle at the point (x, y), and s is the arc length from the point $(1, 0)$ to (x, y), then:

$$\sin \theta = \frac{y}{r} = \frac{y}{1} = \sin s$$

$$\cos \theta = \frac{x}{r} = \frac{x}{1} = \cos s$$

$$\tan \theta = \frac{y}{x} = \tan s$$

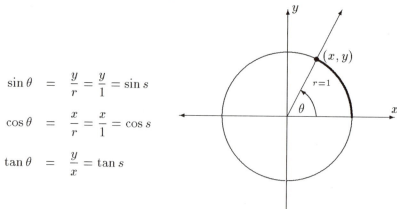

□ Figure 2.46

Chapter 2 The Trigonometric Functions

Similarly:

$$\sec \theta = \sec s, \qquad \csc \theta = \csc s, \qquad \cot \theta = \cot s$$

This means that every relationship between trigonometric functions of a central angle in this book is also true for circular functions of a real number s.

Geometry Reminder

Symmetry of a Circle

Two points with the same y-coordinate are said to be symmetrical about the y-axis if their x-coordinates have the same absolute value.

□ Figure 2.47

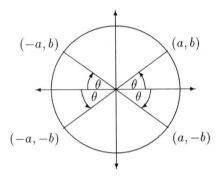

(a, b) and $(-a, b)$ are symmetrical about the y-axis.

(a, b) and $(a, -b)$ are symmetrical about the x-axis.

By using congruent triangles, it is possible to prove that the coordinates of the four points pictured differ only in sign.

Here is an example of how we can use the symmetry of the circle to find the values of the circular functions when s is in the second quadrant.

Example 26 □ Use the symmetry of the circle to find the coordinates of the points for $s = \dfrac{3}{4}\pi, \dfrac{5}{4}\pi, \dfrac{7}{4}\pi$.

□ Figure 2.48

$\left(-\dfrac{\sqrt{2}}{2}, \dfrac{\sqrt{2}}{2}\right)$

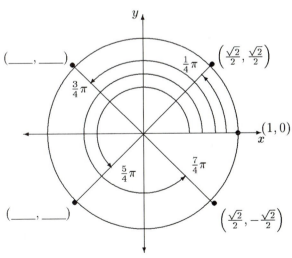

$\left(-\dfrac{\sqrt{2}}{2}, -\dfrac{\sqrt{2}}{2}\right)$

□

2.6 Circular Functions

65

Using a calculator or using the special angles in section 2.1 and the fact that the central angle in radians of a unit circle is equal to the arc length, we can develop the chart below.

Circular Functions of Common Numbers

(x, y)	s	$\cos s$	$\sin s$	$\tan s$
$(1, 0)$	0	1	0	0
$\left(\dfrac{\sqrt{3}}{2}, \dfrac{1}{2}\right)$	$\dfrac{1}{6}\pi$	$\dfrac{\sqrt{3}}{2}$	$\dfrac{1}{2}$	$\dfrac{\sqrt{3}}{3}$
$\left(\dfrac{\sqrt{2}}{2}, \dfrac{\sqrt{2}}{2}\right)$	$\dfrac{1}{4}\pi$	$\dfrac{\sqrt{2}}{2}$	$\dfrac{\sqrt{2}}{2}$	1
$\left(\dfrac{1}{2}, \dfrac{\sqrt{3}}{2}\right)$	$\dfrac{1}{3}\pi$	$\dfrac{1}{2}$	$\dfrac{\sqrt{3}}{2}$	$\sqrt{3}$
$(0, 1)$	$\dfrac{1}{2}\pi$	0	1	Undefined
$(-1, 0)$	π	-1	0	0

Pointer for Less Memory Work

Using your knowledge of the symmetry of a circle, you only need to know the coordinates of two points on a unit circle to write the coordinates of twenty different points of a unit circle.

The x and y coordinates of the point on a unit circle at $\dfrac{1}{4}\pi$ are equal. Therefore (x, y) at $\dfrac{1}{4}\pi$ can be thought of as (a, a).

Using the equation of a circle

$$
\begin{aligned}
x^2 + y^2 &= r^2 \\
\text{becomes} \quad a^2 + a^2 &= 1^2 \\
2a^2 &= 1 \\
a &= \dfrac{1}{\sqrt{2}} \\
a &= \dfrac{\sqrt{2}}{2} \qquad \text{after rationalizing}
\end{aligned}
$$

Hence the coordinates of the point at $s = \dfrac{1}{4}\pi$ are $\left(\dfrac{\sqrt{2}}{2}, \dfrac{\sqrt{2}}{2}\right)$.

Chapter 2 The Trigonometric Functions

Using symmetry and $\left(\dfrac{\sqrt{2}}{2}, \dfrac{\sqrt{2}}{2} \right)$ as coordinates of a point at $s = \dfrac{1}{4}\pi$, we can divide the circle into 8 equal arcs.

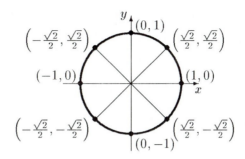

Unit Circle Divided into 8 Equal Arcs

At $\dfrac{1}{6}\pi$ the y coordinate is $\dfrac{1}{2}$ of r.

$$
\begin{aligned}
x^2 + y^2 &= r^2 & \text{equation of a circle} \\
x^2 + \left(\frac{1}{2}\right)^2 &= 1^2 & \text{substituting} \\
x^2 &= \frac{3}{4} \\
x &= \frac{\sqrt{3}}{2}
\end{aligned}
$$

Which means the coordinates for a point at $s = \dfrac{1}{6}\pi$ are $\left(\dfrac{\sqrt{3}}{2}, \dfrac{1}{2} \right)$.

Using symmetry and knowing the coordinates of the point at $s = \dfrac{1}{6}\pi$ are $\left(\dfrac{\sqrt{3}}{2}, \dfrac{1}{2} \right)$, we can divide the circle into 12 equal arcs.

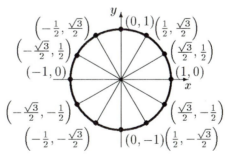

Unit Circle Divided into 12 Equal Arcs

Values of the circular functions for multiples of these numbers can be found by using the symmetry of the unit circle.

Example 27 □ Find $\sin\dfrac{7}{6}\pi$, $\cos\dfrac{7}{6}\pi$, and $\tan\dfrac{7}{6}\pi$.

At $s=\dfrac{7}{6}\pi$ the coordinates of (x,y) are symmetrical to the coordinates of the point at $s=\dfrac{1}{6}\pi$ which are $\left(\dfrac{\sqrt{3}}{2},\dfrac{1}{2}\right)$. The coordinates of the point at $s=\dfrac{7}{6}\pi$ are $\left(-\dfrac{\sqrt{3}}{2},-\dfrac{1}{2}\right)$.

□ Figure 2.51

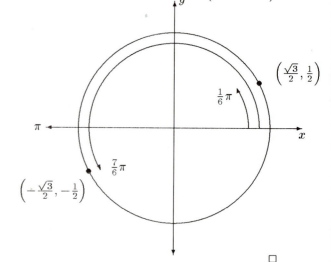

Therefore:

$$\sin\frac{7}{6}\pi \;=\; -\frac{1}{2}$$

$$\cos\frac{7}{6}\pi \;=\; -\frac{\sqrt{3}}{2}$$

$$\tan\frac{7}{6}\pi \;=\; \frac{-\frac{1}{2}}{-\frac{\sqrt{3}}{2}}$$

$$\;=\; \frac{\sqrt{3}}{3}$$

□

Example 28 □ Find all positive and negative numbers in the interval $-2\pi < s < 2\pi$ with coordinates $\left(-\dfrac{\sqrt{3}}{2},\dfrac{1}{2}\right)$ on the unit circle.

From the pointer on pages 66–67 we see that $\dfrac{1}{6}\pi$ maps to $\left(\dfrac{\sqrt{3}}{2},\dfrac{1}{2}\right)$. The point $\left(-\dfrac{\sqrt{3}}{2},\dfrac{1}{2}\right)$ is symmetric to this point. By inspection:

□ Figure 2.52

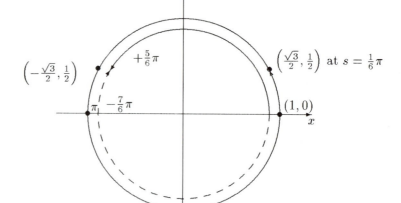

Value of s in positive direction is $\dfrac{5}{6}\pi$

Value of s in negative direction is $-\dfrac{7}{6}\pi$

□

Chapter 2 The Trigonometric Functions

Example 29 □ Find the values of the six trigonometric functions of s at the point on the unit circle with coordinates $(0.8, -0.6)$.

$$\sin s = y$$
$$= -0.6$$
$$\csc s = \frac{1}{y}$$
$$= \frac{1}{-0.6}$$
$$\approx -1.66$$

$$\cos s = x$$
$$= 0.8$$
$$\sec s = \frac{1}{x}$$
$$= \frac{1}{0.8}$$
$$= 1.25$$

$$\tan s = \frac{y}{x}$$
$$= \frac{-0.6}{0.8}$$
$$= -0.75$$
$$\cot s = \frac{x}{y}$$
$$= \frac{0.8}{-0.6}$$
$$\approx -1.33$$

□

Example 30 □ Given that $\sin \frac{1}{6}\pi = \frac{1}{2}$, find $\cos \frac{1}{6}\pi$.

The circular functions are defined on a unit circle whose equation is $x^2 + y^2 = 1$. Because $\cos s = x$ and $\sin s = y$, we can rewrite the equation of the unit circle as:

$$\cos^2 s + \sin^2 s = 1$$

> Note: $\cos^2 s$ means $(\cos s)(\cos s)$.

To find $\cos \frac{1}{6}\pi$ from $\sin \frac{1}{6}\pi$, we substitute $\frac{1}{6}\pi$ for s.

$$\cos^2 \frac{1}{6}\pi + \sin^2 \frac{1}{6}\pi = 1$$
$$\cos^2 \frac{1}{6}\pi + \left(\frac{1}{2}\right)^2 = 1$$
$$\cos^2 \frac{1}{6}\pi = \frac{3}{4}$$
$$\cos \frac{1}{6}\pi \stackrel{?}{=} \pm\frac{\sqrt{3}}{2}$$

> Do I use the positive or negative value for $\pm\frac{\sqrt{3}}{2}$?

> There are two points with $\sin s = \frac{1}{2}$. But at $s = \frac{1}{6}\pi$, $\cos s$ is positive. Therefore, $\cos \frac{1}{6}\pi = +\frac{\sqrt{3}}{2}$.

□

Example 31 □ Using values from the pointer on pages 66–67 and the symmetry of the circle, evaluate $\cos^2 \frac{1}{3}\pi - \sin\left(\frac{3}{2}\pi\right)\left(\sin^2 \frac{1}{3}\pi\right)$.

$$\cos^2 \frac{1}{3}\pi - \left(\sin \frac{3}{2}\pi\right)\left(\sin^2 \frac{1}{3}\pi\right) =$$
$$\left(\frac{1}{2}\right)^2 - (-1)\left(\frac{\sqrt{3}}{2}\right)^2 =$$
$$\frac{1}{4} - (-1)\left(\frac{3}{4}\right) = 1$$

□

2.6 Circular Functions

Example 32 □ If $\sin s = \dfrac{5}{6}$ and $\dfrac{1}{2}\pi < s < \pi$, find $\cos s$.

From Example 30 we know:

$$\cos^2 s + \sin^2 s = 1$$

Substituting:

$$\cos^2 s + \left(\dfrac{5}{6}\right)^2 = 1$$

$$\cos^2 s + \dfrac{25}{36} = 1$$

$$\cos^2 s = \dfrac{11}{36}$$

$$\cos s \overset{?}{=} \pm\dfrac{\sqrt{11}}{6}$$

Because the given domain of s is in the second quadrant, we choose the negative value for $\cos s$.

$$\cos s = -\dfrac{\sqrt{11}}{6}$$

□

Problem Set 2.6

Using the pointer on pages 66–67 and the symmetry of the circle, find each of the following:

1. $\sin \dfrac{3}{4}\pi$, $\cos \dfrac{3}{4}\pi$, and $\tan \dfrac{3}{4}\pi$

2. $\sin \dfrac{5}{6}\pi$, $\cos \dfrac{5}{6}\pi$, and $\tan \dfrac{5}{6}\pi$

3. $\sin \dfrac{5}{3}\pi$, $\cos \dfrac{5}{3}\pi$, and $\tan \dfrac{5}{3}\pi$

4. $\sin \dfrac{7}{4}\pi$, $\cos \dfrac{7}{4}\pi$, and $\tan \dfrac{7}{4}\pi$

5. $\sin \dfrac{4}{3}\pi$, $\cos \dfrac{4}{3}\pi$, and $\tan \dfrac{4}{3}\pi$

6. $\sin \dfrac{1}{2}\pi$, $\sin \pi$, and $\sin \dfrac{3}{2}\pi$

7. $\cos^2 \dfrac{1}{6}\pi - (\cos \pi)\left(\sin^2 \dfrac{1}{6}\pi\right)$

8. $\sin^2 \dfrac{1}{4}\pi - \left(\sin \dfrac{3}{2}\pi\right)\left(\cos^2 \dfrac{1}{4}\pi\right)$

9. $\tan^2 \dfrac{1}{6}\pi \left(\tan \dfrac{1}{4}\pi\right) + \cos^2 \dfrac{1}{4}\pi$

10. $\left(\tan^2 \dfrac{1}{4}\pi\right)\left(\tan^2 \dfrac{1}{6}\pi\right)$

11. $\left(\tan \dfrac{1}{3}\pi\right)\left(\cos \dfrac{1}{6}\pi\right) - \sin \dfrac{1}{6}\pi$

12. $\left(\tan \dfrac{1}{6}\pi\right)\left(\sin \dfrac{1}{3}\pi\right) + \cos^2 \dfrac{1}{3}\pi$

Find the values of all positive and negative arc lengths s in the interval $-2\pi < s < 2\pi$ that correspond to the following ordered pairs on the unit circle.

13. $\left(-\dfrac{1}{2}, -\dfrac{\sqrt{3}}{2}\right)$

14. $\left(\dfrac{\sqrt{2}}{2}, -\dfrac{\sqrt{2}}{2}\right)$

15. $\left(-\dfrac{\sqrt{3}}{2}, -\dfrac{1}{2}\right)$

16. $\left(-\dfrac{\sqrt{2}}{2}, \dfrac{\sqrt{2}}{2}\right)$

17. $(0, -1)$

18. $(-1, 0)$

Find the six trigonometric functions of s at the point on the unit circle with the given coordinates.

19. $\left(-\dfrac{3}{5}, \dfrac{4}{5}\right)$ **20.** $\left(-\dfrac{4}{5}, -\dfrac{3}{5}\right)$ **21.** $\left(-\dfrac{12}{13}, -\dfrac{5}{13}\right)$

22. $\left(\dfrac{5}{13}, -\dfrac{12}{13}\right)$ **23.** $\left(\dfrac{8}{17}, \dfrac{15}{17}\right)$ **24.** $\left(-\dfrac{15}{17}, \dfrac{8}{17}\right)$

Use the identity $\sin^2 s + \cos^2 s = 1$ to solve the following:

25. Given that $\sin \dfrac{1}{3}\pi = \dfrac{\sqrt{3}}{2}$, find $\cos \dfrac{1}{3}\pi$.   **26.** Given that $\cos \dfrac{1}{4}\pi = \dfrac{\sqrt{2}}{2}$, find $\sin \dfrac{1}{4}\pi$.

27. Given that $\sin \dfrac{5}{6}\pi = \dfrac{1}{2}$, find $\cos \dfrac{5}{6}\pi$.   **28.** Given that $\cos \dfrac{7}{6}\pi = -\dfrac{\sqrt{3}}{2}$, find $\sin \dfrac{7}{6}\pi$.

29. Given that $\sin \dfrac{11}{6}\pi = -\dfrac{1}{2}$, find $\tan \dfrac{11}{6}\pi$.   **30.** Given that $\cos \dfrac{5}{3}\pi = \dfrac{1}{2}$, find $\tan \dfrac{5}{3}\pi$.

31. If $\sin s = -\dfrac{1}{3}$ and $\pi < s < \dfrac{3}{2}\pi$, find $\cos s$.   **32.** If $\cos s = \dfrac{3}{4}$ and $\dfrac{3}{2}\pi < s < 2\pi$, find $\sin s$.

33. If $\cos s = -\dfrac{2}{3}$ and $\dfrac{1}{2}\pi < s < \pi$, find $\tan s$.   **34.** If $\sin s = -\dfrac{5}{6}$ and $\pi < s < \dfrac{3}{2}\pi$, find $\tan s$.

Group Writing Activity

The definitions of the circular functions are based upon s being the length of an arc on a unit circle. What are the definitions if a circle other than a unit circle is used? What are the coordinates of a point on a unit circle with a central angle θ, and the coordinates of a point on a circle with radius r with a central angle θ?

Chapter 2	Key Ideas

2.1 The six trigonometric functions are:

$$\sin \theta = \frac{y}{r} \qquad\qquad \csc \theta = \frac{r}{y}$$

$$\cos \theta = \frac{x}{r} \qquad\qquad \sec \theta = \frac{r}{x}$$

$$\tan \theta = \frac{y}{x} \qquad\qquad \cot \theta = \frac{x}{y}$$

2.2 An angle formed by the terminal side and the x-axis is called a reference angle. If you know the coordinates of a point on a circle with a radius of one at $30°$, $45°$, and $60°$, you can give the trigonometric functions of many angles symmetric to these angles.

The coordinates at $30°$ are $\left(\dfrac{\sqrt{3}}{2}, \dfrac{1}{2}\right)$.

The coordinates at $60°$ are the reverse of the coordinates of $30°$, they are $\left(\dfrac{1}{2}, \dfrac{\sqrt{3}}{2}\right)$.

The coordinates at $45°$ are $\left(\dfrac{\sqrt{2}}{2}, \dfrac{\sqrt{2}}{2}\right)$.

2.3 The six trigonometric functions for right triangles are:

$$\sin \theta = \frac{\text{opposite side}}{\text{hypotenuse}} \qquad \csc \theta = \frac{\text{hypotenuse}}{\text{opposite side}}$$

$$\cos \theta = \frac{\text{adjacent side}}{\text{hypotenuse}} \qquad \sec \theta = \frac{\text{hypotenuse}}{\text{adjacent side}}$$

$$\tan \theta = \frac{\text{opposite side}}{\text{adjacent side}} \qquad \cot \theta = \frac{\text{adjacent side}}{\text{opposite side}}$$

The six trigonometric functions for right triangles are applied to many practical situations. Use your calculator for solving right triangles.

2.4 Geometry Reminders

1. If parallel lines are cut by a transversal, the corresponding angles are congruent, alternate interior angles are congruent, and alternate exterior angles are congruent.

2. The sum of the interior angles of a triangle is 180°.

3. Right triangles can be solved using the six trigonometric functions.

4. To solve a right triangle, find all three angles and all three sides of the triangle.

2.5 1. The angle between the observer's line of sight to the object and a horizontal line is called either the angle of elevation or angle of depression.

2. Surveyors and sailors express direction by referring to the angle from a north-south line toward east or west, for example, N30°W or S25°E.

2.6 1. A unit circle is a circle with radius equal to one unit centered at the origin.

2. Symmetry of a circle.

Two points with the same y-coordinates are said to be symmetrical about the y-axis if their x-coordinates have the same absolute value.

3. Circular functions of common numbers $0, \frac{1}{6}\pi, \frac{1}{4}\pi, \frac{1}{3}\pi, \frac{1}{2}\pi, \pi$ on the unit circle can be evaluated using the coordinates of the end of the arc. The common arc lengths and the corresponding coordinates of a point at that arc length on a unit circle are:

Arc length	Coordinates of corresponding points
0	$(1, 0)$
$\frac{1}{6}\pi$	$\left(\frac{\sqrt{3}}{2}, \frac{1}{2}\right)$
$\frac{1}{4}\pi$	$\left(\frac{\sqrt{2}}{2}, \frac{\sqrt{2}}{2}\right)$
$\frac{1}{3}\pi$	$\left(\frac{1}{2}, \frac{\sqrt{3}}{2}\right)$
$\frac{1}{2}\pi$	$(0, 1)$
π	$(-1, 0)$

Chapter 2 Review Test

Find the values of the six trigonometric functions of an angle θ in standard position for the following coordinates of points on the terminal side of the angle. **(2.1)**

1. $(-12, 5)$
2. $(-4, -4)$

Find the values of the remaining trigonometric functions of θ if: **(2.1)**

3. $\sin \theta = -\dfrac{4}{5}$ and $\cos \theta$ is negative
4. $\cos \theta = \dfrac{1}{3}$ and $\tan \theta$ is negative

Find the values of the six trigonometric functions of the following angles given in degrees. **(2.2)**

5. $240°$
6. $315°$

Find values of the six trigonometric functions of the following angles given in radians. **(2.2)**

7. $\dfrac{2}{3}\pi$
8. $\dfrac{15}{4}\pi$

Use your calculator to find the following. Use radian measure. **(2.3)**

9. $\arcsin(0.8291)$
10. $\arcsin(0.7415)$
11. $\arctan(0.2000)$

Solve each of the following right triangles. In each case $\gamma = 90°$. **(2.4)**

12. $a = 4.6$, $\quad b = 6.2$
13. $\alpha = 36.4°$, $\quad a = 0.469$

14. $a = 6.84$, $\quad c = 9.86$
15. $\beta = 42.69°$, $\quad a = 5.648$

16. $b = 92.84$, $\quad c = 128.4$
17. $\alpha = 32.6°$, $\quad b = 9.47$

Solve each of the following:

18. During a severe windstorm, a telephone pole broke several feet above the ground. The top part of the pole bent at the break, with the top touching the ground 48.3 feet from the base of the pole. If the top part of the pole makes an angle of $26.6°$ with the ground, how tall was the original telephone pole to the nearest tenth of a foot? **(2.3)**

19. A scissors has blades that are 8.0 inches from the pivot to the point of the scissors. If the scissors are held open so the points are 3.4 inches apart, what is the angle between the blades to the nearest tenth of a degree? **(2.3)**

20. A boy flying a kite let out 180 feet of string. Assuming that the kite string is straight and that the kite is 70 feet above the ground, to the nearest degree what angle does the kite string make with the ground? **(2.3)**

21. To determine the height of a tree standing on a hillside, a conservationist walked down the slope of the hill 75 feet. From this point he found the angles of elevation of the base and top of the tree to be 26° and 58° respectively. Find the height of the tree to the nearest foot. **(2.5)**

22. Two observers, exactly a mile apart, observe a glider over a line directly between them. At precisely the same time, their measures of the angles of elevation are 48.2° and 36.8° respectively. What is the altitude of the glider (to the nearest foot)? **(2.5)**

23. A surveyor, standing on a 1650-foot cliff, observed two streams of water. While looking directly east, he measured the angles of depression of the nearer stream to be 35.6° and of the farther stream to be 26.7°. To the nearest 10 feet, how far apart were the streams at this particular line of sight? **(2.5)**

24. A fire was observed by two observers in the San Gabriel Mountains. Lookout station A is 15.0 kilometers due east from station B. The bearing from B to the fire directly north of A was N34.2°E. To the nearest 0.1 km how far is the fire from each of the stations? **(2.5)**

Using the pointer on pages 66–67 and the symmetry of the circle, find each of the following. **(2.6)**

25. $\sin \frac{5}{4}\pi$, $\cos \frac{2}{3}\pi$, $\tan \frac{11}{6}\pi$

Find all positive and negative arc lengths s in the interval $-2\pi < x < 2\pi$ that correspond to the following ordered pairs on the unit circle. **(2.6)**

26. $\left(-\frac{\sqrt{2}}{2}, -\frac{\sqrt{2}}{2}\right)$ **27.** $\left(\frac{1}{2}, -\frac{\sqrt{3}}{2}\right)$

Find the values of the six trigonometric functions of s at the point on the unit circle with the given coordinates. **(2.6)**

28. $\left(-\frac{3}{5}, -\frac{4}{5}\right)$ **29.** $\left(-\frac{5}{13}, \frac{12}{13}\right)$

Use the relation $\sin^2 s + \cos^2 s = 1$ to solve the following. **(2.6)**

30. Given that $\sin \frac{4}{3}\pi = -\frac{\sqrt{3}}{2}$, find $\cos \frac{4}{3}\pi$.

31. Given that $\cos \frac{5}{6}\pi = -\frac{\sqrt{3}}{2}$, find $\tan \frac{5}{6}\pi$.

Using the pointer on pages 66–67, evaluate the following. **(2.6)**

32. $\left(\sin \frac{1}{3}\pi\right)\left(\cos \frac{1}{6}\pi\right) - \tan^2 \frac{1}{3}\pi$ **33.** $\left(\cos^2 \frac{1}{2}\pi\right)\left(\tan^2 \frac{1}{4}\pi\right) - \sin^2 \frac{1}{6}\pi$

34. If $\cos s = -\frac{1}{3}$ and $\frac{1}{2}\pi < s < \pi$, find $\sin s$.

Chapter 3

Graphs of the Trigonometric Functions

Contents

Preview

Chapter 2 indicated that triangles are important applications of the trigonometric functions, but triangles are not essential to defining the trigonometric functions. In Chapter 2 we defined these functions in terms of a circle. Defining trigonometric functions in terms of a circle makes them very useful for describing circular motion. The graphs of trigonometric functions give us a picture of circular motion. They can tell us the voltage output of a generator or the position of an overhead cam in an automobile engine.

We will first draw the graphs of generic trigonometric functions; then we will examine how changing the constants of a trigonometric function shifts or expands its graph. The technology of the graphing calculator will be used to explore these changes. Then we will show you a technique called addition of ordinates that lets you draw the graph of a sum of functions from the graphs of its component functions.

We will use the graphing calculator to efficiently implement this idea. We will then turn our attention to the graphs of trigonometric functions where the generic box has no upper or lower bound. However, you will see that the coefficients of tangent, cotangent, secant, and cosecant functions have the same effects as the coefficients of sine and cosine functions, both when graphing by hand or using the graphing calculator.

3.1   Graphing Generic Sine and Cosine Functions

The trigonometric functions are periodic functions; that is, they repeat themselves. Therefore, if we graph each function up to the point where it repeats itself, we will know what shape to repeat for the entire curve. The basic shape of each function is what we call the generic curve.

Just as we graph other functions by plotting points, we can use points to graph the generic curve for a sine function. To get points to plot we use the values we developed in Chapters 1 and 2, along with the symmetry of the circle to find other values.

Here is a plot of $y = \sin x$.

□ Figure 3.1

x	Approximate Value of Sin x
0	0
$\frac{1}{6}\pi$	0.500
$\frac{1}{4}\pi$	0.707
$\frac{1}{3}\pi$	0.866

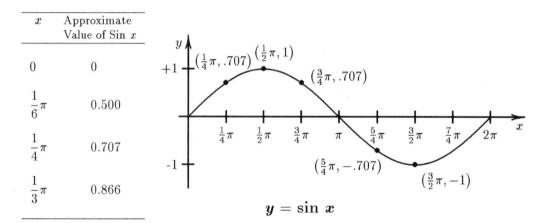

$$y = \sin x$$

Use a calculator to find additional values of $\sin x$.

What about values of x greater than 2π or less than 0?

These are values where the generic sine curve repeats itself. In section 1.1 we defined coterminal angles. Use them to find the sine of an angle outside the domain $0 \leq x < 2\pi$.

Chapter 3 Graphs of the Trigonometric Functions

Using Your Graphing Calculator

To get an appropriate graph on the screen of your graphing calculator, you must set the limits of the screen. The four corners of the screen are defined by $X_{\min}, X_{\max}, Y_{\min}$, and $Y_{\max}$. These values should be selected so that the graph you are working on fits nicely in the screen.

To graph the generic box of $y = \sin x$ make the settings below on your calculator.

Graph   Range

$Y = \sin x$

$$X_{\min} = 0 \qquad Y_{\min} = -1.5$$
$$X_{\max} = 6.28 \qquad Y_{\max} = 1.5$$
$$X_{\text{scl}} = .785 \qquad Y_{\text{scl}} = .5$$

□ Figure 3.2

After you execute the graph you should get this graph on your viewing screen.

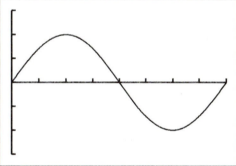

Next we will look at how a graph of $y = \sin x$ repeats itself.

Example 1 □ Find the sine of $\dfrac{13}{4}\pi$.

Because $\dfrac{13}{4}\pi$ is larger than 2π, we first find a coterminal angle less than 2π.

$$\frac{13}{4}\pi - 2\pi = \frac{5}{4}\pi$$

$\dfrac{5}{4}\pi$ is in the 3rd quadrant. It is also symmetrical to $\dfrac{1}{4}\pi$. Drawing a sketch may help you to see that $\sin \dfrac{5}{4}\pi$ is the negative of the $\sin \dfrac{1}{4}\pi$.

$$\sin \frac{13}{4}\pi = -\frac{\sqrt{2}}{2}$$
$$\approx -0.7071 \text{ (to four decimal places)}$$

□

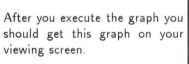

In radian mode, my calculator does that automatically.

Yes. It first reduces any angle to a value $0 \le x < 2\pi$. Then it calculates the $\sin x$.

There are an infinite number of values of x where $\sin x = -\dfrac{\sqrt{2}}{2}$, each 2π apart.

In general, $\sin(x + 2k\pi) = \sin x$.

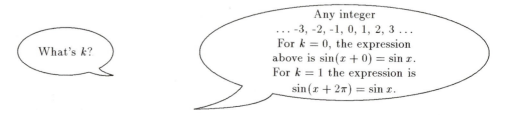

What's k?

Any integer
... -3, -2, -1, 0, 1, 2, 3 ...
For $k = 0$, the expression
above is $\sin(x + 0) = \sin x$.
For $k = 1$ the expression is
$\sin(x + 2\pi) = \sin x$.

Definition 3.1 Period of a Function

$$
\begin{aligned}
f(x) &= f(x + kp) \\
\text{where } x &= \text{any real number} \\
p &= \text{a specific positive real number} \\
k &= \text{any integer}
\end{aligned}
$$

The period of a periodic function is p.

In other words, the value of $\sin x$ repeats itself every 2π radians or $360°$. Because $\sin x = \sin(x + 2k\pi), \sin x$ is a periodic function with a period of 2π radians or $360°$.

□ Figure 3.3

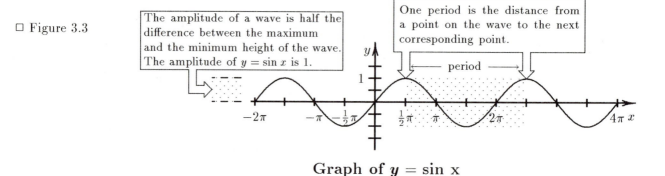

The amplitude of a wave is half the difference between the maximum and the minimum height of the wave. The amplitude of $y = \sin x$ is 1.

One period is the distance from a point on the wave to the next corresponding point.

period

Graph of $y = \sin x$

Notice that x can be any real number, but the value of $\sin x$ is between -1 and $+1$.

Generic Box for a Sine Curve

A sine curve is a series of repetitions of a basic curve. This basic curve, like all generic items, can be enclosed in a simple box. To build the box, locate a few critical points to define the shape of the function. These critical points are the points where the curve crosses the x-axis, the minimum and the maximum values of the function.

Chapter 3 Graphs of the Trigonometric Functions

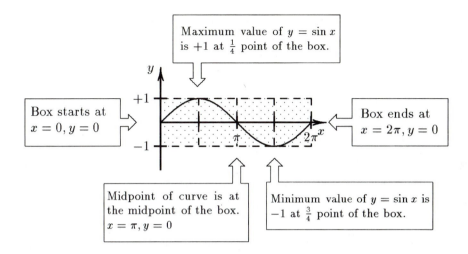

We call the points labeled above critical points.

Each box contains one period of the sine function. We call the curve graphed in the generic box the principal cycle of the function. To graph more than one period, just duplicate the principal cycle and label the critical points.

Generic Box for a Cosine Curve

If we set r equal to one unit, $\cos x$ is given by the x-coordinate of a point on the unit circle. Plotting these values yields:

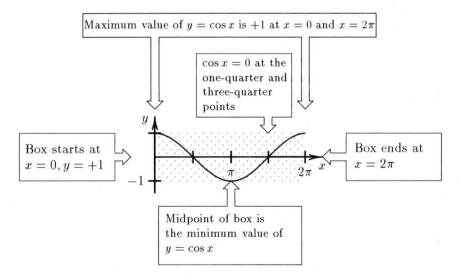

Repeating the generic cosine curve or principal cycle gives the graph of as many periods as we want. The actual graph of $y = \cos x$ contains all of the cycles.

3.1 Graphing Generic Sine and Cosine Functions

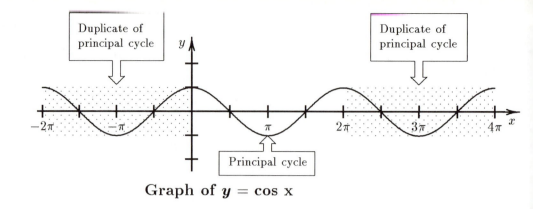

Duplicate of principal cycle

Principal cycle

Graph of $y = \cos x$

Graphing Trigonometric Functions in Engineering and Technology

When graphing functions, we generally use the real numbers as domain and range. For the trigonometric functions this includes the use of radians for the domain.

In fields of technology, some functions relate quantities of different units of measurement. The domain might be time, pressure, date, or degrees. Some examples are distance as a function of time, volume as a function of pressure, or temperature as a function of date. Pistons in automobile engines are connected to crankshafts which turn in circles much like a bicycle pedal or pencil sharpener crank to make the piston go up and down.

Example 2 □ Plot the height of the outer edge of a 5-cm crankshaft above the horizontal as a function of its angle of rotation.

□ Figure 3.7

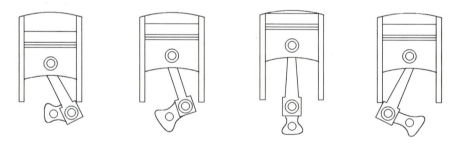

$$h = 5\sin 0° \qquad h = 5\sin 45° \qquad h = 5\sin 90° \qquad h = 5\sin 150°$$
$$= 0 \qquad\quad \approx 3.54 \text{ cm} \qquad = 5 \text{ cm} \qquad\quad = 2.5 \text{ cm}$$

□

If we plot these points and add a few others, we get the following graph.

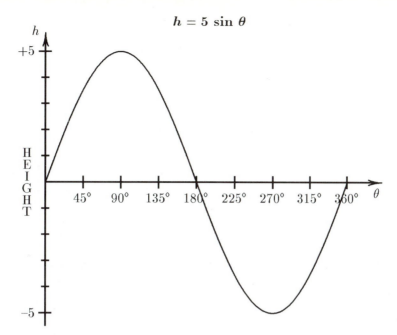

$$h = 5 \sin \theta$$

Graph of Crankshaft Height vs Angle of Rotation

□ Figure 3.8

Changing the Amplitude of a Sine or Cosine Function

Notice what multiplying a basic sine curve by a constant does.

$$y = A \sin x$$

It produces a curve with the same period as $y = \sin x$, but the amplitude of the curve expands or contracts by a factor of A.

□ Figure 3.9

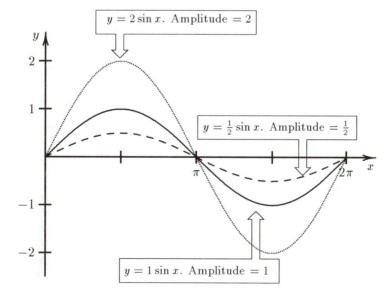

A negative value of A produces the reflection about the x-axis of the same curve with a positive value of A.

Using Your Graphing Calculator

To reproduce the graph above, set the range of values of your graphing calculator to the following:

Graph		Range	
$Y_1 = \sin x$	$X_{\min} = 0$	$Y_{\min} = -2.1$	
$Y_2 = 2\sin x$	$X_{\max} = 6.28$	$Y_{\max} = 2.1$	
$Y_3 = .5\sin x$	$X_{\text{scl}} = .785$	$Y_{\text{scl}} = .5$	

□ Figure 3.10

Execute the graph.
You should get this graph on
your viewing screen.

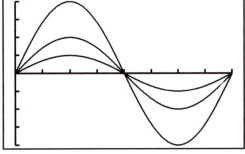

Next use your graphing calculator with the same range to plot $y = 2\sin x$ and $y = -2\sin x$ on the same screen.

□ Figure 3.11

After executing the graph
you should get this graph on
your viewing screen.

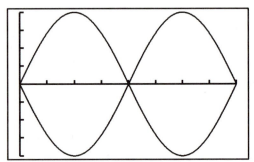

What happened? The graph of $y = -2\sin x$ is a reflection of $y = 2\sin x$ about the x-axis.

The effect of A on the graph of $y = A \cos x$ is illustrated in the graph below.

□ Figure 3.12

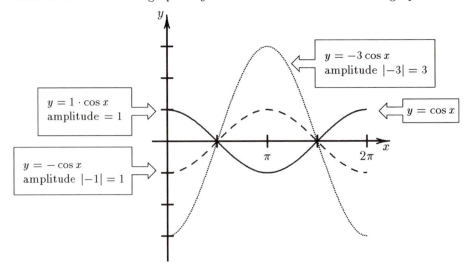

$y = -3\cos x$
amplitude $|-3| = 3$

$y = 1 \cdot \cos x$
amplitude $= 1$

$y = \cos x$

$y = -\cos x$
amplitude $|-1| = 1$

The Effect of A on the Generic Sine Curve Box

□ Figure 3.13

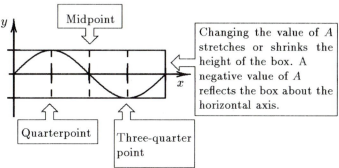

Changing the value of A stretches or shrinks the height of the box. A negative value of A reflects the box about the horizontal axis.

Changing the value of A does not affect the locations of the end, mid-, or quarter points of the generic box. Next we will examine how to stretch or shrink the distance between the critical points.

The Effect of B on Generic Sine and Cosine Curves

One more step toward generalization of the graph of $y = \sin x$ is to consider the effect of multiplying x, the argument of the function, by a constant. We have examined $y = A \sin Bx$. For $B = 1$, this curve has a period of 2π or $360°$.

Let's make a table of values for $y = \sin 2x$; in this case $B = 2$ and $A = 1$.

x	0	$\frac{1}{6}\pi$	$\frac{1}{4}\pi$	$\frac{1}{3}\pi$	$\frac{1}{2}\pi$	$\frac{2}{3}\pi$	$\frac{3}{4}\pi$	$\frac{5}{6}\pi$	π
$2x$	0	$\frac{1}{3}\pi$	$\frac{1}{2}\pi$	$\frac{2}{3}\pi$	π	$\frac{4}{3}\pi$	$\frac{3}{2}\pi$	$\frac{5}{3}\pi$	2π
$\sin 2x$	0	0.86	1	0.86	0	-0.86	-1	-0.86	0

While x has gone through only half a period, $2x$ has gone through a complete period.

□ Figure 3.14

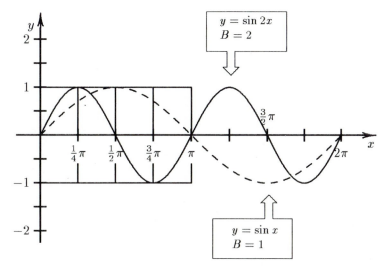

$y = \sin 2x$
$B = 2$

$y = \sin x$
$B = 1$

Setting $B = 2$ shrinks the width of the box from 2π to π; hence two periods will now fit between zero and 2π.

Changing the value of B will stretch or shrink the generic curve box along the x-axis. This will increase or decrease the period.

Consider what happens when $B = \dfrac{1}{3}$.

Example 3 □ Graph $y = 2\sin\dfrac{1}{3}x$

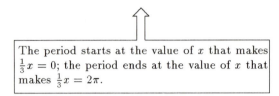

The period starts at the value of x that makes $\frac{1}{3}x = 0$; the period ends at the value of x that makes $\frac{1}{3}x = 2\pi$.

To find the starting point or left edge of the box, set the argument of the sine function to zero.

$$\frac{1}{3}x = 0$$
$$x = 0 \quad \Leftarrow \boxed{\text{Left side of box}}$$

To find the right edge of the box, set the argument of the sine function to 2π.

$$\frac{1}{3}x = 2\pi$$
$$x = 6\pi \quad \Leftarrow \boxed{\text{Right side of the box}}$$

Now we can sketch $y = 2\sin\dfrac{1}{3}x$. Next divide the box into 4 quarters each $\dfrac{6}{4}\pi = \dfrac{3}{2}\pi$ wide and draw the graph.

□ Figure 3.15

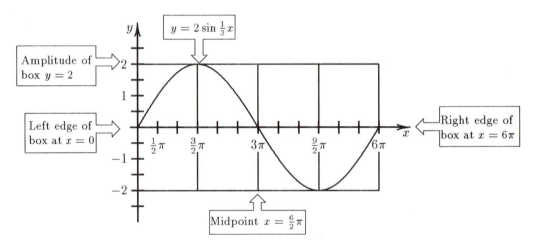

This time, with $B = \dfrac{1}{3}$, instead of taking 2π to complete one period in the generic box, it took 6π to complete a period. Another way to say this is that only $\dfrac{1}{3}$ of a period fit between zero and 2π.

In effect, B tells us how many periods will fit in an interval of 2π.

□

Chapter 3 Graphs of the Trigonometric Functions

$$B = \frac{2\pi}{\text{(length of period)}} \quad \longleftarrow \boxed{\text{length of generic period}}$$

Therefore for a period of 6π

$$B = \frac{2\pi}{6\pi}$$
$$B = \frac{1}{3}$$

Or if we want a sine function with a period of $\frac{2}{3}\pi$

$$B = \frac{2\pi \quad \longleftarrow \boxed{\text{length of generic period}}}{\frac{2}{3}\pi \quad \longleftarrow \boxed{\text{length of desired period}}}$$
$$B = 3$$

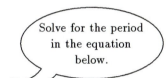

If I know B, how do I find the period?

Solve for the period in the equation below.

$$B = \frac{2\pi \quad \longleftarrow \boxed{\text{length of generic period}}}{P \quad \longleftarrow \boxed{\text{length of a particular function}}}$$
$$P = \frac{2\pi}{B}$$

Example 4 □ Find the period of $y = 2\cos\frac{3}{4}x$.

This time $B = \frac{3}{4}$.

The period of $y = 2\cos\frac{3}{4}x$ is

$$P = \frac{2\pi}{\frac{3}{4}}$$
$$P = 2\pi \cdot \frac{4}{3}$$
$$P = \frac{8}{3}\pi$$

This function will require an x interval of $\frac{8}{3}\pi$ to

complete one cycle. □

Example 5 □ Given the function $y_1 = \sin x$, find a function y_2 that is 3 times as high as y_1 and has half as many periods between 0 and 4π as y_1.

You must find A and B in the equation $y = A \sin Bx$ so that the above criteria are met.

What do A & B mean again?

$|A|$ is the amplitude. B is the number of generic boxes you can fit between 0 and 2π.

For the given function $y_1 = \sin x$, $A_1 = 1$ which means the amplitude is 1. Also, one generic box fits between zero and 2π because $B_1 = 1$. For the curve to be 3 times as high as y_1, the amplitude of y_1 must be multiplied times 3.

$$A_2 = 3 \cdot 1$$

For half as many periods to occur in the interval 0 to 4π, half as many must also occur from 0 to 2π so

$$B_2 = \frac{1}{2} \cdot 1$$

Hence, the equation is $y_2 = 3 \sin \frac{1}{2}x$. □

 Using Your Graphing Calculator

Plot equations y_1 and y_2 in example 5 on your graphing calculator from 0 to 4π to compare them.

Hints to find an appropriate value for the X_{scl}.
1. If the B coefficient of one function is a multiple of the other, try π divided by 2 times the largest value of B.
2. If the B coefficient of one function is not a multiple of the other, for example $y = \sin 2x$ and $y = \sin 3x$, start with X_{scl} equal to π divided by twice the product of the B coefficients.

□ Figure 3.16

$$y_1 = \sin x$$
$$y_2 = 3 \sin \frac{1}{2}x$$

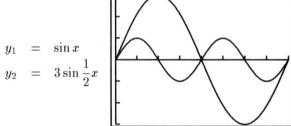

Example 6 □ Which positive x value closest to zero yields the maximum value of the function $y = 2 \sin 3x$?

We know that the maximum value of $y = \sin x$ first occurs at the quarter point of the generic box.

Left Edge	Right Edge	Period
$3x = 0$	$3x = 2\pi$	$\frac{2}{3}\pi - 0$
$x = 0$	$x = \frac{2}{3}\pi$	$\frac{2}{3}\pi$

The quarter point of the graph is one fourth of the way through the period.

$$\text{Quarter point} = \frac{2}{3}\pi \cdot \frac{1}{4}$$
$$= \frac{1}{6}\pi$$

Therefore, the first positive maximum value of $y = 2 \sin 3x$ occurs at $\frac{1}{6}\pi$. □

Using Your Graphing Calculator

You can verify this by using the trace feature of your graphing calculator. By using the arrow buttons, move the trace indicator to the first maximum point to the right of the y-axis. The decimal value of x closely approximates $\frac{1}{6}\pi$.

□ Figure 3.17

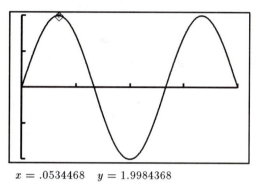

$x = .0534468 \quad y = 1.9984368$

Simple Harmonic Motion

Have you ever noticed when you remove a melon from a grocery scale, the scale springs back beyond the resting point several times until it comes to rest?

□ Figure 3.18

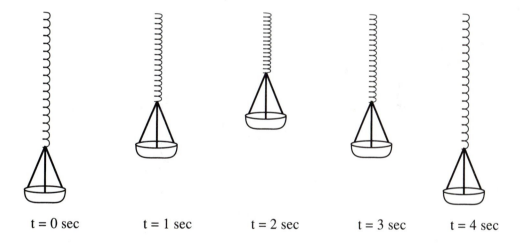

$t = 0$ sec $t = 1$ sec $t = 2$ sec $t = 3$ sec $t = 4$ sec

This motion approximates what is called damped simple harmonic motion.

Some examples of simple harmonic motion are circular in nature such as the pendulum in a grandfather's clock going back and forth. Others are linear in nature such as a child on a pogo stick, a bouncing ball, or a bottle in an ocean wave. These motions can be described by the equations

$$d = A \sin \omega t \quad \text{or} \quad d = A \cos \omega t$$

where d is the displacement from the point of equilibrium.

3.1 Graphing Generic Sine and Cosine Functions

87

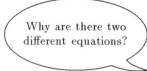

Why are there two different equations?

If the harmonic motion starts at the midline, we usually use the sine function. If it starts in a displaced position, we usually use the cosine function.

Example 7 □ A weight on the end of a spring is pulled to 3 inches below a spring's equilibrium point and then released. The position of the weight with respect to the equilibrium point at time t is given by:

$$y = -3\cos\frac{3\pi}{2}t \qquad\qquad t \text{ in seconds}$$

Find the period, the highest point it will reach, and the frequency (the number of up and down movements in a second).

To find the period, we need to find both sides of the generic box.

Left Edge	Right Edge	Period
$\dfrac{3\pi}{2}t = 0$	$\dfrac{3\pi}{2}t = 2\pi$	$\dfrac{4}{3} - 0$
$t = 0$	$t = \dfrac{4}{3}$	$\dfrac{4}{3}$

One period occurs every $\dfrac{4}{3}$ seconds. The period is the number of seconds per cycle. The number of periods per second is called the frequency. The frequency is the reciprocal of the period.

$$\text{frequency} = \frac{3}{4}\frac{\text{cycles}}{\text{sec}}$$

Because this is a cosine function with a negative value of A, the highest point occurs at the midpoint of each period.

$$\frac{3\pi}{2}t = \pi \qquad \Longleftarrow \boxed{\text{midpoint of generic cosine box at } \theta = 0}$$

$$t = \frac{2}{3}$$

Hence the highest point occurs at $t = \dfrac{2}{3}$ sec. and every $\dfrac{4}{3}$ seconds thereafter. Substituting $t = \dfrac{2}{3}$ in $y = -3\cos\dfrac{3\pi}{2}t$ yields $y = 3$ inches above the resting point for the highest point.

□

Problem Set 3.1

Use your knowledge of symmetry and the coordinates of a point at an angle of $0, \frac{1}{6}\pi, \frac{1}{4}\pi, \frac{1}{3}\pi$, or $\frac{1}{2}\pi$ to provide the following functional values.

1. $\sin \frac{2}{3}\pi$ **2.** $\sin \frac{3}{4}\pi$ **3.** $\cos \frac{4}{3}\pi$ **4.** $\cos \frac{5}{6}\pi$

5. $\sin \frac{5}{3}\pi$ **6.** $\sin \frac{3}{2}\pi$ **7.** $\cos \pi$ **8.** $\cos \frac{7}{6}\pi$

9. $\sin \frac{11}{6}\pi$ **10.** $\cos \frac{2}{3}\pi$ **11.** $\sin \frac{4}{3}\pi$ **12.** $\cos \frac{11}{6}\pi$

Using your knowledge of symmetry, find approximations to two decimal places of $y = \sin x$ and $y = \cos x$ for each multiple of $\frac{1}{6}\pi$ and $\frac{1}{4}\pi$ between 0 and 2π.

Use these values to plot approximations of the principal cycle of the following graphs:

13. $y = \sin x$ **14.** $y = \cos x$

Use your graphing calculator to plot the following curves. Use the given information to set the range of your graphing calculator so the graph fits the viewing screen nicely and the critical points occur at the tick marks. Record these values. List the values you choose for $X_{\min}, X_{\max}, X_{\text{scl}}, Y_{\min}, Y_{\max}$, and Y_{scl}. Compare your choices to the answers in the back. Remember your choices don't have to match ours exactly but they should be reasonably close.

15. $y = \sin x$ Graph one period starting at $x = 0$

16. $y = \cos x$ Graph one period ending at $x = 2\pi$

17. $y = 2\sin x$ Graph two periods ending at $x = 2\pi$

18. $y = 2\cos x$ Graph two periods starting at $x = -2\pi$

19. $y = \sin 2x$ Graph three periods starting at $x = -2\pi$

20. $y = \frac{1}{2}\cos 2x$ Graph three periods ending at $x = 4\pi$

21.   Graph the reflection about the x-axis of the function in problem 19.

22.   Graph the reflection about the x-axis of the function in problem 20 with the addition that the new graph has an amplitude four times that of the graph in problem 20.

For what positive x value closest to zero do the minimums of the following functions occur?

23. $y = 3\sin \frac{1}{3}x$ **24.** $y = \frac{1}{4}\cos 7x$

For the following functions how many periods occur in the given domain?   **25.** $y = \frac{1}{2}\sin 5x$ $3\pi \leq x \leq 7\pi$

What is the length of the period for the following functions?

27. $y = -3\sin \frac{1}{3}x$ **28.** $y = \frac{2}{3}\cos 4x$

29. $y = 3\sin \frac{1}{2}x$ **30.** $y = -\cos 3x$

For the following functions y_1, give the equation of the function y_2 that meets the given criteria:

31. $y_1 = \cos x$ y_2 is twice as high, with the same period as y_1

32. $y_1 = 3\sin 2x$ y_2 is half as high, with twice as many periods between 3π and 8π as y_1

33. The graph of the amplitude of the sound from a tuning fork is a sine curve. The shorter the period, the higher the frequency, producing a higher pitch. The graph that approximates the sound of the tuning fork is given by the equation:

$$y = .06 \sin 512\pi t \qquad t \text{ in seconds}$$

What is the period of this function? What is the frequency? (The frequency is the number of periods per second.) The frequency of middle C is 264 cycles per second. How does the sound of the tuning fork compare to middle C?

34. If the position of a weight on a spring is given by

$$y = 10 \cos \frac{5\pi}{4} t \qquad t \text{ in seconds}$$

Find
a) the period
b) the lowest point it would reach
c) the frequency (the number of up and down movements in a unit of time)

Group Writing Activity

On page 85 we developed the formula $P = \dfrac{2\pi}{B}$ for the length of a period of a sine or cosine function. How could you use this information to graph a quick sketch of two periods of sine or cosine functions given an arbitrary starting point?

3.2 Shifting Generic Curves Right/Left or Up/Down

Phase Shift

Sine and cosine waves are excellent representations of many physical phenomena; for example, electric voltage, electric current, light waves, and water waves. Each of these waves can be shifted forward or backward. For example, the peak value of an electric voltage is shifted as an alternating current is passed through a capacitor. This shift is called phase shift. Mathematically, phase shift is induced by adding a constant to the argument of a periodic function.

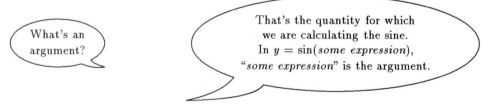

In $y = A \sin (Bx + C)$, the effect of C is to shift the sine wave along the x-axis.

A generic sine wave begins its period when the argument is equal to zero. The wave ends its period when the argument is equal to 2π.

Example 8 □ Graph one period of $y = \sin\left(x + \frac{1}{3}\pi\right)$.

The principal cycle of the graph will not start at $x = 0$. We need to calculate the left and right edges of the generic box.

The left edge is the value of x that makes the argument $\left(x + \frac{1}{3}\pi\right)$ equal to zero.

$$
\begin{aligned}
x + \frac{1}{3}\pi &= 0 \\
x &= -\frac{1}{3}\pi \quad \Longleftarrow \boxed{\text{Left edge of box}}
\end{aligned}
$$

The right edge of the generic box occurs when the argument of the function equals 2π.

$$
\begin{aligned}
x + \frac{1}{3}\pi &= 2\pi \\
x &= \frac{5}{3}\pi \quad \Longleftarrow \boxed{\text{Right edge of box}}
\end{aligned}
$$

The period is the width of the box.

$$
\frac{5}{3}\pi - \left(-\frac{1}{3}\pi\right) = 2\pi \quad \Longleftarrow \boxed{\text{Period}}
$$

Now that the period is known, we can calculate the quarter points.

Quarter point	Midpoint	Three-quarter point
Left side $+\ \dfrac{1}{4}$ period	Left side $+\ \dfrac{1}{2}$ period	Left side $+\ \dfrac{3}{4}$ period
$-\dfrac{1}{3}\pi + \dfrac{1}{4}\cdot 2\pi$	$-\dfrac{1}{3}\pi + \dfrac{1}{2}\cdot 2\pi$	$-\dfrac{1}{3}\pi + \dfrac{3}{4}\cdot 2\pi$
$-\dfrac{1}{3}\pi + \dfrac{1}{2}\pi$	$-\dfrac{1}{3}\pi + \pi$	$-\dfrac{1}{3}\pi + \dfrac{3}{2}\pi$
$\dfrac{1}{6}\pi$	$\dfrac{2}{3}\pi$	$\dfrac{7}{6}\pi$

□ Figure 3.19

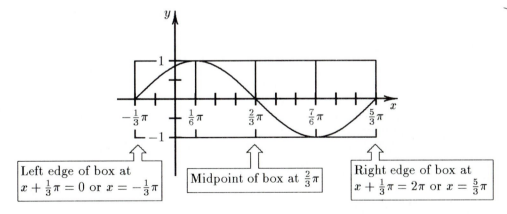

Left edge of box at $x + \frac{1}{3}\pi = 0$ or $x = -\frac{1}{3}\pi$

Midpoint of box at $\frac{2}{3}\pi$

Right edge of box at $x + \frac{1}{3}\pi = 2\pi$ or $x = \frac{5}{3}\pi$

□

In the example above, the phase shift $-\frac{1}{3}\pi$ is the amount the box is moved from its normal starting point.

Example 9 □ Graph one period of $y = \cos\left(2x - \dfrac{1}{3}\pi\right)$.

To find the left edge of the box, set the argument, $2x - \dfrac{1}{3}\pi$, equal to the normal starting point of a generic cosine curve.

$$2x - \frac{1}{3}\pi = 0 \qquad \boxed{\text{Cosine curves normally go from 0 to } 2\pi}$$
$$2x = \frac{1}{3}\pi$$
$$x = \frac{1}{6}\pi \qquad \boxed{\text{Left edge of box}}$$

We have just found the value of x that makes the argument $2x - \dfrac{1}{3}\pi$ equal to zero. This gives us the value of x where the generic curve begins.

The combined effect of the coefficient of x being 2 and $C = -\dfrac{1}{3}\pi$ is to shift the generic box to the right $\dfrac{1}{6}\pi$.

To find the right side of the box, set the argument to the endpoint of one period of a generic cosine function.

$$2x - \frac{1}{3}\pi = 2\pi$$
$$2x = \frac{6}{3}\pi + \frac{1}{3}\pi$$
$$2x = \frac{7}{3}\pi$$
$$x = \frac{7}{6}\pi \qquad \boxed{\text{Right edge of box}}$$

The period is the difference between the left and right edge of the box.

$$\text{period} = \frac{7}{6}\pi - \frac{1}{6}\pi \qquad \boxed{\text{Length of box}}$$
$$= \pi \qquad \boxed{\text{Period}}$$

Now that we know the starting point of the principal cycle and the period, we can find the other critical points.

Quarter point		Midpoint		Three-quarter point	
left side $+$	$\dfrac{1}{4}$ period	left side $+$	$\dfrac{1}{2}$ period	left side $+$	$\dfrac{3}{4}$ period
$\dfrac{1}{6}\pi$ $+$	$\dfrac{1}{4}\pi$	$\dfrac{1}{6}\pi$ $+$	$\dfrac{1}{2}\pi$	$\dfrac{1}{6}\pi$ $+$	$\dfrac{3}{4}\pi$
$\dfrac{2}{12}\pi$ $+$	$\dfrac{3}{12}\pi$	$\dfrac{1}{6}\pi$ $+$	$\dfrac{3}{6}\pi$	$\dfrac{2}{12}\pi$ $+$	$\dfrac{9}{12}\pi$
$\dfrac{5}{12}\pi$		$\dfrac{2}{3}\pi$		$\dfrac{11}{12}\pi$	

Chapter 3 Graphs of the Trigonometric Functions

Now we can sketch the graph of $y = \cos\left(2x - \dfrac{1}{3}\pi\right)$

First draw the box:

□ Figure 3.20

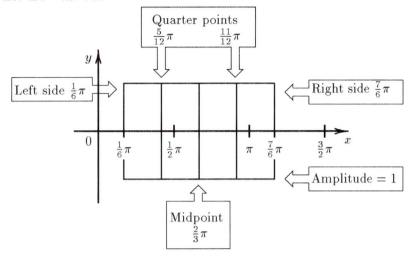

Next fill in the curve:

□ Figure 3.21

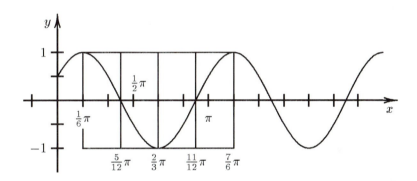

□

Using Your Graphing Calculator

To draw this graph using your graphing calculator, make the following settings:

$$Y_{\min} = -1.2 \qquad X_{\min} = 0$$
$$Y_{\max} = 1.2 \qquad X_{\max} = 6.807$$
$$Y_{\text{scl}} = 0.5 \qquad X_{\text{scl}} = 0.262$$

Where did you get 0.262 for x scale?

That's a decimal approximation for $\frac{1}{12}\pi$. I wanted the tick marks at the critical points.

The next page gives you one way to determine a good value for x scale.

Using Your Graphing Calculator

To determine the value to use for X_{scl} for the equation $y = A \sin(Bx + C)$ or $y = A \cos(Bx + C)$ so that the critical points of the function occur at the tick marks use

$$X_{\text{scl}} = \frac{C}{B}$$

If the critical points do not occur at the tick marks, double the denominator.

$$X_{\text{scl}} = \frac{C}{2B}$$

If some critical points still do not have tick marks, continue doubling the denominator until all the critical points occur at the tick marks. For example,

$$y = \frac{1}{2} \sin\left(2x - \frac{1}{3}\pi\right)$$

$$\text{use } X_{\text{scl}} = \frac{\frac{1}{3}\pi}{2}$$

$$= \frac{1}{6}\pi$$

□ Figure 3.22

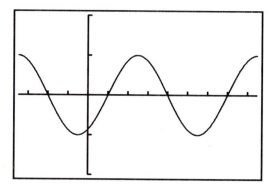

All the critical points do not occur at the tick marks, so divide X_{scl} by 2. The new X_{scl} becomes

$$\frac{\frac{1}{6}\pi}{2} = \frac{1}{12}\pi$$

□ Figure 3.23

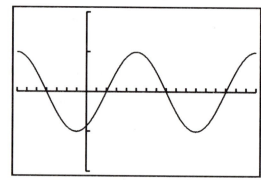

We have done it.

Chapter 3 Graphs of the Trigonometric Functions

Pointer to Find Phase Shift

Above we found the starting point of the generic box by finding the value of x that made the argument of the function $y = \cos\left(2x - \frac{1}{3}\pi\right)$ equal to zero. We can follow the same procedure to find the phase shift of any trigonometric function.

To find the phase shift of

$$y = A\sin\left(Bx + C\right) \text{ or } y = A\cos\left(Bx + C\right)$$

set $Bx + C$ equal to zero and solve for x.

$$Bx + C = 0$$
$$x = -\frac{C}{B}$$

$x = -\dfrac{C}{B}$ is the value of x where the argument of the function is zero and the principal cycle begins. This means we can look at the equation of a trigonometric function and immediately determine its period and phase shift.

$$\text{Period} = \frac{2\pi}{B}$$
$$\text{Phase Shift} = -\frac{C}{B}$$

This means that $y = \cos\left(2x - \frac{1}{3}\pi\right)$ has

$$\text{Period} = \frac{2\pi}{2}$$
$$= \pi$$
$$\text{Phase Shift} = -\frac{-\frac{1}{3}\pi}{2}$$
$$= \frac{1}{6}\pi$$

> Is there an easier way to find phase shift?

> There is a different way, if we write the general equation in a slightly different form.

We can write the general equation of a sine function in the form

$$y = A\sin\left[N(x + s)\right]$$

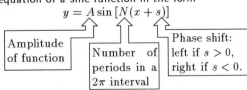

Amplitude of function

Number of periods in a 2π interval

Phase shift: left if $s > 0$, right if $s < 0$.

Consider the funciton above, $y = \cos\left(2x - \frac{1}{3}\pi\right)$. If we factor the 2 from the argument we get

$$y = \cos\left[2\left(x - \frac{1}{6}\pi\right)\right]$$

For this function the amplitude is 1, there are 2 periods in a 2π interval, therefore the length of a period is π. The phase shift is $\frac{1}{6}\pi$ to the right. Notice $+\frac{1}{6}\pi$ is the value of x that makes the argument of the function zero. Therefore, $+\frac{1}{6}\pi$ is the starting point of the principal cycle.

Example 10 □ Graph $y = \sin\left(\dfrac{1}{2}x + \dfrac{1}{3}\pi\right)$ for one period.

The left edge of the box is the value of x that makes the argument equal 0.

$$\frac{1}{2}x + \frac{1}{3}\pi = 0$$

$$x = -\frac{2}{3}\pi \quad \longleftarrow \boxed{\text{Left edge of box}}$$

The right edge of the box is the value of x that makes the argument equal 2π.

$$\frac{1}{2}x + \frac{1}{3}\pi = 2\pi$$

$$\frac{1}{2}x = \frac{5}{3}\pi$$

$$x = \frac{10}{3}\pi \quad \longleftarrow \boxed{\text{Right edge of box}}$$

The period of the function is the difference between the right and left edges of the box.

$$\text{Period} = \frac{10}{3}\pi - \left(-\frac{2}{3}\pi\right) \text{ which is } 4\pi.$$

Now that we know the period, we can find the other critical points of the principal cycle.

Quarter point	Midpoint	Three-quarter point
Left side $\ +\ \dfrac{1}{4}$ period	Left side $\ +\ \dfrac{1}{2}$ period	Left side $\ +\ \dfrac{3}{4}$ period
$-\dfrac{2}{3}\pi \ +\ \dfrac{1}{4}\cdot 4\pi$	$-\dfrac{2}{3}\pi \ +\ \dfrac{1}{2}\cdot 4\pi$	$-\dfrac{2}{3}\pi \ +\ \dfrac{3}{4}\cdot 4\pi$
$-\dfrac{2}{3}\pi \ +\ \pi$	$-\dfrac{2}{3}\pi \ +\ 2\pi$	$-\dfrac{2}{3}\pi \ +\ 3\pi$
$\dfrac{1}{3}\pi$	$\dfrac{4}{3}\pi$	$\dfrac{7}{3}\pi$

Now sketch the curve $y = \sin\left(\dfrac{1}{2}x + \dfrac{1}{3}\pi\right)$ for one period.

□ Figure 3.24

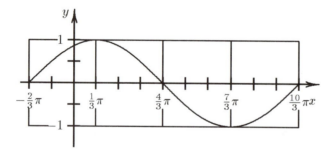

□

Frequently, in applications such as using an oscilloscope in electronics or tuning an automobile, adjustments need to be made that shift an existing curve right or left. To find the equation of the modified functions, relate the original starting point of the generic box to the shifted starting point.

Chapter 3 Graphs of the Trigonometric Functions

Example 11 □ If $y_1 = \sin\left(2x - \dfrac{1}{6}\pi\right)$, find the function y_2 that has the same amplitude and period but is shifted $\dfrac{1}{3}\pi$ to the right or y_1.

First find the starting point of the generic box for $y = \sin\left(2x - \dfrac{1}{6}\pi\right)$. This is an equation of the form $y = A\sin(Bx + C)$. The starting point occurs when $Bx + C = 0$.

$$2x - \frac{1}{6}\pi = 0$$

$$2x = \frac{1}{6}\pi$$

$$x = \frac{1}{12}\pi \qquad \text{starting point of original generic box}$$

Next find the x-value of the new starting point, which is $\dfrac{1}{3}\pi$ to the right, by adding $\dfrac{1}{3}\pi$ to the original starting point, $\dfrac{1}{12}\pi$.

$$\frac{1}{12}\pi + \frac{1}{3}\pi = \frac{5}{12}\pi$$

Now look for the value of C so that $(Bx + C)$ will evaluate to zero when x is at the new starting point $\dfrac{5}{12}\pi$. We keep $B = 2$ because we do not want to change the period of the function.

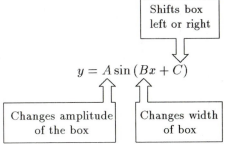

$$2x + C = 0$$

$$2\left(\frac{5}{12}\pi\right) + C = 0$$

$$C = -\frac{10}{12}\pi$$

$$C = -\frac{5}{6}\pi$$

This is the x value of the new starting point.

Value of C that makes the function cross the x-axis at $x = \frac{5}{12}\pi$

The new equation becomes

$$y_2 = \sin\left(2x - \frac{5}{6}\pi\right)$$

Shifting a Curve Up or Down

We now know the effects of three constants on the generic box for a trigonometric function.

Shifts box left or right

$$y = A\sin(Bx + C)$$

Changes amplitude of the box

Changes width of box

Adding a constant to a trigonometric function has the same effect as adding a constant to any function. It shifts the graph of the function up or down.

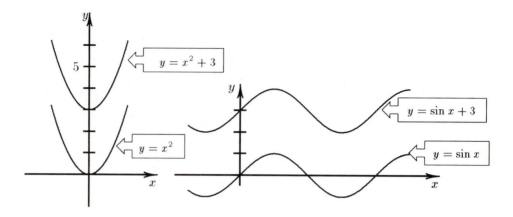

In the case of $\sin x$:

$$y = A \sin (Bx + C) + k$$

k controls the vertical displacement. It shifts graph up or down.

Example 12 □ Graph $y = \cos x + 2$.

Is that the same as $y = \cos (x + 2)$?

No, $y = \cos (x + 2)$ would shift $y = \cos x$ left. $y = \cos x + 2$ shifts the curve up.

We include a brief table of values to verify that the graph is correct.

□ Figure 3.26

x	$\cos x$	$y = \cos x + 2$
0	1	3
$\frac{1}{4}\pi$	0.707	2.707
$\frac{1}{2}\pi$	0	2
π	-1	1
$\frac{3}{2}\pi$	0	2
2π	1	3

Notice the third column of this table is the second column plus 2. □

Chapter 3 Graphs of the Trigonometric Functions

In summary, for any sine or cosine function,

$$y = A \sin(Bx + C) + k \text{ or } y = A \cos(Bx + C) + k$$

1. The amplitude is $|A|$ and if $A < 0$, the sine or cosine graph is reflected about the x-axis.

2. The period is $\dfrac{2\pi}{B}$ which is the width of the generic box or of a cycle.

3. The phase shift is $-\dfrac{C}{B}$. This is the value of x that makes the argument of the function equal to zero and is where the left edge of the generic box is located.

4. The constant k yields a vertical shift of the function $y = A \sin(Bx + C)$ or $y = A \cos(Bx + C)$ upward if $k > 0$ and downward if $k < 0$.

Problem Set 3.2

Find the starting point, quarter point, midpoint, three-quarter point, and endpoint of a generic box. Then sketch two periods of the following graphs.

1. $y = \sin\left(x - \dfrac{1}{4}\pi\right)$
2. $y = \cos(x - \pi)$

3. $y = \sin\left(x + \dfrac{1}{4}\pi\right)$
4. $y = \cos(x + \pi)$

5. $y = -2\cos\left(x - \dfrac{1}{3}\pi\right)$
6. $y = -3\sin\left(x + \dfrac{1}{9}\pi\right)$

7. $y = \dfrac{1}{2}\sin\left(\dfrac{1}{2}x - \dfrac{1}{6}\pi\right)$
8. $y = 4\cos\left(2x + \dfrac{3}{4}\pi\right)$

Use your graphing calculator to graph two periods of the following functions. Use the given starting point or ending point of the curve to set the range of your graphing calculator so that the graph fits the viewing screen nicely. List the values you chose for $X_{\min}$, $X_{\max}$, X_{scl}, $Y_{\min}$, $Y_{\max}$, and Y_{scl}.

9. $y = \cos\left(x + \dfrac{1}{2}\pi\right)$ starting at $x = -\pi$

10. $y = \sin\left(x + \dfrac{1}{2}\pi\right)$ ending at $x = 3\pi$

11. $y = \cos\left(x - \dfrac{1}{2}\pi\right)$ ending at $x = \dfrac{11}{3}\pi$

12. $y = \sin\left(x - \dfrac{1}{2}\pi\right)$ starting at $x = -\dfrac{1}{3}\pi$

13. $y = -2\sin\left(2x + \dfrac{1}{4}\pi\right)$ starting at $x = -\dfrac{1}{2}\pi$

14. $y = \cos\left(\dfrac{1}{2}x - \dfrac{1}{8}\pi\right)$ ending at $x = \dfrac{33}{4}\pi$

Graph a 2-period segment of each of the functions below starting or ending at the point indicated. Within each segment you have graphed, identify the coordinates of the point requested.

15. $y = -2 \sin \left(2x - \dfrac{1}{4}\pi\right)$ ending at $x = \dfrac{7}{4}\pi$
identify the first maximum

16. $y = 3 \cos \left(\dfrac{1}{2}x + \dfrac{1}{8}\pi\right)$ starting at $x = -\dfrac{1}{3}\pi$
identify the first x-intercept

17. $y = 2 \sin \left(2x - \dfrac{1}{3}\pi\right)$ starting at $x = -\dfrac{1}{4}\pi$
identify the first minimum

18. $y = -\dfrac{3}{4} \cos \left(\dfrac{2}{3}x - \dfrac{1}{3}\pi\right)$ ending at $x = \dfrac{11}{4}\pi$
identify the first x-intercept

19. $y = -3 \cos \left(\dfrac{1}{2}x + \dfrac{1}{4}\pi\right)$ ending at $x = \dfrac{29}{4}\pi$
identify the last maximum

20. $y = \dfrac{1}{2} \sin \left(2x - 2\pi\right)$ starting at $x = -\dfrac{1}{2}\pi$
identify the last maximum

For the following functions, how many periods occur in the given domain? Use your graphing calculator to verify the results.

21. $y = -3 \sin \left(\dfrac{1}{2}x + \dfrac{1}{2}\pi\right) \ -\pi \le x \le 7\pi$ **22.** $y = 4 \cos \left(\dfrac{1}{3}x - \dfrac{2}{3}\pi\right) \ 0 \le x \le 12\pi$

What is the length of one period for the following functions? Use your graphing calculator to verify the results.

23. $y = \dfrac{1}{2} \cos \left(3x + \dfrac{1}{2}\pi\right)$ **24.** $y = -2 \sin \left(\dfrac{1}{6}x - \dfrac{1}{6}\pi\right)$

Using the same trigonometric function as y_1, give the equation of the function y_2 that meets the given criteria:

25. $y_1 = \cos x$ y_2 is $\dfrac{1}{2}$ as high as y_1 and the graph is shifted to the right of y_1 by $\dfrac{1}{4}\pi$

26. $y_1 = 2 \sin 3x$ y_2 is twice as high as y_1 with $\dfrac{1}{2}$ as many repetitions in any interval as y_1

27. $y_1 = -\dfrac{3}{4} \sin 2x$ y_2 is twice as high as y_1 with twice as many repetitions in any interval as y_1

28. $y_1 = -\dfrac{2}{3} \cos \dfrac{1}{2}x$ y_2 is the reflection about the x-axis of a curve that is twice as high as y_1 with three times as many periods in any interval as y_1

29. $y_1 = \dfrac{1}{2} \sin \left(3x - \dfrac{1}{2}\pi\right)$ y_2 is $\dfrac{2}{3}$ as high as y_1 and the graph is shifted to the left of y_1 by $\dfrac{1}{4}\pi$

30. $y_1 = 2 \cos \left(\dfrac{1}{2}x + \dfrac{1}{6}\pi\right)$ y_2 is twice as high as y_1 and the graph is shifted to the right of Y_1 by $\dfrac{1}{2}\pi$

Graph the following functions by hand.

31. $y = \cos x - 2$ **32.** $y = \sin x - 3$

33. $y = 3 \cos x + 2$ **34.** $y = 2 \sin x + 2$

Use your graphing calculator to graph 2 periods of each of the following functions. Set your range to best fit the viewing screen. List those values.

35. $y = \dfrac{1}{2}\sin 2x - 1$ **36.** $y = \dfrac{3}{2}\cos 2x + 1$

37. $y = -2\sin 3x - 3$ **38.** $y = -3\cos 2x - 2$

39. $y = |\sin x|$ **40.** $y = |\cos x|$

Group Writing Activity

Below are four copies of the graphs of a function $y_1 = f(x)$.
1. By referring to any copy of the graph of $y_1 = f(x)$ determine the following:
   a. The value of y_1 at $x = 1$.
   b. The value of y_1 at $x = 2$.
   c. $f(1)$ d. $f(2)$ e. $f(4)$ f. $f(0)$

2. On each copy of the graph of $y_1 = f(x)$ sketch the graphs of the functions y_2–y_5 as determined by the rule above each graph.

(a) $y_2 = 2f(x)$

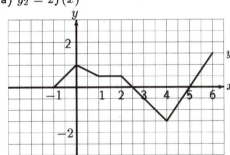

(b) $y_3 = f(2x)$

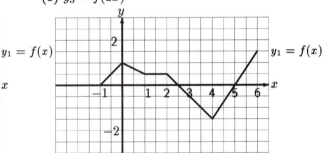

(c) $y_4 = -f(x)$

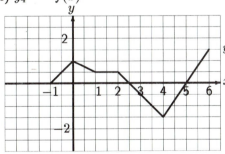

(d) $y_5 = f(x + 2)$

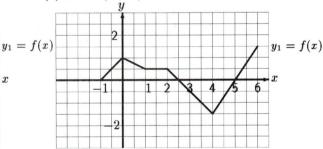

3. Using your results from above, explain what happens to the graph when you:
   a. multiply the function by a constant.
   b. multiply the argument of a function by a constant.
   c. add a constant to the argument of a function.

3.3 Using the Graphing Calculator to Graph Functions by Addition of Ordinates

The ordinate of a point is its y-coordinate. In the previous example, the y-coordinate of each point to be plotted was found by adding $+2$ to the value of $\cos x$ for each value of x. It is possible to consider $y = \cos x + 2$ the sum of the two functions so $y_1 = \cos x$ and $y_2 = +2$; then $y = y_1 + y_2$. We could do this addition graphically.

Why would you want to do that?

Frequently we can use graphical methods to simplify drawing the curve that is the sum of simpler functions.

Example 13 □ Graph $y = \sin x + \cos x$.

Let's write this table of values sideways:

x	0	$\frac{1}{6}\pi$	$\frac{1}{4}\pi$	$\frac{1}{3}\pi$	$\frac{1}{2}\pi$	$\frac{2}{3}\pi$	$\frac{3}{4}\pi$	$\frac{5}{6}\pi$	π
$f_1 = \sin x$	0	.500	.707	.866	1	.866	.707	.500	0
$f_2 = \cos x$	1	.866	.707	.500	0	$-.500$	$-.707$	$-.866$	-1
$f_1 + f_2$	1	1.366	1.414	1.366	1	.366	0	$-.366$	-1

Notice each of the values in the third row is the sum of the values in the two rows above it. In other words the value of $y = \sin x + \cos x$ can be found by adding the values of $y = \sin x$ and $y = \cos x$ together for any given value of x.

Before graphing calculators, mathematicians frequently used this technique to draw complicated graphs by simply adding the ordinates or y values of each simpler component function of the complicated graph.

Texas Instruments' TI-81, -82, and -85 graphing calculators make it particularly easy to use this technique.

Define the following functions on your calculator.

$$Y_1 = \sin x \qquad Y_2 = \cos x$$

Now to define Y_3 as $Y_1 + Y_2$ do the following:

1. Go to $Y_3 =$ on the screen

2. Press $\boxed{\text{2ND}}$, $\boxed{\text{VARS}}$

3. Press $\boxed{1}$ to select the first function

4. Enter $\boxed{+}$

5. Press $\boxed{\text{2ND}}$, $\boxed{\text{VARS}}$

6. Press $\boxed{2}$ to select the 2nd function.

Your screen should now display

$$
\begin{aligned}
Y_1 &= \sin x \\
Y_2 &= \cos x \\
Y_3 &= Y_1 + Y_2 \\
Y_4 &=
\end{aligned}
$$

Now press $\boxed{\text{GRAPH}}$ and the calculator will first graph $Y_1 = \sin x$ then graph $Y_2 = \cos x$ and finally graph Y_3, the sum of $Y_1 + Y_2$.

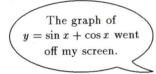
The graph of $y = \sin x + \cos x$ went off my screen.

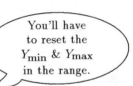
You'll have to reset the $Y_{\min}$ & $Y_{\max}$ in the range.

□

Example 14 □ Use addition of ordinates to graph $y = \sin x + \dfrac{1}{3}\sin 3x$.

Use the edge of a piece of paper to add the ordinate of y_2 to the ordinate of y_1 at each grid line.

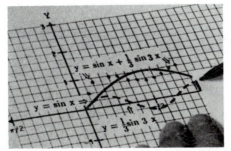

Mark the distance that y_2 is above the x-axis.

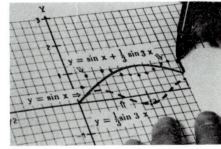

Slide the mark so its bottom is at y_1 and mark $y_1 + y_2$ at the top of the mark.

If y_2 is negative, you subtract its ordinate from the value of y_1.

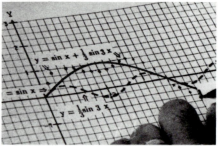

Mark the distance that y_2 is below the x-axis.

Because y_2 is negative, mark that distance below y_1.

Figure 3.27 is the result of this process.

□ Figure 3.27

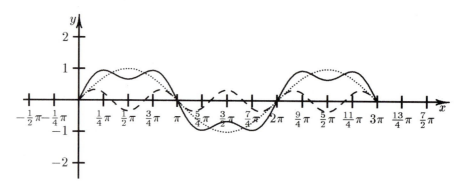

It is no accident that the graph of $y = \sin x + \dfrac{1}{3}\sin 3x$ is beginning to take a rectangular shape. We have just drawn the graph of the first two terms of the Fourier series for a square wave. Electrical and aerodynamical engineers use Fourier analysis because many differently shaped curves can be represented as a sum of sine and cosine waves. A Fourier series for a square wave is $y = \sin x + \dfrac{1}{3}\sin 3x + \dfrac{1}{5}\sin 5x + \dfrac{1}{7}\sin 7x + \cdots$ Below is a graph of the first 3 terms of this series.

□ Figure 3.28

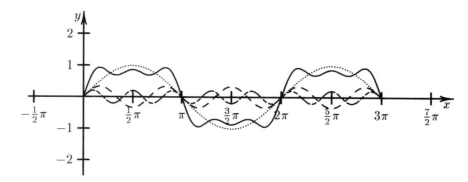

□

Chapter 3 Graphs of the Trigonometric Functions

Using Your Graphing Calculator

Experiment with your graphing calculator by graphing the first seven terms of the Fourier series for a square wave above.

Problem Set 3.3

Sketch a graph of each of the following functions using the method of addition of ordinates.

1. $y = \dfrac{1}{3}x + \sin x$

2. $y = \dfrac{1}{2}x + \cos x$

3. $y = \cos x - \dfrac{1}{2}x$

4. $y = \sin x - \dfrac{1}{3}x$

5. $y = \sin x + \sin 2x$

6. $y = \cos x + \cos 2x$

7. $y = 2\cos x - \sin 2x$

8. $y = 3\sin x - \cos 2x$

9. $y = \cos x + \cos \dfrac{1}{2}x$

10. $y = \sin x + \sin \dfrac{1}{2}x$

11. $y = \sin x + \dfrac{1}{2}\cos \dfrac{1}{2}x$

12. $y = \cos x + \dfrac{1}{2}\sin \dfrac{1}{2}x$

13. $y = |\sin x| + |\cos x|$

14. $y = |\cos x| - |\sin x|$

Sketch the graphs below without using a calculator. To sketch a graph of $y = \cos^2 x$ by hand, you may need to make a table of values.

15. $y = x + \cos^2 x$

16. $y = \cos^2 x + \sin^2 x$

3.4   Graphing the Tangent and Cotangent Functions

In the previous sections we were able to keep the graph of $\sin x$ and $\cos x$ inside a generic box. In the next sections we will deal with trigonometric functions that do not have upper or lower limits. Hence our generic box will have neither a top nor a bottom. It will, however, have sides that serve as asymptotes for the curves.

There is one other major difference between sine or cosine curves and tangent or cotangent curves. It is possible to draw the graph of a sine or cosine function through as many periods as you desire without ever lifting your pen from the paper. This is one test to determine if a function is continuous. The other four trigonometric functions all have values where there are breaks or discontinuities in the graph of those functions.

A Visualization of the Trigonometric Functions

Next we will develop a visualization of $\sin x$, $\cos x$, and $\tan x$. This visualization involves the use of points (x, y) on and off of a circle. Therefore we shall substitute the Greek letter θ (theta) for x as a measure of angle to avoid confusion between x the coordinate and x the angle. Hence we will deal with $\sin\theta$, $\cos\theta$, and $\tan\theta$.

Recall the definition of the trigonometric functions from Chapter 2.

$$\sin\theta = \frac{y}{r}, \qquad \cos\theta = \frac{x}{r}, \qquad \tan\theta = \frac{y}{x}$$

If you choose $r = 1$ unit, these definitions become

$$\sin\theta = y, \qquad \cos\theta = x, \qquad \tan\theta = \frac{y}{x}$$

Recall from geometry that a line that touches a circle at exactly one point is a tangent line.

Below is a drawing of a circle with a radius of one unit. A line tangent to the circle has been drawn through the point (1,0).

□ Figure 3.29

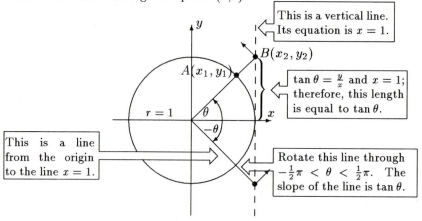

Point A is on the circle of radius 1 unit.

$$\cos\theta = x_1$$
$$\sin\theta = y_1$$

Point B is at the intersection of the terminal side of angle θ and the line tangent to the circle at the x-axis.

$$\tan\theta = \frac{y_2}{x_2}$$
$$\tan\theta = y_2 \qquad \text{because } x_2 = 1 \text{ unit}$$

The distance from the x-axis to point B is y_2. It is a good visualization of tangent θ. Visualize how this distance changes as θ increases.

Chapter 3 Graphs of the Trigonometric Functions

Consider this figure as θ varies through values greater than $-\frac{1}{2}\pi$ to less than $+\frac{1}{2}\pi$.

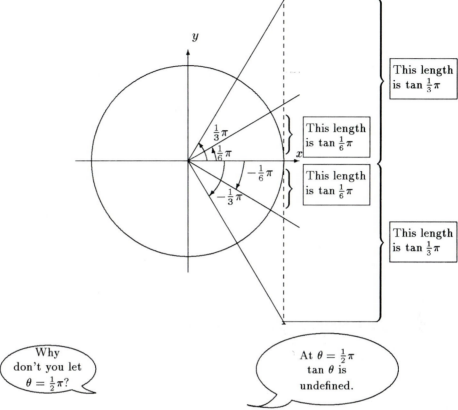

□ Figure 3.30

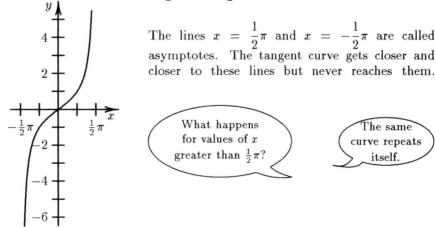

As θ approaches $-\frac{1}{2}\pi$, the tangent of θ is a large negative number. As θ increases toward zero, the absolute value of $\tan\theta$ decreases toward zero. At $\theta = 0$, $\tan\theta$ is 0. Then, as θ increases, the value of $\tan\theta$ increases dramatically as θ gets closer and closer to $\frac{1}{2}\pi$.

Because $\tan\theta = \frac{y}{x}$ and at $\theta = \frac{1}{2}\pi$, $x = 0$, any attempt to calculate $\tan\theta$ at $\theta = \frac{1}{2}\pi$ involves division by zero. Hence $\tan\frac{1}{2}\pi$ is undefined.

A graph of $y = \tan x$ for $-\frac{1}{2}\pi < x < \frac{1}{2}\pi$ looks like this:

□ Figure 3.31

The lines $x = \frac{1}{2}\pi$ and $x = -\frac{1}{2}\pi$ are called asymptotes. The tangent curve gets closer and closer to these lines but never reaches them.

3.4 Graphing the Tangent and Cotangent Functions

As θ reaches π, $\tan\theta$ becomes zero. Beyond π, $|\tan\theta|$ increases until θ reaches $\dfrac{3}{2}\pi$, where x is again zero and $\tan\theta$ is undefined.

□ Figure 3.32

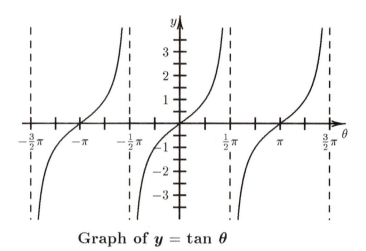

Graph of $y = \tan\theta$

Notice that $\tan\theta$ is undefined at $\theta = \ldots -\dfrac{3}{2}\pi, -\dfrac{1}{2}\pi, \dfrac{1}{2}\pi \ldots$ The asymptotes of the curve are at these values of θ. To see why the tangent curve repeats, look at a unit circle again and consider rotating a line for values of θ greater than $\dfrac{1}{2}\pi$ and less than $\dfrac{3}{2}\pi$.

□ Figure 3.33

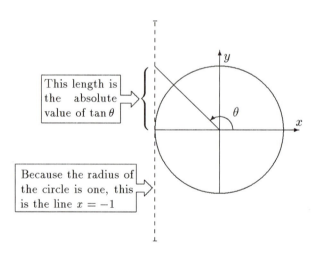

This length is the absolute value of $\tan\theta$

Because the radius of the circle is one, this is the line $x = -1$

At values of θ slightly greater than $\dfrac{1}{2}\pi$, $\tan\theta$ is a large negative number.

$$\tan\theta = \frac{+y}{-1}$$

x is negative in the third quadrant

Chapter 3 Graphs of the Trigonometric Functions

$y = \tan\,\theta$

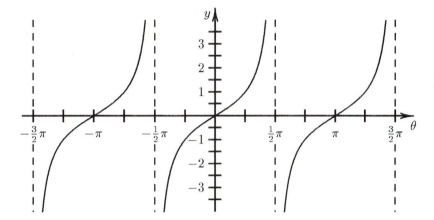

□ Figure 3.34

Graph of $y = $ Cotangent θ

The next graph we will examine is $y = $ cotangent θ. It is very similar to the graph of $y = $ tangent θ.

The definitions of $\tan\theta$ and $\cot\theta$ are

$$\tan\theta = \frac{y}{x} \qquad \cot\theta = \frac{x}{y}$$

From this it follows that

$$\cot\theta = \frac{1}{\tan\theta} \qquad \text{if } \tan\theta \neq 0$$

The $\cot\theta$ is the reciprocal of the $\tan\theta$. We use this relationship to find values of $\cot\theta$ on a calculator.

Example 15 □ Use your calculator to find $\cot 30°$ to four decimal places.

There are two ways to do this

1. Enter $1 \div \tan 30°$ then press EXE.

or **2.** Enter $(\tan 30°)$ and press x^{-1} then press EXE.

In either case we get $\cot 30° = 1.7321$. □

We could use the technique in the previous example to build a table of values for $\cot\theta$. However, since we are interested only in the general shape of the curve, there is a way to reduce calculation.

Example 16 □ Sketch a graph of $y = \cot\theta$.

First sketch the graph of $y = \tan\theta$ on the graph.

□ Figure 3.35

When $\tan\theta$ is	$\cot\theta$ is $\dfrac{1}{\tan\theta}$
1	1
$\dfrac{1}{2}$	2
$\dfrac{1}{4}$	4
0	undefined
−1	−1
$-\dfrac{1}{2}$	−2

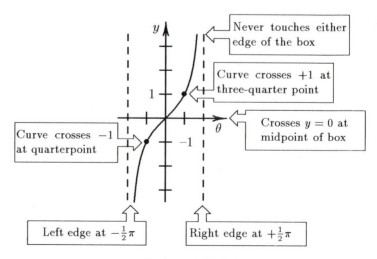

Graph of $y = \cot\theta$

□

 Using Your Graphing Calculator

Because there is no ⌐COT⌐ key on your calculator, to plot a graph of $\cot x$ you will need to plot $y = \dfrac{1}{\tan x}$. You may also use $y = (\tan x)^{-1}$.

A Generic Box for Tangent and Cotangent

Because the values of tangent and cotangent each have no upper or lower limit, our box cannot have a top or a bottom. The critical points for a tangent box are the midpoint where the curve crosses the axis, the points where the value of $\tan\theta$ is 1, and the two asymptotes which are $\theta = -\dfrac{1}{2}\pi$ and $\theta = \dfrac{1}{2}\pi$.

□ Figure 3.36

$y = \tan\theta$

Never touches either edge of the box

Curve crosses +1 at three-quarter point

Crosses $y = 0$ at midpoint of box

Curve crosses −1 at quarterpoint

Left edge at $-\frac{1}{2}\pi$

Right edge at $+\frac{1}{2}\pi$

Chapter 3 Graphs of the Trigonometric Functions

$y = \cot \theta$

The asymptotes of $\cot \theta$ are $\theta = 0$ and $\theta = \pi$. The curve has a period of π.

□ Figure 3.37

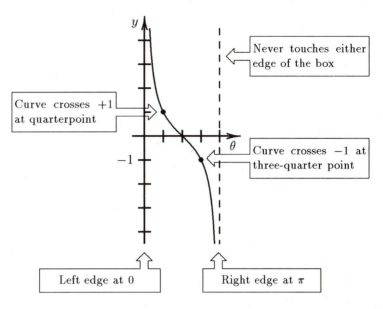

Never touches either edge of the box

Curve crosses $+1$ at quarterpoint

Curve crosses -1 at three-quarter point

Left edge at 0

Right edge at π

Example 17 □ Use your graphing calculator to sketch a graph of three periods of $y = \cot 2x$.

To find the left side of the box, set the argument equal to the normal asymptote of $\cot x$.

We need to set $X_{\min}$ and $X_{\max}$ so that three periods will be included. First find the left and right edges of the principal cycle in the generic box.

Left Edge	Right Edge	Period
$2\theta \ = \ 0$	$2\theta \ = \ \pi$	$\dfrac{1}{2}\pi - 0$
$\theta \ = \ 0$	$\theta \ = \ \dfrac{1}{2}\pi$	$\dfrac{1}{2}\pi$

Therefore, to plot three periods we need to include at least $3\left(\dfrac{1}{2}\pi\right) - 0$ between $X_{\min}$ and $X_{\max}$. The authors chose

$$
\begin{aligned}
X_{\min} &= 0 \\
X_{\max} &= 4.72 \\
X_{\text{scl}} &= \frac{1}{4}\pi
\end{aligned}
\qquad
\begin{aligned}
Y_{\min} &= -4 \\
Y_{\max} &= +4 \\
Y_{\text{scl}} &= .5
\end{aligned}
$$

Remember to plot $y = \dfrac{1}{\tan 2x}$ to get a graph of $y = \cot 2x$.

3.4 Graphing the Tangent and Cotangent Functions 111

$$y = \cot 2x$$

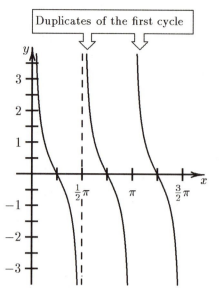

Duplicates of the first cycle

□ Figure 3.38

□

Example 18 □ Graph one period of $y = \tan\left(2x + \dfrac{1}{6}\pi\right)$.

Find the edges of the box and the period by setting the argument equal to the normal asymptotes of $y = \tan x$.

Left Edge	Right Edge	Period
$2x + \dfrac{1}{6}\pi = -\dfrac{1}{2}\pi$	$2x + \dfrac{1}{6}\pi = \dfrac{1}{2}\pi$	Right Edge $-$ Left Edge
$2x = -\dfrac{1}{2}\pi - \dfrac{1}{6}\pi$	$2x = \dfrac{1}{2}\pi - \dfrac{1}{6}\pi$	$\dfrac{1}{6}\pi - \left(-\dfrac{1}{3}\pi\right)$
$2x = -\dfrac{4}{6}\pi$	$2x = \dfrac{2}{6}\pi$	$\dfrac{1}{2}\pi$
$x = -\dfrac{1}{3}\pi$	$x = \dfrac{1}{6}\pi$	

Next we find the quarter points

Quarter point	Midpoint	Three-quarter point
Left Edge $+\ \dfrac{1}{4}$ period	Left Edge $+\ \dfrac{1}{2}$ period	Left Edge $+\dfrac{3}{4}$ period
$-\dfrac{1}{3}\pi + \dfrac{1}{4}\cdot\dfrac{1}{2}\pi$	$-\dfrac{1}{3}\pi + \dfrac{1}{2}\cdot\dfrac{1}{2}\pi$	$-\dfrac{1}{3}\pi + \dfrac{3}{4}\cdot\dfrac{1}{2}\pi$
$-\dfrac{1}{3}\pi + \dfrac{1}{8}\pi$	$-\dfrac{1}{3}\pi + \dfrac{1}{4}\pi$	$-\dfrac{1}{3}\pi + \dfrac{3}{8}\pi$
$-\dfrac{5}{24}\pi$	$-\dfrac{1}{12}\pi$	$\dfrac{1}{24}\pi$

With these five points we can make a good sketch.

Chapter 3 Graphs of the Trigonometric Functions

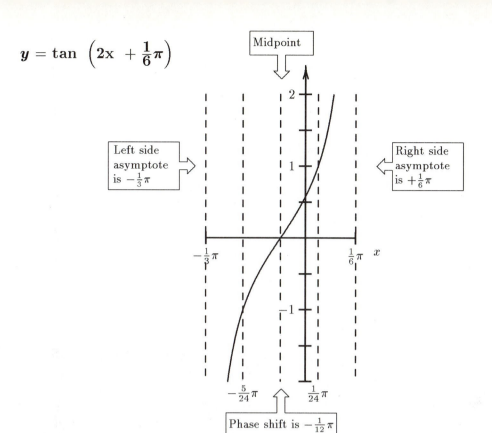

$$y = \tan\left(2x + \frac{1}{6}\pi\right)$$

Midpoint

2

Left side asymptote is $-\frac{1}{3}\pi$

1

Right side asymptote is $+\frac{1}{6}\pi$

□ Figure 3.39

$-\frac{1}{3}\pi$ $\frac{1}{6}\pi$ x

-1

$-\frac{5}{24}\pi$ $\frac{1}{24}\pi$

Phase shift is $-\frac{1}{12}\pi$ □

Note: As with the sine and cosine functions, the period of $y = \tan(Bx + C)$ or $y = \cot(Bx + C)$ is

$$\frac{\text{period of } y = \tan x \ (\text{ or } \cot x)}{B} = \frac{\pi}{B}$$

The phase shift can be found, as for sine and cosine functions, using

$$\text{Phase Shift} = -\frac{C}{B}$$

Problem Set 3.4

State the period for the following functions.

1. $y = \tan x$ 2. $y = \tan 2x$ 3. $y = \cot x$

4. $y = \cot 3x$ 5. $y = \tan \frac{1}{4}x$ 6. $y = \cot \frac{1}{2}$

7. $y = \cot 4x$ 8. $y = \tan \frac{1}{3}$ 9. $y = \cot \frac{1}{4}$

10. $y = \tan \frac{1}{2}$ 11. $y = \tan 4x$ 12. $y = \cot \frac{1}{3}$

Give the left and right asymptotes of the generic box of the following functions.

13. $y = \tan x$ 14. $y = \tan 2x$ 15. $y = \cot x$

16. $y = \cot 2x$ 17. $y = \cot \frac{1}{2}$ 18. $y = \cot \frac{1}{3}$

19. $y = \tan \frac{1}{2}$ 20. $y = \tan \frac{1}{3}$ 21. $y = \tan\left(x + \frac{1}{3}\pi\right)$

22. $y = \tan\left(x - \frac{1}{6}\pi\right)$ 23. $y = \cot\left(x + \frac{1}{4}\pi\right)$ 24. $y = \cot\left(x + \frac{1}{3}\pi\right)$

Use your graphing calculator to plot the following curves. Use the given information to set the range of your graphing calculator so the graph fits the viewing screen nicely, and determine where the graph first crosses the x-axis within the segment graphed. Record this value. List the values you choose for $X_{\min}, X\max, X_{\text{scl}}, Y_{\min}, Y\max,$ and Y_{scl}.

25. $y = \tan 2x$ Graph three periods starting at $x = 0$

26. $y = \tan \dfrac{1}{2}x$ Graph three periods ending at $x = 3\pi$

27. $y = \cot \dfrac{1}{2}x$ Graph three periods ending at $x = 2\pi$

28. $y = \cot 3x$ Graph four periods starting at $x = -\dfrac{1}{3}$

29. $y = \tan \left(x - \dfrac{1}{3}\pi \right)$ Graph three periods starting at $x = -\dfrac{1}{6}\pi$

30. $y = \tan \left(2x + \dfrac{1}{3}\pi \right)$ Graph four periods ending at $x = \dfrac{7}{4\pi}$

31. $y = \cot \left(x - \dfrac{1}{6}\pi \right)$ Graph three periods ending at $x = \dfrac{17}{6}\pi$

32. $y = \cot \left(\dfrac{1}{2}x + \dfrac{1}{4}\pi \right)$ Graph three periods starting at $-\dfrac{3}{2}\pi$

33. $y = \tan \left(\dfrac{1}{2}x + \dfrac{1}{6}\pi \right)$ Graph four periods starting at $x = -\dfrac{1}{2}\pi$

34. $y = \tan \left(\dfrac{1}{3}x - \dfrac{1}{12}\pi \right)$ Graph three periods ending at $x = \dfrac{101}{12}\pi$

For the following functions, how many periods occur in the domain $-\pi \le x \le 9\pi$?

35. $y = \cot \left(\dfrac{1}{2}x - \dfrac{1}{6}\pi \right)$ **36.** $y = \tan \left(2x + \dfrac{1}{2}\pi \right)$

Find the smallest positive value of x where each graph below crosses the x-axis.

37. $y = \tan \left(\dfrac{1}{2}x - \dfrac{1}{4}\pi \right)$ **38.** $y = \cot \left(2x + \dfrac{1}{3}\pi \right)$

39. $y_1 = \tan \left(\dfrac{1}{3}x - \dfrac{1}{12}\pi \right)$ Write a tangent function y_2 that has the same period as y_1 but is shifted $\dfrac{1}{3}\pi$ to the left.

40. $y_1 = \dfrac{1}{2}\cot 4x$ Define a cotangent function y_2 that has half as many periods in any interval as y_1.

3.5   Graphing the Secant and Cosecant Functions

The two remaining trigonometric functions that we have not graphed are $\sec x$ and $\csc x$. We will use the graphs of $\cos x$ and $\sin x$ to generate these graphs.

Because $\cos \theta = \dfrac{x}{r}$ and $\sec \theta = \dfrac{r}{x}$ it follows that $\sec \theta = \dfrac{1}{\cos \theta}$.

To find a value for $\sec x$ on your calculator, find $\cos x$ then use the reciprocal key.

Example 19 □ Find $\sec \dfrac{1}{3}\pi$.

$$\cos \dfrac{1}{3}\pi = 0.5$$

$$\sec \dfrac{1}{3}\pi = \dfrac{1}{\cos \frac{1}{3}\pi}$$

$$= 2.0$$

□

Example 20 □ Sketch a graph of $y = \sec x$.

We can use a sketch of $y = \cos x$ to draw $y = \sec x$.

□ Figure 3.40

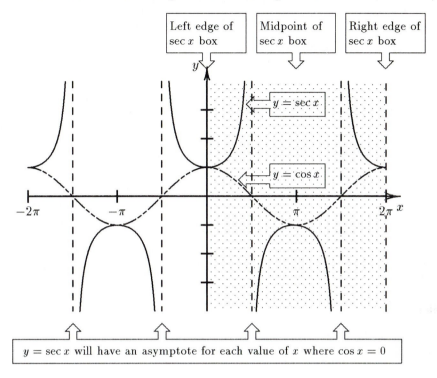

| Left edge of sec x box | Midpoint of sec x box | Right edge of sec x box |

$y = \sec x$

$y = \cos x$

$y = \sec x$ will have an asymptote for each value of x where $\cos x = 0$

□

A Generic Secant Box

When

$\cos x$ is	$\sec x$ is
0	undefined
$\frac{1}{4}$	4
$\frac{1}{2}$	2
1	1
-1	-1
$-\frac{1}{2}$	-2

□ Figure 3.41

$y = \sec x$

$y = \cos x$

| Left edge $x = 0$ | Midpoint $x = \pi$ | Right edge $x = 2\pi$ |

A $y = \sec x$ box contains two half segments that open up and a segment that opens down.

Example 21 □ Sketch a graph of $y = \csc x$.

Since $\sin x = \dfrac{y}{r}$ and $\csc x = \dfrac{r}{y}$, it follows $\csc x = \dfrac{1}{\sin x}$. We use the same technique to graph $y = \csc x$ as we use to graph $y = \sec x$.

□ Figure 3.42

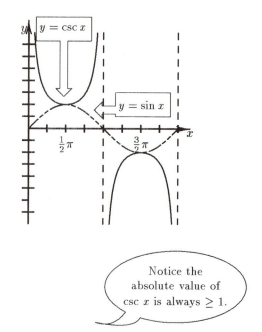

The asymptotes occur when $\sin x = 0$.

Notice the absolute value of $\csc x$ is always ≥ 1.

□

Example 22 □ Sketch a graph of $y = 2 \csc \left(2x - \dfrac{1}{6}\pi \right)$.

The left edge of the box is at the value of x that makes the argument equal zero. The right edge is the value of x that makes the argument equal 2π.

Left Edge	Right Edge	Period
$2x - \dfrac{1}{6}\pi = 0$	$2x - \dfrac{1}{6}\pi = 2\pi$	$\dfrac{13}{12}\pi - \dfrac{1}{12}\pi$
$x = \dfrac{1}{12}\pi$	$x = \dfrac{13}{12}\pi$	π

The midpoint of the box is halfway between the left and right edges at $\dfrac{1}{12}\pi + \dfrac{1}{2}\pi = \dfrac{7}{12}\pi$. Now we can sketch the graph of $y = 2 \csc \left(2x - \dfrac{1}{6}\pi \right)$.

Chapter 3 Graphs of the Trigonometric Functions

□ Figure 3.43

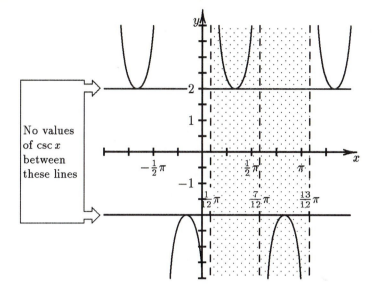

No values
of csc x
between
these lines

The coefficient of 2
means that the curve
has values $|y| \geq 2$.

It also
makes the curve
steeper.

Problem Set 3.5

Evaluate the following functions.

1. $\sec \dfrac{1}{6}\pi$ **2.** $\sec \dfrac{4}{3}\pi$ **3.** $\csc \dfrac{5}{3}\pi$ **4.** $\csc \dfrac{3}{4}\pi$

5. $\sec \dfrac{1}{3}\pi$ **6.** $\sec \dfrac{7}{6}\pi$ **7.** $\csc \dfrac{1}{4}\pi$ **8.** $\csc \dfrac{5}{4}\pi$

9. $\sec \dfrac{5}{6}\pi$ **10.** $\sec \dfrac{7}{4}\pi$ **11.** $\csc \dfrac{2}{3}\pi$ **12.** $\csc \dfrac{11}{6}\pi$

On the same set of axes, graph the corresponding sine or cosine function first, then graph:

13. $y = 2\sec \dfrac{1}{2}x$ **14.** $y = \dfrac{1}{2}\csc 3x$

Use your graphing calculator to graph two periods of the following functions. Use the given information to set the range on your graphing calculator. List the range values from the graphing calculator that give you the best fit on the viewing screen. The tick marks should occur at the critical points.

15. $y = 2\sec x$ starting point at $-\pi$ **16.** $y = 2\csc x$ ending point at π

17. $y = \csc 3x$ ending point at $\dfrac{1}{2}\pi$ **18.** $y = \csc \dfrac{1}{2}x$ starting point at $-\dfrac{1}{2}\pi$

19. $y = \sec\left(x + \dfrac{1}{4}\pi\right)$ starting point at $-\dfrac{1}{2}\pi$

20. $y = \sec\left(2x + \dfrac{1}{3}\pi\right)$ ending point at $\dfrac{1}{3}\pi$

21. $y = \csc\left(x + \dfrac{1}{6}\pi\right)$ ending point at π

22. $y = \csc\left(\dfrac{1}{2}x - \dfrac{1}{4}\pi\right)$ starting point at $-\dfrac{1}{2}\pi$

23. $y = \dfrac{1}{2}\sec\left(2x + \dfrac{1}{3}\pi\right)$ starting point at $-\dfrac{1}{6}\pi$

24. $y = 2\sec\left(\dfrac{1}{2}x - \dfrac{1}{4}\pi\right)$ ending point at 3π

25. $y = \dfrac{1}{2}\csc\left(2x + \dfrac{1}{3}\pi\right)$ ending point at $\dfrac{1}{4}\pi$

26. $y = 2\csc\left(\dfrac{1}{2}x - \dfrac{1}{3}\pi\right)$ starting point at $-\dfrac{1}{3}\pi$

27. $y_1 = 2\sec\left(\dfrac{1}{2}x - \dfrac{1}{6}\pi\right)$ Write a secant function y_2 that has the same period as y_1 but is shifted $\dfrac{1}{4}\pi$ to the right.

28. $y_1 = -\dfrac{1}{2}\csc\left(2x + \dfrac{1}{3}\pi\right)$ Define a cosecant function y_2 that has the same period as y_1 but is shifted $\dfrac{1}{2}\pi$ to the left.

3.6   Qualitative Analysis of Trigonometric Functions

In applications such as shifting the scale on an electronic piano, we sometimes know the amplitude and pitch we desire to achieve. Then we need to find the equation of a function with the desired qualities.

If we recognize the general shape of a curve as a particular function, we can modify the coefficients to make the function pass through specified points.

□ Figure 3.44

Example 23 □ What is the equation of the curve below?

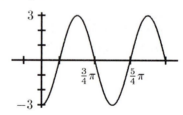

We recognize this as a sine function or a cosine function. First we will write a sine function that matches this curve.

In general, a sine function is

$$y = A\sin(Bx + C) + k$$

In this case $k = 0$ because the curve is equally above and below the x-axis. We know $A = 3$ because the minimum y value is -3 and the maximum y value is $+3$.

Determining the period and phase shift takes more detective work. From the graph we know the midpoint and right edge of the generic box. So, half the period must be the difference between the endpoint and midpoint.

$$\frac{5}{4}\pi - \frac{3}{4}\pi \;=\; \frac{2}{4}\pi \quad \Longleftarrow \boxed{\text{half the period}}$$

$$2\left(\frac{2}{4}\right)\pi \;=\; \pi \quad \Longleftarrow \boxed{\text{full period}}$$

Chapter 3 Graphs of the Trigonometric Functions

Because the period π is half the normal period of a sine function, we know $B = 2$.

To find the phase shift observe that the midpoint of the generic box occurs at $x = \frac{3}{4}\pi$. The midpoint of the generic box of a sine function is at $x = \pi$. Therefore, we want a value of C that makes the argument $(Bx + C)$ equal to π when $x = \frac{3}{4}\pi$. Knowing $B = 2$ we can solve for C.

$$
\begin{aligned}
Bx + C &= \pi \qquad \Longleftarrow \boxed{\text{at midpoint}} \\
2\left(\frac{3}{4}\pi\right) + C &= \pi \\
\frac{3}{2}\pi + C &= \frac{2}{2}\pi \\
C &= -\frac{1}{2}\pi
\end{aligned}
$$

The equation of this curve is

$$
y = 3\sin\left(2x - \frac{1}{2}\pi\right)
$$

We could also use a cosine function to represent this curve.

We know a generic cosine function is

$$
y = A\cos(Bx + C) + k
$$

We know the vertical displacement is 0, the amplitude is 3, and the period is π. So the equation of this graph as a cosine function is

$$
y = 3\cos(2x + C) + 0
$$

If we observe that in a generic cosine graph the curve first crosses the x-axis when the argument equals $\frac{1}{2}\pi$, we can solve for the value of C that makes a cosine curve cross the x-axis at $x = \frac{3}{4}\pi$.

$$
\begin{aligned}
Bx + C &= \frac{1}{2}\pi \\
2\left(\frac{3}{4}\pi\right) + C &= \frac{1}{2}\pi \\
\frac{3}{2}\pi + C &= \frac{1}{2}\pi \\
C &= -\pi
\end{aligned}
$$

As a cosine function this is a graph of

$$
y = 3\cos(2x - \pi)
$$

You can test that both of the functions produce the same graph with your graphing calculator. $\qquad\square$

Write the equations of two functions that will produce each of the graphs below.

1.

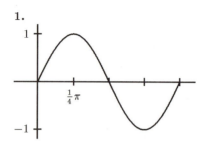

2.

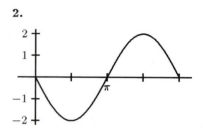

3.

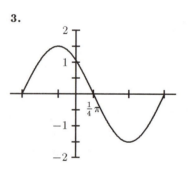

4.

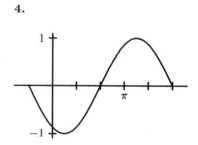

5.

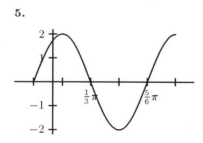

6.

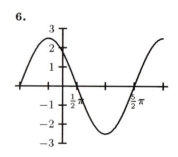

7.

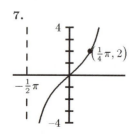

$\left(\frac{1}{4}\pi, 2\right)$

8.

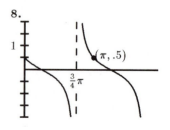

$(\pi, .5)$

3.1 **1.** Coordinates of points corresponding to angles that are multiples of $\frac{1}{6}\pi, \frac{1}{4}\pi, \frac{1}{3}\pi$, $\frac{1}{2}\pi$, and π are found using the symmetry of a circle.

2. Values of the trigonometric functions $\frac{1}{6}\pi, \frac{1}{4}\pi, \frac{1}{3}\pi, \frac{1}{2}\pi$, and π along with their multiples are used in graphing the sine and cosine graphs.

3. The period of the sine and cosine graphs is 2π radians.

4. The generic box is used in graphing trigonometric functions.

5. When graphing $y = A \sin Bx$ or $y = A \cos Bx$,

A is the amplitude and

B determines the period.

3.2 When graphing $y = A \sin(Bx + C) + k$ and $y = A \cos(Bx + C) + k$,

A is the amplitude,

B determines the period. (The value of x that makes $Bx + C = 0$ is the left edge of the generic box. The value of x that makes $Bx + C = 2\pi$ is the right edge of the generic box.) The period is the difference between the right and left edges of the box. This difference always yields the period $= \frac{2\pi}{B}$.

C causes a shift of the graph along the x-axis, phase shift $= -\frac{C}{B}$.

k causes a vertical shift of the generic box.

3.3 It is possible to sketch the graph of a function that is the sum of simpler functions by graphing each simpler function and then adding the ordinates at selected points.

3.4 **1.** When sketching a graph of $y = A \tan(Bx + C) + k$, edges of the generic box occur at values of x that make the argument of the function equal to $-\frac{1}{2}\pi$ or $+\frac{1}{2}\pi$.

2. The period of the tangent function is π radians.

3. When sketching a graph of $y = A \cot(Bx + C) + k$, edges of the generic box occur at values of x that make the argument of the function equal to zero or π.

4. The period of the cotangent function is π radians.

5. To find the value of the cotangent functions with a calculator, we use $\cot x = \frac{1}{\tan x}$.

6. The generic box of the tangent and cotangent graphs open at the top and bottom.

7. The generic tangent box is $-\frac{1}{2}\pi < x < \frac{1}{2}\pi$ along the x-axis.

8. The generic cotangent box is $0 < x < \pi$ along the x-axis.

9. When sketching $y = \tan(Bx + C)$ and $y = \cot(Bx + C)$:

B determines the period which is $\frac{\pi}{B}$, and C causes a shift along the x-axis equal to $-\frac{C}{B}$.

3.5 **1.** To find the value of $\sec x$ and $\csc x$ functions, use: $\sec x = \dfrac{1}{\cos x}$ and $\csc x = \dfrac{1}{\sin x}$.

      2. Use $\sec x = \dfrac{1}{\cos x}$ and $\csc x = \dfrac{1}{\sin x}$ to graph the secant and cosecant functions.

      3. The generic secant and cosecant boxes are: $0 \leq x \leq 2\pi$ along the x-axis and open at the top and bottom.

      4. The absolute value of $\sec x$ and $\csc x$ is always greater than or equal to one.

      5. When sketching $y = A\sec(Bx + C) + k$ and $y = A\csc(Bx + C) + k$:

A determines the amplitude of the graph.

The period of the graph is $\dfrac{2\pi}{B}$.

C causes a shift of the graph along the x-axis equal to $-\dfrac{C}{B}$.

k causes a vertical shift of the graph.

3.6 **1.** To determine the equation of the trigonometric functions sine and cosine from the graph, determine its period, then determine its amplitude by examining the maximum or minimum points on the graph, and then determine its phase shift from the position of the critical points.

      2. For graphs of tangent, cotangent, secant, and cosecant you will need to know at least one point on the graph where the value of the function is not zero to determine the amplitude.

Chapter 3 Review Test

Find the following functional values.

1. $\sin \dfrac{7}{4}\pi$ **(3.1)**

2. $\cos \dfrac{4}{3}\pi$ **(3.1)**

3. $\sec \dfrac{1}{3}\pi$ **(3.5)**

4. $\csc \dfrac{2}{3}\pi$ **(3.5)**

State the period for each of the following functions.

5. $y = \tan 3x$ **(3.4)**

6. $y = \cot \dfrac{1}{3}x$ **(3.4)**

7. $y = \sin \left(3x - \dfrac{1}{2}\pi\right)$ **(3.2)**

8. $y = \sec \left(\dfrac{1}{4}x + \dfrac{1}{6}\pi\right)$ **(3.5)**

Determine the endpoints, quarter point, midpoint, and three-quarter point of the generic box for each curve below. Then sketch by hand one period of the following. **(3.2)**

9. $y = 2\cos \left(2x + \dfrac{1}{3}\pi\right)$

10. $y = \dfrac{1}{2}\sin \left(3x - \dfrac{1}{2}\pi\right)$

Sketch the graphs of the following functions in one period; then duplicate one period on each side on the generic box. **(3.4)**

11. $y = \tan \left(2x + \dfrac{1}{3}\pi\right)$

12. $y = \cot \left(\dfrac{1}{2}x - \dfrac{1}{2}\pi\right)$

Use your graphing calculator to plot the following curves. Use the given information to set the range on your graphing calculator so the graph fits the screen nicely, and the critical points occur at the tick marks. Record the range values from your calculator.

13. $y = 2\cos x$ Graph three periods starting at $x = -\pi$ **(3.1)**

14. $y = -\dfrac{1}{2}\sin x$ Graph two periods ending at $x = 3\pi$ **(3.1)**

15. $y = 3\sin \dfrac{1}{3}x$ Graph two periods ending at $x = 3\pi$ **(3.1)**

16. $y = 2\cos \dfrac{1}{2}x$ Graph two periods starting at $x = -\dfrac{1}{2}\pi$ **(3.1)**

17. $y = \tan 2x$ Graph three periods starting at $x = -\dfrac{1}{4}\pi$ **(3.4)**

18. $y = \sec 2x$ Graph three periods ending at $x = 2\pi$ **(3.5)**

19. If the position of a weight on a spring is given by $\qquad$ **(3.1)**

$$y = 3\sin\frac{2\pi}{3}t \qquad y \text{ in cm and } t \text{ in sec}$$

Find
a) the period
b) the difference in height between the highest and lowest points of the weight's travel
c) the number of complete up and down cycles the weight will make in one second

Using the same trigonometric function as y_1, give the equation of the function y_2 that meets the given criteria.

20. $y_1 = \dfrac{2}{3}\cos x$ $\qquad$ y_2 has the same period and amplitude as y_1 but is reflected about the x-axis **(3.1)**

21. $y_1 = 3\sin\dfrac{1}{3}x$ $\qquad$ y_2 is $\dfrac{1}{2}$ as high as y_1 with twice as many repetitions in any interval **(3.1)**

22. $y_1 = \tan\dfrac{1}{2}x$ $\qquad$ y_2 has a period half as long as the period of y_1 **(3.4)**

23. $y_1 = 2\sec\left(\dfrac{1}{2}x - \dfrac{1}{12}\pi\right)$ $\qquad$ y_2 has the same period as y_1 but is shifted to the right $\dfrac{1}{3}\pi$ **(3.5)**

24. $y_1 = -\dfrac{1}{2}\cos\left(2x - \dfrac{1}{4}\pi\right)$ $\qquad$ y_2 has the same period as y_1 but is shifted to the left $\dfrac{1}{6}\pi$ **(3.2)**

Find two functions that the following graphs represent. **(3.6)**

25.

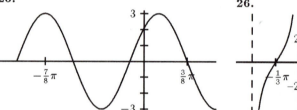

26.

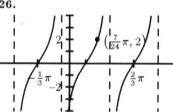

Sketch a graph of each of the following functions using the method of addition of ordinates. **(3.3)**

27. $y = -\sin 2x + 3$ $\qquad$ **28.** $y = 2\sin x + \cos 2x$ $\qquad$ **29.** $y = \cos x + \dfrac{1}{3}\cos 3x$

Chapter 4

Inverse Trigonometric Functions

Contents

Preview

One key use of mathematics is to describe a physical situation so well that we can predict what will happen next. When the value of one variable is uniquely determined by the value of another variable, the first variable is said to be a function of the second. For example, the distance traveled by a car is a function of the number of turns made by its wheels. The height of a piston in a cylinder is a function of the rotation of the crankshaft.

In this chapter we will study relations, functions, and their inverses in general. In the second half of the chapter we will turn our attention to the inverses of the trigonometric functions. An automotive engineer, for example, would use the inverse of $\sin x$ to find the amount of rotation of a crankshaft needed to displace a piston a specific distance up a cylinder.

4.1 Relations, Functions, and Their Inverses

Many people who use mathematics use it because mathematics helps describe a relationship between two variables. A variable that may assume any one of a set of values is referred to as an independent variable. The quantity that changes as a result of a change in the independent variable is referred to as a function of the independent variable.

It may also be called the dependent variable.

In functional notation the independent variable is enclosed in parentheses.

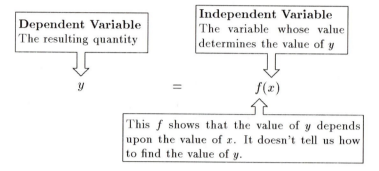

Definition 4.1A Relation

A relation is a connection between two variables. The connections may be defined with a set of ordered pairs, a table, or an equation.

Definition 4.1B Domain and Range

The domain of a relation is the set of possible replacements for the independent variable. Generally we let x represent the independent variable.

The range of a relation is the set of possible values for the dependent variable. y is usually used to represent the dependent variable.

Unless otherwise specified, the domain of a relation is generally considered to be the set of all real numbers except those numbers where the relationship is undefined.

The range is the set of possible y values that result as each value of the domain is substituted in the relationship.

Example 1 □ Give the range and domain of the following relationships.

(a) $y = x^2$

Domain is all real numbers because x^2 is defined for all real numbers.

Range is $y \geq 0$ because there is no way to get a negative value from x^2.

(b) $y^2 = x$

Domain is $x \geq 0$ because y^2 is always positive and x must be equal to y^2.

Range is $y \in$ reals (pronounced y is an element of the reals) because any real numbers can be substituted for y. For instance $(9, -3)$ is part of this relation because $(-3)^2 = 9$ is true.

□ Figure 4.1

(c) $\dfrac{x^2}{9} + \dfrac{y^2}{25} = 1$

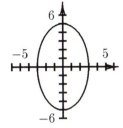

This is the ellipse on the right.

Domain: $-3 \le x \le 3$ because values of x greater than 3 or less than negative 3 do not yield real values for y.

Range: $-5 \le y \le +5$ because the acceptable values of x produce values of y in this range.

□

Using Your Graphing Calculator

Note

To graph equations, you must first solve for y as an explicit function of x. Sometimes this will yield a radical. For instance $x^2 + y^2 = 9$ yields $y = \pm\sqrt{9 - x^2}$. To graph equations involving $\pm$ radicals, you must enter two separate functions. Enter

$$y = \boxed{\sqrt{}}\,(9 - x^2) \text{ to graph the top of the circle.}$$

Then enter

$$y = \boxed{-}\,\boxed{\sqrt{}}\,(9 - x^2) \text{ to graph the lower half of the circle.}$$

This gives us the entire graph of

$$x^2 + y^2 = 9$$

Why didn't I get a round circle on the display?

You probably need to adjust the range for x and y. Read below.

Most graphing calculators have screens wider in the x direction than in the y direction. If you try to graph a circle with radius 4 by setting $X_{\min} = -5$, $X_{\max} = 5$, $Y_{\min} = -5$, $Y_{\max} = 5$, the graph will appear squashed vertically because the distance between the x ticks is less than the distance between the y tick marks. To compensate on a TI calculator try changing to $X_{\min} = -7.5$, $X_{\max} = 7.5$. TI calculators have a feature on the ZOOM menu called "Square" which will automatically adust the x and y axes so that the grid you are graphing on consists of squares. If your calculator does not have this feature, try making $(X_{\max} - X_{\min}) = 1.5\,(Y_{\max} - Y_{\min})$.

Definition 4.1C Function

A function is a relation where for each permissible replacement of the independent variable, there is exactly one value for the dependent variable. Another way to express this idea is that a function is a relation where no two ordered pairs have the same first component.

The key idea of a function is predictability. For each value of the independent variable there must be exactly one value of the dependent variable. All functions are relations, but not all relations are functions.

In the example above $y = x^2$ is a function because each possible value from the domain of x produces exactly one value for y.

$y^2 = x$ is a relation but not a function because each x value has two possible y values that can be associated with it. For example, $(9, 3)$ and $(9, -3)$ are both part of the relationship $y^2 = x$.

$\dfrac{x^2}{9} + \dfrac{y^2}{25} = 1$ is also a relation but not a function because each x value of the domain, except $x = 3$ or $x = -3$, has more than one y value associated with it.

Example 2 □ Tell if $y = \dfrac{1}{x}$ is a function and give its domain and range.

It is a function because for each x value there is only one y value. Its domain is $x \in$ reals, $x \neq 0$ because division by zero is undefined. Its range is $y \in$ reals, $y \neq 0$ because all y values are possible except $y = 0$ because $\dfrac{1}{x} = 0$ has no solution.

□

Example 3 □ What is the range and domain of the function $y = \sin x$?

Because the domain is not specified, x may be any real number. The range of $\sin x$ is $-1 \leq \sin x \leq +1$.

When an unbroken graph represents a function, the coordinates of every one of the infinite number of ordered pairs on the curve is a member of the function. □

Vertical Line Test for a Function

One test for a function is the vertical line test. A vertical line will cross the graph of a function only once.

□ Figure 4.2

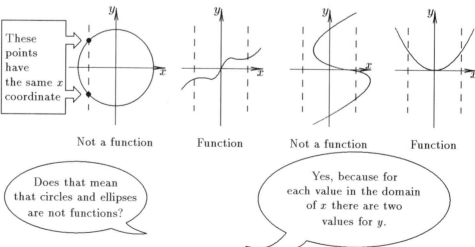

There are obviously many important relations that are not functions. Functions are a subset of relations. A special kind of function is a one-to-one function.

Chapter 4 Inverse Trigonometric Functions

Definition 4.1D One-to-One Function

A one-to-one function is a function in which each value of y has exactly one x value.

$y = x^3$ is an example of a one-to-one function. $y = x^2$ is a function but it is not a one-to-one function because each value in the range (y) has two values in the domain (x) associated with it.

If a graph can pass both the vertical and a similar horizontal line test, it is a one-to-one function.

□ Figure 4.3

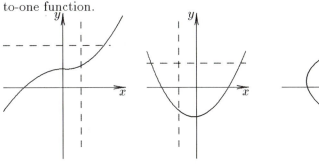

One-to-one
function

Function
not one-to-one

Not a function

One-to-one
function

$y = \sin x$ is an example of a function that is not a one-to-one function.

Example 4 □ Give the domain and range of $y = \csc x$.

□ Figure 4.4

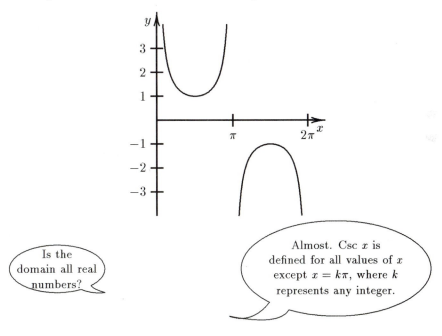

Is the
domain all real
numbers?

Almost. Csc x is
defined for all values of x
except $x = k\pi$, where k
represents any integer.

Therefore we say the domain is $x \in$ reals, $x \neq k\pi$.

In the case of the range of csc s, notice that $|\csc x|$ is always greater than or equal to 1.

We must exclude $-1 < y < +1$ from the range of $y = \csc x$.

The range is $|y| \geq 1$. Is $y = \csc x$ a function? _____

Yes

No

Is $y = \csc x$ a one-to-one function? _____

□

Inverse of a Relation

If we interchange x and y in a relation it has the effect of interchanging the first and second components of the ordered pairs that define the relation; we thus produce another relation called its inverse. The domain of the original relation becomes the range of its inverse and vice versa.

Example 5 □ Find the inverse relation of the relation below and graph it.

□ Figure 4.5

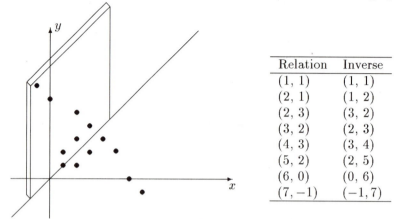

Relation	Inverse
(1, 1)	(1, 1)
(2, 1)	(1, 2)
(2, 3)	(3, 2)
(3, 2)	(2, 3)
(4, 3)	(3, 4)
(5, 2)	(2, 5)
(6, 0)	(0, 6)
(7, −1)	(−1, 7)

Notice, if you placed a mirror on the line $y = x$, the points of the inverse relation would be the reflections of the points of the original relation.

In general, the inverse of any relation is a reflection of that relation across the line $y = x$. You find any inverse of a relation by interchanging the coordinates (x, y) of each point to get (y, x). Therefore, a way to write the equation of the inverse of a relation is to interchange x and y in the original equation and solve for y.

□

Example 6 □ Write the equation of the inverse of the relation $y = x^2 + 1$.

To write the equation of the inverse, interchange x and y.

□ Figure 4.6

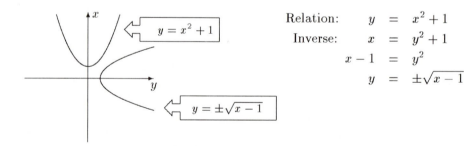

$$
\begin{aligned}
\text{Relation:} \quad y &= x^2 + 1 \\
\text{Inverse:} \quad x &= y^2 + 1 \\
x - 1 &= y^2 \\
y &= \pm\sqrt{x - 1}
\end{aligned}
$$

□

Which is the inverse?

Each equation is the inverse of the other equation.

Chapter 4 Inverse Trigonometric Functions

Example 7 □ Write the equation of the inverse of $y = x^3$.

Original equation: $\quad y \;=\; x^3$
Inverse: $\qquad\qquad\; x \;=\; y^3$
$\qquad\qquad\qquad\quad\; y \;=\; \sqrt[3]{x} \qquad$ (Solving for y)

□

Yes

Is the original equation a function? _____

How do you tell by looking at an equation if it's a function?

Normally, you'll have to sketch the graph and apply the vertical line test.

Yes

Is the inverse a function? _____

Is the inverse of a function always a function?

No, look at example 6. $y = \pm\sqrt{x - 1}$ does not pass the vertical line test.

The inverse of a one-to-one function, however, is always a function.

Pointer for Better Understanding:

Relations and Their Inverses

A relation, as the name implies, is a relationship between the first and second components of an ordered pair. Relationships, for example father to child, also imply a relationship in the opposite direction. Each father has a child and each child has a father. Notice that each child has exactly one father, analogous to a mathematical function. But, each father does not necessarily have one child, analogous to a mathematical relation. Father ↔ child, therefore, is not a one-to-one relationship.

An example of a one-to-one relationship is husband and wife.

Problem Set 4.1

By examining the graphs below and asking if there is exactly one y value for each x value, determine if the graph represents a function.

1.

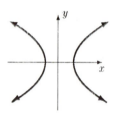

2.

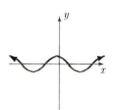

3.

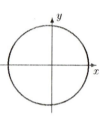

4.

5.

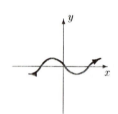

6.

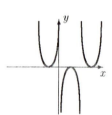

7.

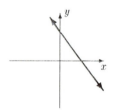

8.

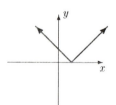

9.

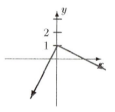

Assume that the following graphs are what they appear to be, that is, if a graph appears to touch the x-axis in one point, it really does. If it appears to cross the y-axis at $y = 3$, it does not cross at $y = 3.01$ or 2.99.

For the following graphs:

    (a) Determine the domain and range.

    (b) Determine if each of these graphs represents a one-to-one function.

10.

11.

12.

13.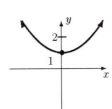

14.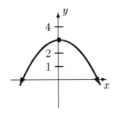

15.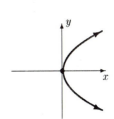

16.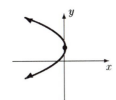

17.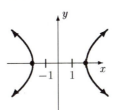

18.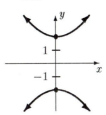

For the following equations:
  (a) Determine the domain and range.
  (b) Determine if the graphs of these equations represent one-to-one functions. (HINT: Use your graphing calculator to sketch a graph.)

19. $y = x^2 + 2$ **20.** $y = x^3$ **21.** $y = \tan x$

22. $y = \sin x$ **23.** $x = y^2$ **24.** $x - 3y = 8$

25. $y = |x|$ **26.** $y = |x - 2|$ **27.** $y = |x| + 2$

28. $y = |x + 3|$ **29.** $y = \cos x$ **30.** $y = \sec x$

For the following equations:
  (a) Write the inverse. Solve for y in terms of x.
  (b) Determine if the original equation is a function.
  (c) Use your graphing calculator to plot the graph of the inverse and apply a vertical line test to see if it is a function.

31. $2x + 3y = 8$ **32.** $y = x^2 + 3$ **33.** $y = x^3 - 3$

34. $y = 2 - x^2$ **35.** $y = 4 - x^3$ **36.** $x = 2y - 1$

37. $x = 2$ **38.** $x = y^2$ **39.** $y = 2$

40. $4x^2 - y^2 = 16$ **41.** $x^2 + 25y^2 = 25$ **42.** $x = y^3 - 3$

Write the inverse of the following equations. Sketch the graph of the original equation and its inverse on the same axis.

43. $y = 2x + 3$ **44.** $y^2 = x$ **45.** $y = x^2 + 2$

46. $3x + 2y = 6$ **47.** $y = 3 - x^2$ **48.** $y = x^2 - 2$

49. $x = 3 - y^2$ **50.** $x = y^3 - 2$

4.2   Inverses of the Trigonometric Functions

In the previous section, we found the inverse of a function by interchanging x and y in the equation that defined the function.

We can use that technique with trigonometric functions. Consider $y = \sin x$. This function is satisfied by ordered pairs of the form (x, y); one example of such a pair is $\left(\frac{1}{6}\pi, 0.5\right)$. Interchange x and y in the original equation. This gives $x = \sin y$. The ordered pairs found by reversing the elements of the original function satisfy the inverse. For example, $\left(\frac{1}{6}\pi, 0.5\right)$ becomes $\left(0.5, \frac{1}{6}\pi\right)$ after reversal which satisfies $x = \sin y$.

However we usually use a different notation for the inverse of a trigonometric function.

4.2A Notation for the Inverse of the Sine Relation

If $y = \sin x$, then its inverse relation is represented by either of the following:

$$y = \arcsin x \qquad \text{read "y is the arc sine of x"}$$
$$y = \sin^{-1} x \qquad \text{read "y is the inverse sine of x"}$$

Recall the definition of radian measure as the ratio of an arc length to the radius.

The expression $y = \arcsin x$ then means y is the arc whose sine is x. For example, if $x = 0.5$, then $y = \frac{1}{6}\pi$. Another way to say exactly the same thing is:

$$y = \sin^{-1} x$$

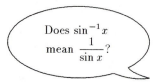

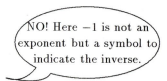

Notice that the symbol -1 is applied to the word **sin** and not the number **sin x**. The reciprocal of $\sin x$ is $(\sin x)^{-1}$:

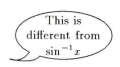

Example 8 □ Express the equation $\sin x = 0.5$, using an inverse trigonometric relation.

In $\sin x = 0.5$, 0.5 is the value of $\sin x$.

In inverse notation we say $\sin^{-1} 0.5 = x$.   □

Example 9 □ Graph $y = \sin x$ and its inverse $y = \sin^{-1} x$.

□ Figure 4.7

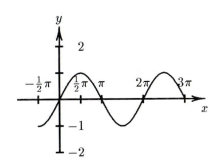

Function $y = \sin x$	
x	y
$-\pi/2$	-1
$-\pi/4$	-0.7
$-\pi/6$	-0.5
0	0
$\pi/6$	0.5
$\pi/4$	0.7
$\pi/2$	1.0
$5\pi/6$	0.5
π	0
$7\pi/6$	-0.5
$3\pi/2$	-1
2π	0

Inverse $y = \sin^{-1}$	
x	y
-1	$-\pi/2$
-0.7	$-\pi/4$
-0.5	$-\pi/6$
0	0
0.5	$\pi/6$
0.7	$\pi/4$
1.0	$\pi/2$
0.5	$5\pi/6$
0	π
-0.5	$7\pi/6$
-1	$3\pi/2$
0	2π

□

Inverse of $y = $ Sine x

The previous section indicated that the inverse of a function may or may not be a function. For example, $y = \sin x$ is a function because for every value of x there is one and only one value for y. $y = \sin^{-1} x$ is a different story; if $x = 0.5$, y could be $\frac{1}{6}\pi, \frac{5}{6}\pi, \frac{13}{6}\pi, \frac{17}{6}\pi$ to name only a few. We would like the inverse of the sine to be a function so that each input produces only one output. To do this we must restrict the range of the inverse sine function.

We say that $y = \mathrm{Sin}^{-1} x$ is defined for value of y only between $-\frac{1}{2}\pi$ and $\frac{1}{2}\pi$.

Why pick that range?

To pick a set of outputs that match every possible valid input for sin x without any duplicate outputs. The most reasonable range is $-\frac{1}{2}\pi \le y \le \frac{1}{2}\pi$.

Graph of $y = \sin^{-1}x$

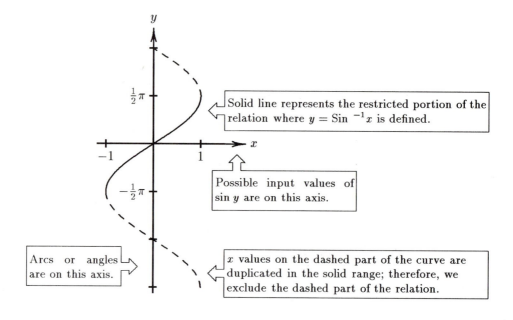

Solid line represents the restricted portion of the relation where $y = \text{Sin}^{-1}x$ is defined.

Possible input values of $\sin y$ are on this axis.

Arcs or angles are on this axis.

x values on the dashed part of the curve are duplicated in the solid range; therefore, we exclude the dashed part of the relation.

4.2B   The Domain and Range of $\text{Sin}^{-1}x$

The function $y = \text{Sin}^{-1}x$ is defined for the domain $-1 \leq x \leq +1$ and for the range $-\frac{1}{2}\pi \leq y \leq +\frac{1}{2}\pi$.

Why do you write $y = \sin^{-1}x$ sometimes and $y = \text{Sin}^{-1}x$ other times? When is it capitalized?

The upper-case indicates that we mean the restricted values of y so that $\text{Sin}^{-1}x$ is a function with values in the defined range $-\frac{1}{2}\pi \leq \text{Sin}^{-1}x \leq \frac{1}{2}\pi$.

Example 10 ☐ Without using your calculator, find the value for $\text{Sin}^{-1}(-0.5)$.

How am I supposed to know that?

We expect you to recognize the coordinates of the points at $\frac{1}{6}\pi, \frac{1}{4}\pi, \frac{1}{3}\pi$ on a unit circle. See the pointer on pages 66–67.

Chapter 4 Inverse Trigonometric Functions

We can rewrite $\text{Sin}^{-1}(-0.5)$ without using inverse function as $\sin\theta = -0.5$. There are two values less than 2π where $\sin\theta = -0.5$.

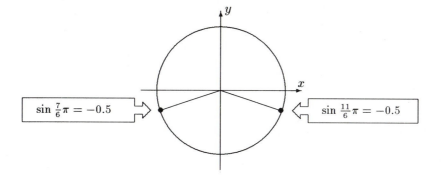

$$\boxed{\sin\tfrac{7}{6}\pi = -0.5}$$ $$\boxed{\sin\tfrac{11}{6}\pi = -0.5}$$

The capital **S** on $\text{Sin}^{-1}(-0.5)$ tells us that the range of $\text{Sin}^{-1}x$ is restricted so that $-\dfrac{1}{2}\pi \le \text{Sin}^{-1}x \le \dfrac{1}{2}\pi$. We will express $\dfrac{11}{6}\pi$ as $-\dfrac{1}{6}\pi$.

Then $\text{Sin}^{-1}(-0.5) = -\dfrac{1}{6}\pi$

☐

Example 11 ☐ Find the exact value of $\text{Sin}^{-1}\left(\dfrac{\sqrt{3}}{2}\right)$.

> These words, **exact value**, mean "do it without a calculator."

> Use your knowledge of the unit circle.

From the unit circle $\sin\left(\dfrac{1}{3}\pi\right) = \dfrac{\sqrt{3}}{2}$.

Therefore $\text{Sin}^{-1}\left(\dfrac{\sqrt{3}}{2}\right) = \dfrac{1}{3}\pi$.

☐

Inverse of $y = \text{Cosine } x$

4.2C Notation for the Inverse of the Cosine Function

If $y = \cos x$, then the inverse relation is represented by the following:

$$y = \arccos x$$
$$y = \cos^{-1}x$$

This still leaves
the problem that each
value of cosine has many
possible angles that could
be paired with it.

That's why we
call it the **inverse relation**.
To make a function we must
restrict the range.

☐ Figure 4.10

Graph of $y = \cos^{-1} x$

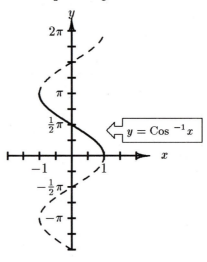

$y = \text{Cos}^{-1} x$

Imagine that you are a member of a committee charged with the task of selecting an appropriate restriction for arccos x. The requirements are:

1. Keep it simple.

2. Insure that arccos x can assume all possible values of x between -1 and $+1$.

3. Guarantee that only one y-value corresponds to each x-value.

To select an appropriate domain for $\cos \theta$, consider a unit circle. Cos θ is the x-coordinate of a point on the circle.

☐ Figure 4.11

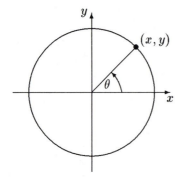

If θ varies from 0 to π, $\cos \theta$ will assume all possible values between -1 and $+1$.

Chapter 4 Inverse Trigonometric Functions

4.2D The Domain and Range of Cos^{-1}x

The function $y = \text{Cos}^{-1}x$ is defined for the domain $-1 \leq x \leq +1$ and for the range $0 \leq y \leq \pi$.

Example 12 □ Without using your calculator, find the value of x so that $x = \text{Cos}^{-1}\left(-\dfrac{\sqrt{2}}{2}\right)$.

First rewrite $x = \text{Cos}^{-1}\left(-\dfrac{\sqrt{2}}{2}\right)$ without using inverse notation. Within the range of the inverse cosine, $x = \text{Cos}^{-1}\left(-\dfrac{\sqrt{2}}{2}\right)$ is equivalent to $\cos x = -\dfrac{\sqrt{2}}{2}$.

Cos x is negative in the second and third quadrants. But the inverse cosine is defined only in the range 0 to π. Therefore we choose x in the second quadrant so that:

$$\cos x = -\frac{\sqrt{2}}{2}$$

Using our knowledge of the unit circle:

$$\theta = \frac{3}{4}\pi$$

□

Inverse of $y = $ Tangent x

The inverse function of $\tan x$ has similar considerations for $\text{Cos}^{-1}x$ and $\text{Sin}^{-1}x$.

4.2E Notation for the Inverse of the Tangent Function

If $y = \tan x$, the inverse function is represented by

$$y = \text{Tan}^{-1}x$$
$$\text{or} \quad y = \arctan x$$

where the domain of x is any real number but the range of y is $-\dfrac{1}{2}\pi < y < \dfrac{1}{2}\pi$.

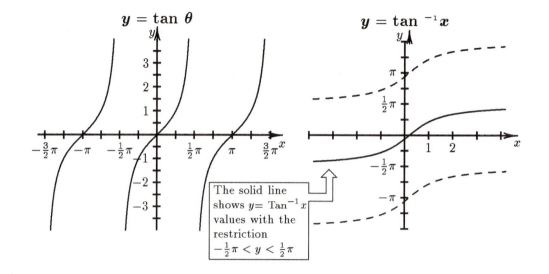

Example 13 □ Find the value of θ so that $\theta = \text{Tan}^{-1}(-\sqrt{3})$ without using your calculator.

Where $\text{Tan}^{-1}x$ is defined:

$$\theta = \text{Tan}^{-1}(-\sqrt{3}) \text{ is equivalent to } \tan\theta = -\sqrt{3}.$$

Tan θ is negative in the second and fourth quadrants but is defined from only $-\frac{1}{2}\pi$ to $\frac{1}{2}\pi$. Therefore we pick the value in the fourth quadrant where:

$$\tan\theta = -\sqrt{3}$$

Because $\tan\theta = \dfrac{y}{x}$, and, on a unit circle at $\theta = -\dfrac{1}{3}\pi, x = \dfrac{1}{2}$ and $y = -\dfrac{\sqrt{3}}{2}$:

□ Figure 4.13

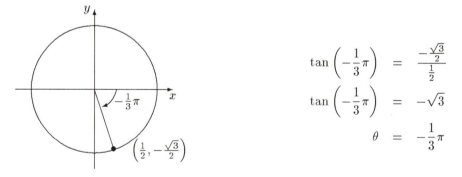

$$\tan\left(-\frac{1}{3}\pi\right) = \frac{-\frac{\sqrt{3}}{2}}{\frac{1}{2}}$$

$$\tan\left(-\frac{1}{3}\pi\right) = -\sqrt{3}$$

$$\theta = -\frac{1}{3}\pi$$

□

It is also possible to define the inverse functions for $y = \sec x, y = \csc x$, and $y = \cot x$. The idea is to pick a range of values for x so that for each possible value of y there will be only one value of x.

Inverses of the Trigonometric Functions

□ Figure 4.14

$$y = \text{Sin}^{-1} x$$

Domain: $-1 \leq x \leq +1$

Range: $-\dfrac{1}{2}\pi \leq y \leq \dfrac{1}{2}\pi$

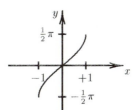

□ Figure 4.15

$$y = \text{Cos}^{-1} x$$

Domain: $-1 \leq x \leq +1$

Range: $0 \leq y \leq \pi$

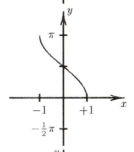

□ Figure 4.16

$$y = \text{Tan}^{-1} x$$

Domain: $x \in \text{reals}$

Range: $-\dfrac{1}{2}\pi < y < \dfrac{1}{2}\pi$

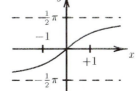

□ Figure 4.17

$$y = \text{Csc}^{-1} x$$

Domain: $1 \leq |x|$

Range: $0 < |y| \leq \dfrac{1}{2}\pi$

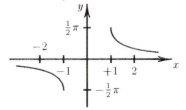

Sometimes the range of $y = \text{Csc}^{-1} x$ is defined as $0 < y \leq \dfrac{1}{2}\pi$ AND $\pi < y \leq \dfrac{3}{2}\pi$. The implications of these choices are usually discussed in a calculus course.

□ Figure 4.18

$$y = \text{Sec}^{-1} x$$

Domain: $1 \leq |x|$

Range: $0 \leq y \leq \pi, y \neq \dfrac{1}{2}\pi$

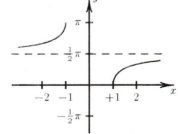

□ Figure 4.19

Sometimes the range of $y = \text{Sec}^{-1} x$ is defined as $0 \leq y < \dfrac{1}{2}\pi$ AND $\pi \leq y < \dfrac{3}{2}\pi$.

$$y = \text{Cot}^{-1} x$$

Domain: $x \in \text{reals}$

Range: $0 < y < \pi$

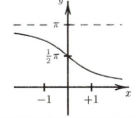

Problem Set 4.2A

Write each of the following relationships without using inverse functions. Do not evaluate.

1. $\mathrm{Cos}^{-1}\left(-\dfrac{\sqrt{3}}{2}\right) = x$

2. $\mathrm{Sin}^{-1}\left(-\dfrac{\sqrt{2}}{2}\right) = x$

3. $\mathrm{Tan}^{-1}(0.001) = x$

4. $\mathrm{Cot}^{-1}(100) = x$

5. $\mathrm{Sec}^{-1}(5) = \theta$

6. $\mathrm{Csc}^{-1}(4) = \theta$

7. $\mathrm{Sin}^{-1}(-0.005) = \theta$

8. $\mathrm{Cos}^{-1}(-1.0) = \theta$

Assume x is within the domain of the appropriate inverse trigonometric function. Write each of the following equations as an inverse trigonometric relationship. Do not evaluate the angle.

9. $\sin x = \dfrac{\sqrt{3}}{2}$

10. $\cos x = \dfrac{\sqrt{2}}{2}$

11. $\tan x = 1.2$

12. $\cot x = 0.002$

Using your knowledge of standard angles, give the exact value of y in radians.

13. $y = \mathrm{Sin}^{-1}\dfrac{1}{2}$

14. $y = \mathrm{Cos}^{-1}\left(-\dfrac{\sqrt{3}}{2}\right)$

15. $y = \mathrm{Tan}^{-1}\sqrt{3}$

16. $y = \mathrm{Cos}^{-1}0$

17. $y = \mathrm{Sin}^{-1}\left(-\dfrac{\sqrt{3}}{2}\right)$

18. $y = \mathrm{Tan}^{-1}1$

19. $y = \mathrm{Cos}^{-1}\left(\dfrac{\sqrt{3}}{2}\right)$

20. $y = \mathrm{Tan}^{-1}\left(\dfrac{\sqrt{3}}{3}\right)$

21. $y = \mathrm{Sin}^{-1}\left(-\dfrac{1}{2}\right)$

22. $y = \mathrm{Cot}^{-1}\sqrt{3}$

23. $y = \mathrm{Sec}^{-1}2$

24. $y = \mathrm{Csc}^{-1}\sqrt{2}$

25. Use your graphing calculator to plot a graph of $y = \mathrm{Sin}^{-1}x$ from $X_{\mathrm{min}} = -6.28$ to $X_{\mathrm{max}} = 6.28$. Why doesn't your calculator put values for points like $x = -6, -4, -2$?

Practice with Inverse Functions

In Chapter 1 you worked problems where you were given the value of a trigonometric function of θ as a fraction. You used that value to find the coordinates of a point on the terminal side of the angle θ. From the coordinates of that point and its distance from the origin you were able to calculate all six trigonometric functions of θ. In the next example we will use a similar process to help develop your understanding of inverse trigonometric functions.

Example 14 □ Find the value of $\sin\left[\mathrm{Cos}^{-1}\left(-\dfrac{3}{5}\right)\right]$ without using a calculator.

The value of $\mathrm{Cos}^{-1}\left(-\dfrac{3}{5}\right)$ is not immediately evident, so we will let $\alpha = \mathrm{Cos}^{-1}\left(-\dfrac{3}{5}\right)$. This can be written

$$\cos \alpha = -\frac{3}{5}$$

$\cos \alpha$ is negative, and $\mathrm{Cos}^{-1}y$ is defined only in the first and second quadrants, therefore the terminal side of the angle α must be in the second quadrant.

Because $\cos \alpha = \dfrac{x}{r} = -\dfrac{3}{5}$, a point on a circle with radius 5 and the x-coordinate of -3 will be on the terminal side of angle α.

We must find the y-coordinate.

□ Figure 4.20

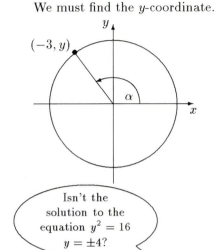

Using the equation of a circle:

$$
\begin{aligned}
x^2 + y^2 &= r^2 \\
(-3)^2 + y^2 &= 5^2 \\
9 + y^2 &= 25 \\
y^2 &= 16 \\
y &= 4
\end{aligned}
$$

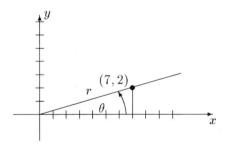

Isn't the solution to the equation $y^2 = 16$ $y = \pm 4$?

Correct, but in the second quadrant y is positive.

Now that we know $x, y,$ and r, we can evaluate $\sin \alpha$ directly.

$$\sin \alpha = \frac{y}{r} = \frac{4}{5}$$

Therefore

$$\sin \left[\text{Cos}^{-1} \left(-\frac{3}{5} \right) \right] = \frac{4}{5}$$

□

Example 15 □ Find the value of $\sec \left[\text{Tan}^{-1} \left(\dfrac{2}{7} \right) \right]$ without using your calculator.

$\text{Tan}^{-1} \left(\dfrac{2}{7} \right) = \theta$ can be rewritten without inverse notation as $\tan \theta = \dfrac{2}{7}$, provided θ is in the range of the inverse tangent. To keep θ in the range of the inverse tangent and have $\tan \theta$ positive, θ must be in the first quadrant.

□ Figure 4.21

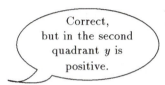

Because $\tan \theta = \dfrac{2}{7} = \dfrac{y}{x}$, there is a point on the terminal side of angle θ with coordinate $(7, 2)$. At this point the value of r can be found from the equation.

$$
\begin{aligned}
x^2 + y^2 &= r^2 \\
(7)^2 + (2)^2 &= r^2 \\
49 + 4 &= r^2 \\
53 &= r^2 \\
\sqrt{53} &= r
\end{aligned}
$$

Knowing $x, y,$ and r, we can write the value of any trigonometric function of θ.

$$
\begin{aligned}
\sec \theta &= \frac{r}{x} \\
&= \frac{\sqrt{53}}{7}
\end{aligned}
$$

□

Problem Set 4.2B

Use the method shown in Example 14 to evaluate the following. Write your answers as fractions. Do not rationalize denominators.

1. $\cos\left[\text{Cos}^{-1}\left(\dfrac{1}{2}\right)\right]$ 2. $\sin\left[\text{Cos}^{-1}\left(-\dfrac{\sqrt{3}}{2}\right)\right]$ 3. $\sin\left[\text{Sin}^{-1}\left(-\dfrac{1}{2}\right)\right]$

4. $\cos\left[\text{Sin}^{-1}\left(\dfrac{4}{5}\right)\right]$ 5. $\sin\left[\text{Cos}^{-1}\left(-\dfrac{12}{13}\right)\right]$ 6. $\tan\left[\text{Sin}^{-1}\left(\dfrac{5}{13}\right)\right]$

7. $\cos\left[\text{Sin}^{-1}\left(-\dfrac{4}{5}\right)\right]$ 8. $\tan\left[\text{Cos}^{-1}\left(-\dfrac{5}{13}\right)\right]$ 9. $\sec\left[\text{Sin}^{-1}\left(\dfrac{3}{5}\right)\right]$

10. $\cot\left[\text{Cos}^{-1}\left(-\dfrac{3}{5}\right)\right]$ 11. $\csc\left[\text{Cot}^{-1}\left(-\dfrac{12}{5}\right)\right]$

For which values of x are the following statements true?

12. $\text{Sin}^{-1}(\sin x) = x$ 13. $\text{Cos}^{-1}(\cos x) = x$ 14. $\text{Tan}^{-1}(\tan x) = x$

4.3 Finding Inverses of Trigonometric Functions Using a Calculator

Finding inverse trigonometric functions is not limited to the few values for which we have memorized coordinates of points on the unit circle. A calculator can provide approximations of any inverse trigonometric function so long as the input is in the domain of the function.

 Using Your Calculator

Use your calculator to find $\text{Sin}^{-1}(0.9205)$.

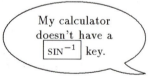 My calculator doesn't have a $\boxed{\text{SIN}^{-1}}$ key.

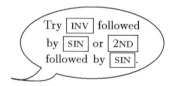 Try $\boxed{\text{INV}}$ followed by $\boxed{\text{SIN}}$ or $\boxed{\text{2ND}}$ followed by $\boxed{\text{SIN}}$.

Before you can get an answer that makes sense, you need to know if you are trying to map real numbers (the value of the sine function) to a real number or to an angle in degrees.

When you set the mode of calculator to DEG or RAD you make this choice.

Choosing RAD mode
$$\text{Sin}^{-1}(0.9205) \approx 1.1693.$$

Example 16 □ Find the value of θ in degrees if $\cos \theta = 0.19$.

In the range of inverse cosine:

$$\cos \theta = 0.19 \text{ is equivalent to } \text{Cos}^{-1}(0.19) = \theta$$

Set your calculator mode to degrees then press $\boxed{2\text{ND}}$ $\boxed{\text{COS}}$ followed by .19 $\boxed{\text{ENTER}}$ to get $\theta \approx 79°$.

□

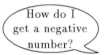

Using Your Calculator

Example 17 □ Use your calculator to find a value of $\text{Sin}^{-1}(-0.80)$ between $-\frac{1}{2}\pi$ and $\frac{1}{2}\pi$.

Because the range of accepted values is expressed in radians rather than degrees, set your calculator to the radian mode. Then enter -0.80.

How do I get a negative number?

Make sure you press the negative key, not the subtraction key.

Press $\boxed{2\text{ND}}$ $\boxed{\text{SIN}}$ $\boxed{(-)}$ $\boxed{.}$ $\boxed{8}$ $\boxed{\text{ENTER}}$. The value -0.9272952 should appear.

Therefore:

$$\text{Sin}^{-1}(-0.80) = -0.9272 \text{ to four places}$$

Notice the calculator provides a value from the domain of $y = \text{Sin}^{-1}x$.

The manufacturer carefully planned that. Otherwise calculators would have to provide an infinite number of values for $\text{Sin}^{-1}(-.80)$.

Example 18 □ Find the value of x so that $\tan x = -1.732$.

In the range of the inverse tangent:

$$\tan x = (-1.732) \text{ is equivalent to } \text{Tan}^{-1}(-1.732) = x.$$

The variable used for the angle is x instead of θ, which indicates that the angle should be expressed in radians. Therefore, set your calculator in radian mode.

Press $\boxed{2\text{ND}}$ $\boxed{\text{TAN}}$ $\boxed{(-)}$ then type 1.732. Press $\boxed{\text{ENTER}}$.
You should get: $x = -1.047$.

□

4.3 Finding Inverses of Trigonometric Functions Using a Calculator

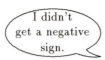

Notice in the example above that the value of x is very close to $-\dfrac{1}{3}\pi$.

Example 19 □ Use a calculator to find $\tan\left[\operatorname{Sin}^{-1}(0.342)\right]$.

$$\operatorname{Sin}^{-1}(0.342) \approx 20°$$
$$\tan 20° \approx 0.3640$$

□

Example 20 □ In radian measure what is $\operatorname{Sin}^{-1}(\sin 0.4)$?

$$\sin 0.4 \approx 0.3894183$$
$$\operatorname{Sin}^{-1}(0.3894183) \approx 0.4$$
$$\text{Therefore: } \operatorname{Sin}^{-1}(\sin 0.4) = 0.4$$

□

Inverse operations have the property of undoing each other. For example the inverse operation of writing on a board is erasing the board.

closing it

The inverse operation of opening a door is _____

Example 21 □ In radian mode find $\cos\left(\operatorname{Cos}^{-1}0\right)$.

$\operatorname{Cos}^{-1}0 = \dfrac{1}{2}\pi$ because $\dfrac{1}{2}\pi$ is the angle with a cosine of 0 and $\cos\dfrac{1}{2}\pi = 0$.

Therefore $\cos\left(\operatorname{Cos}^{-1}0\right) = 0$ □

This is a little like asking, "Who is Mary's daughter's mother?"

Mary. Each child has only one mother.

Chapter 4 Inverse Trigonometric Functions

Example 22 □ In degree mode, find $\mathrm{Csc}^{-1}(2)$.

Because calculators don't have a key for csc, rewrite this in terms of $\sin\theta$.

$$
\begin{aligned}
\text{Let } \theta &= \mathrm{Csc}^{-1}(2). \\
\text{Then } \csc\theta &= 2 \\
\text{but } \csc\theta &= \frac{1}{\sin\theta} \\
2 &= \frac{1}{\sin\theta} \\
2\sin\theta &= 1 \\
\sin\theta &= \frac{1}{2} \\
\theta &= 30° \\
\text{So } \mathrm{Csc}^{-1}(2) &= 30°
\end{aligned}
$$

□

Example 23 □ Use your calculator to find $\mathrm{Cot}^{-1}(5)$.

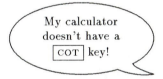

My calculator doesn't have a $\boxed{\text{COT}}$ key!

Most don't. So express this as a tangent.

If $\theta = \mathrm{Cot}^{-1}(5)$ then $\cot\theta = 5$;

$$
\begin{aligned}
\text{Because } \tan\theta &= \frac{1}{\cot\theta} \\
\tan\theta &= \frac{1}{5} \qquad \text{or, in brief, } \mathrm{Cot}^{-1}(5) = \mathrm{Tan}^{-1}\left(\frac{1}{5}\right)
\end{aligned}
$$

Now use your calculator:

$$
\begin{aligned}
\tan\theta &= 0.2000 \\
\theta &\approx 11.3° \\
\text{so } \mathrm{Cot}^{-1}(5) &\approx 11.3°
\end{aligned}
$$

□

Using Your Calculator

Most calculators do not have a key for $\sec x, \csc x, \cot x$ or their inverses. We can still evaluate these functions if we recall the following:

$$\csc \theta = \frac{1}{\sin \theta}$$

$$\sec \theta = \frac{1}{\cos \theta}$$

$$\cot \theta = \frac{1}{\tan \theta}$$

Therefore, if we wish to evaluate $\mathrm{Csc}^{-1}(2)$, we recall that it is the same angle as $\mathrm{Sin}^{-1}\left(\frac{1}{2}\right)$. In general,

$$\mathrm{Csc}^{-1}x = \mathrm{Sin}^{-1}\left(\frac{1}{x}\right)$$

$$\mathrm{Sec}^{-1}x = \mathrm{Cos}^{-1}\left(\frac{1}{x}\right)$$

$$\mathrm{Cot}^{-1}x = \mathrm{Tan}^{-1}\left(\frac{1}{x}\right)$$

Group Writing Activity

Using Example 22 as a model of a specific case, derive the general formulas above.

Problem Set 4.3

Find the value of x in radians between 0 and 1.5708 for the following. Give your result correct to four decimal places.

1. $\sin x = 0.3019$ **2.** $\cos x = 0.9002$ **3.** $\csc x = 3.0015$

4. $\sec x = 4.1602$ **5.** $\tan x = 8.0102$ **6.** $\cot x = 0.8012$

Use your calculator to find the value of the following correct to the nearest degree.

7. $\mathrm{Sin}^{-1}(0.7219)$ **8.** $\mathrm{Cos}^{-1}(0.6015)$ **9.** $\mathrm{Tan}^{-1}(2.8906)$

10. $\mathrm{Sin}^{-1}(-0.5621)$ **11.** $\mathrm{Tan}^{-1}(-0.8712)$ **12.** $\mathrm{Cos}^{-1}(-0.9284)$

13. $\mathrm{Sec}^{-1}(3.9012)$ **14.** $\mathrm{Tan}^{-1}(-0.1862)$

Use your calculator to find the value of the following in radians correct to two decimal places.

15. $\mathrm{Cos}^{-1}(0.5291)$ **16.** $\mathrm{Sin}^{-1}(0.1286)$ **17.** $\mathrm{Tan}^{-1}(0.8294)$

18. $\mathrm{Cos}^{-1}(-0.9586)$ **19.** $\mathrm{Sin}^{-1}(-0.7291)$ **20.** $\mathrm{Sec}^{-1}(-1.3481)$

21. $\mathrm{Csc}^{-1}(1.8432)$ **22.** $\mathrm{Cot}^{-1}(0.5398)$

Evaluate each of the following. If the answer is an angle, give your answer to the nearest degree. If the answer is a number, round to four decimal places.

23. $\tan\left[\text{Sin}^{-1}(0.6598)\right]$ **24.** $\cos\left[\text{Tan}^{-1}(-2.6932)\right]$

25. $\cot\left[\text{Tan}^{-1}(2.8600)\right]$ **26.** $\text{Cos}^{-1}(\cos 75°)$

27. $\text{Sin}^{-1}(\tan 30°)$ **28.** $\text{Tan}^{-1}(\cot 26°)$

29. $\text{Cot}^{-1}(\sec 40°)$ **30.** $\text{Csc}^{-1}(\cot 15°)$

Chapter 4   Key Ideas

4.1 **1.** The independent variable may assume any one of the values in the replacement set.

2. The dependent variable is the resulting value after a value has been assigned to the independent variable.

3. A relation is a connection between two variables.

4. The domain of a relation is the set of all possible replacements of the independent variable.

5. The range of a relation is the set of all possible values of the dependent variable.

6. A function is a relation in which for each value of the independent variable there is one and only one value of the dependent variable.

7. One test for a function is the vertical line test. A vertical line will cross the graph of a function only once.

8. A one-to-one function is a function in which each value of the dependent variable has exactly one value of the independent variable associated with it.

9. The inverse of a relation is a relation in which the first and second components of the ordered pairs that define the relation have been interchanged.

10. The inverse of a function is not necessarily a function.

4.2 **1.** The inverse of $y = \sin x$ is $y = \sin^{-1} x$ or $y = \arcsin x$.

2. The inverse of $y = \cos x$ is $y = \cos^{-1} x$ or $y = \arccos x$.

3. The inverse of $y = \tan x$ is $y = \tan^{-1} x$ or $y = \arctan x$.

4. The domain of $y = \text{Sin}^{-1} x$ is $-1 \le x \le 1$.

5. The range of $y = \text{Sin}^{-1} x$ is $-\frac{1}{2}\pi \le y \le \frac{1}{2}\pi$.

6. The domain of $y = \text{Cos}^{-1} x$ is $-1 \le x \le 1$.

7. The range of $y = \text{Cos}^{-1} x$ is $0 \le y \le \pi$.

8. The domain of $y = \text{Tan}^{-1} x$ is $x \in R$.

9. The range of $y = \text{Tan}^{-1} x$ is $-\frac{1}{2}\pi < y < \frac{1}{2}\pi$.

10. $\text{Csc}^{-1} x = \text{Sin}^{-1}\left(\dfrac{1}{x}\right)$

11. $\text{Sec}^{-1} x = \text{Cos}^{-1}\left(\dfrac{1}{x}\right)$

12. $\text{Cot}^{-1} x = \text{Tan}^{-1}\left(\dfrac{1}{x}\right)$

Chapter 4   Review Test

For the following graphs:
  (a) Determine the domain and range.
  (b) Determine if the graph represents a function. **(4.1)**

1.   **2.**   **3.**   **4.**

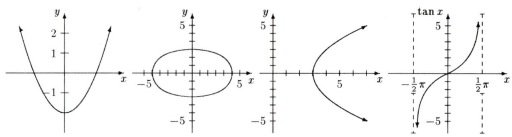

For the following equations:
  (a) Determine the domain and range.
  (b) Determine if they are one-to-one. **(4.1)**

5. $y = x^2 - 1$ **6.** $x = y^2 + 2$ **7.** $x + 3y = 6$

For the following equations:
  (a) Write the inverse. Solve for y in terms of x.
  (b) Determine if the original equation is a function.
  (c) Determine if the inverse is a function. **(4.1)**

8. $y = 3x - 6$ **9.** $x^2 - 4y^2 = 1$ **10.** $y = x^3 + 4$ **11.** $y = x^2 - 3$

Give the exact value of the following expressions in radians. **(4.2)**

12. $\text{Sin}^{-1}\left(-\dfrac{1}{2}\right)$ **13.** $\text{Cos}^{-1}(-1)$ **14.** $\text{Tan}^{-1}(-\sqrt{3})$

15. $\text{Cot}^{-1}(\sqrt{3})$ **16.** $\text{Sec}^{-1}(\sqrt{2})$ **17.** $\text{Csc}^{-1}\left(\dfrac{2\sqrt{3}}{3}\right)$

Use your calculator to find the value of the following correct to the nearest degree. **(4.3)**

18. $\text{Sin}^{-1}(-0.8692)$ **19.** $\text{Cos}^{-1}(-0.4395)$ **20.** $\text{Sec}^{-1}(2.6845)$

Use your calculator to find the value of the following in radians correct to three decimal places. **(4.2)**

21. $\text{Tan}^{-1}(-3.68)$ **22.** $\text{Csc}^{-1}(2.84)$ **23.** $\text{Sin}^{-1}(-0.829)$

Evaluate each of the following correct to the nearest degree if the answer is an angle, and to four decimal places if your answer is a number. **(4.2)**

24. $\tan\left[\text{Cos}^{-1}(-0.2651)\right]$ **25.** $\text{Cos}^{-1}(\sin 57°)$ **26.** $\sec\left[\text{Tan}^{-1}(1.53)\right]$

Evaluate the following. **(4.2)**

27. $\tan\left[\text{Cos}^{-1}\left(-\dfrac{4}{5}\right)\right]$ **28.** $\csc\left[\text{Tan}^{-1}\left(\dfrac{3}{4}\right)\right]$ **29.** $\sin\left[\text{Sec}^{-1}\left(\dfrac{5}{3}\right)\right]$

Chapter 5

Basic Trigonometric Identities

Contents

Preview

Just as factoring helps us to simplify algebraic expressions, trigonometric relations help us to simplify and rewrite trigonometric expressions. When trigonometry is applied to real world applications, additions, subtractions, multiplications, and divisions can lead to rather imposing expressions. Trigonometric identities help us reduce these complex expressions to more convenient forms.

As you work with trigonometric identities, you will need to apply many of the skills of manipulation for algebraic expressions you learned in algebra. If you find this chapter difficult, we suggest that you review the chapter on factoring and rational expressions in your algebra text.

5.1 Fundamental Identities

Because identities are a special kind of equation, we will start this chapter by defining some terms that help us talk about equations and identities.

5.1A Definition of A Conditional Equation

A conditional equation is a statement that is true on condition that the variable is replaced with the proper value.

Examples of a conditional equation are:

$$x + 3 \;=\; 5 \quad \text{true if } x \;=\; 2$$
$$2x \;=\; 8 \quad \text{true if } x \;=\; 4$$

5.1B Definition of an Identity

An identity is an equation that is true for all valid replacements of the variable.

Examples of an identity are:

$$x + x \;=\; 2x \qquad \text{true for all } x$$

$$\frac{x}{x} \;=\; 1 \qquad \text{true for all } x, x \neq 0$$

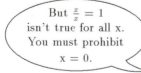

But $\frac{x}{x} = 1$ isn't true for all x. You must prohibit x = 0.

True. That's why we say all **valid** replacements of the variable. In trigonometric identities we need to be careful with this.

The definitions of the six trigonometric functions are repeated below for reference. Bear in mind that these definitions are valid only when the denominator is not equal to zero. Although we will not continue to state the restriction each time, it still applies.

5.1C Definition of the Trigonometric Functions

If the terminal side of an angle is rotated through an angle θ, then for any point on the rotating side at a distance r from the vertex

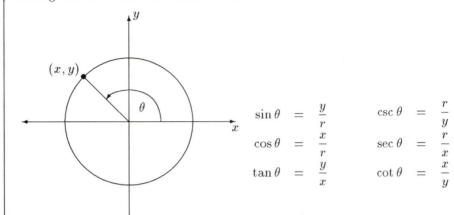

□ Figure 5.1

$$\sin \theta = \frac{y}{r} \qquad \csc \theta = \frac{r}{y}$$

$$\cos \theta = \frac{x}{r} \qquad \sec \theta = \frac{r}{x}$$

$$\tan \theta = \frac{y}{x} \qquad \cot \theta = \frac{x}{y}$$

Six trigonometric identities that follow immediately from the definitions above are called the reciprocal identities.

5.1D The Reciprocal Identities

$$(1) \quad \sin \theta = \frac{1}{\csc \theta} \qquad\qquad (4) \quad \csc \theta = \frac{1}{\sin \theta}$$

$$(2) \quad \cos \theta = \frac{1}{\sec \theta} \qquad\qquad (5) \quad \sec \theta = \frac{1}{\cos \theta}$$

$$(3) \quad \tan \theta = \frac{1}{\cot \theta} \qquad\qquad (6) \quad \cot \theta = \frac{1}{\tan \theta}$$

Example 1 □ List the value of θ where $\csc \theta = \dfrac{1}{\sin \theta}$ is undefined.

This identity is undefined when $\sin \theta = 0$ because that would involve division by zero. Therefore, $\csc \theta = \dfrac{1}{\sin \theta}$ where $\theta \neq k\pi$.

□

What's $k\pi$?

The k stands
for any integer
$\ldots$-2, -1, 0, +1, +2 $\ldots$
so $k\pi$ represents
$\ldots$-2π, -π, 0, π, 2π $\ldots$

Product Identities   The other reciprocal identities have similar restrictions about division by zero. Three other product identities follow directly from the definitions of the trigonometric functions.

$$(7) \qquad \sin \theta \cdot \csc \theta = 1$$
$$\frac{y}{r} \cdot \frac{r}{y} = 1$$
$$1 = 1$$

$$(8) \qquad \cos \theta \cdot \sec \theta = 1$$
$$\frac{x}{r} \cdot \frac{r}{x} = 1$$
$$1 = 1$$

A third product identity is

$$(9) \qquad \tan \theta \cdot \cot \theta = 1$$

Ratio Identities   From the definitions of the trigonometric functions, we can also derive two ratio identities.

The first ratio identity follows from the definition of $\tan \theta$.

$$\tan \theta = \frac{y}{x}$$

To get the new identity, divide both the numerator and denominator by r.

$$\tan \theta = \frac{\dfrac{y}{r}}{\dfrac{x}{r}}$$

$$(10) \qquad \tan \theta = \frac{\sin \theta}{\cos \theta}$$

The second ratio identity follows from the definition of $\cot \theta$:

$$\cot \theta = \frac{x}{y}$$

Dividing numerator and denominator by r

$$\cot \theta = \frac{\dfrac{x}{r}}{\dfrac{y}{r}}$$

$$(11) \qquad \cot \theta = \frac{\cos \theta}{\sin \theta}$$

Chapter 5 Basic Trigonometric Identities

In trigonometry, we are talking about a point (x, y) rotating around a circle with radius r. The equation of a circle is

$$x^2 + y^2 = r^2.$$

□ Figure 5.2

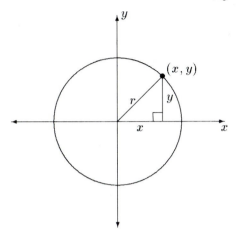

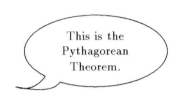

This is the Pythagorean Theorem.

Pythagorean Identities

The equation of a circle leads to three identities called the Pythagorean identities.

$$
\begin{aligned}
x^2 + y^2 &= r^2 \\
\frac{x^2}{r^2} + \frac{y^2}{r^2} &= \frac{r^2}{r^2} \qquad \text{dividing by } r^2 \\
\left(\frac{x}{r}\right)^2 + (\underline{})^2 &= 1
\end{aligned}
$$

$\frac{y}{r}$

$$(12) \qquad (\cos \theta)^2 + (\sin \theta)^2 = 1$$

To simplify writing, we use $\sin^2\theta$ to represent $(\sin \theta)^2$.

Now let's divide both sides of the equation of a circle by x^2:

$$
\begin{aligned}
x^2 + y^2 &= r^2 \\
\frac{x^2}{x^2} + \frac{y^2}{x^2} &= \frac{r^2}{x^2} \\
\underline{} + \left(\frac{y}{x}\right)^2 &= \left(\frac{r}{x}\right)^2
\end{aligned}
$$

1

$$(13) \qquad 1 + \tan^2\theta = \sec^2\theta$$

Dividing both sides of the equation of a circle by y^2 yields the third Pythagorean identity:

$$
\begin{aligned}
x^2 + y^2 &= r^2 \\
\frac{x^2}{y^2} + \underline{} &= \frac{r^2}{y^2}
\end{aligned}
$$

$\frac{y^2}{y^2}$

$$(14) \qquad \cot^2\theta + 1 = \csc^2\theta$$

These fundamental identities are extremely useful. They will be used throughout the remainder of this book, they occur frequently in calculus, and they are essential to electronics, mechanics, optics and all areas that use trigonometry. Therefore you should memorize them.

The identities above have been developed as a function of an angle θ which can be measured in degrees or radians.

5.1 Fundamental Identities

On a unit circle with $r = 1$, the measure of central angle θ in radians is exactly equal to the arc s, therefore all these identities are also true for trigonometric functions of real numbers.

The fundamental identities are collected below for your convenience. In this collection the variable u can stand for an angle θ, an arc length on a unit circle, a real number, or any algebraic or trigonometric expression that calculates to a real number.

5.1E The Fundamental Trigonometric Identities

Reciprocal Identities

(1) $\sin u = \dfrac{1}{\csc u}$

(2) $\csc u = \dfrac{1}{\sin u}$

(3) $\cos u = \dfrac{1}{\sec u}$

(4) $\sec u = \dfrac{1}{\cos u}$

(5) $\tan u = \dfrac{1}{\cot u}$

(6) $\cot u = \dfrac{1}{\tan u}$

Product Identities

(7) $\sin u \cdot \csc u = 1$

(8) $\cos u \cdot \sec u = 1$

(9) $\tan u \cdot \cot u = 1$

Ratio Identities

(10) $\tan u = \dfrac{\sin u}{\cos u}$

(11) $\cot u = \dfrac{\cos u}{\sin u}$

Pythagorean Identities

Identity

(12) $\cos^2 u + \sin^2 u = 1$

(13) $1 + \tan^2 u = \sec^2 u$

(14) $\cot^2 u + 1 = \csc^2 u$

Alternate Forms

$\sin^2 u = 1 - \cos^2 u$

$\cos^2 u = 1 - \sin^2 u$

$1 = \sec^2 u - \tan^2 u$

$\tan^2 u = \sec^2 u - 1$

$1 = \csc^2 u - \cot^2 u$

$\cot^2 u = \csc^2 u - 1$

Do I have to memorize all these?

If you memorize (1), (3), and (5), just a little algebra will produce (2), (4), (6), (7), (8), and (9) any time you want them.

Reductions of Trigonometric Expressions

Frequently scientists and mathematicians find themselves staring at complicated trigonometric expressions. They use identities and algebraic manipulation to reduce these expressions to less complicated expressions.

         Chapter 5 Basic Trigonometric Identities

Example 2 □ Reduce $\sin\theta\cot\theta$ to $\cos\theta$.

$$\sin\theta\cot\theta \;=\; \sin\theta\cot\theta$$

The reflexive property of equality says that everything is equal to itself.

$$=\; \sin\theta\cdot\frac{\cos\theta}{\sin\theta}$$

identity (6)

$$=\; \cos\theta$$

□

Example 3 □ Reduce $(1+\cos u)(1-\cos u)$ to $\sin^2 u$.

$$(1+\cos u)(1-\cos u) \;=\; (1+\cos u)(1-\cos u)$$

reflexive property of equality

$$=\; 1-\cos^2 u$$

multiply binomials

$$=\; \sin^2 u$$

identity (12)

□

But identity (12) is $\cos^2 u + \sin^2 u = 1$.

Look at the alternate forms of this identity.

Example 4A □ Reduce $\dfrac{1+\tan^2\theta}{\tan^2\theta(1+\cot^2\theta)}$ to 1.

$$\frac{1+\tan^2\theta}{\tan^2\theta(1+\cot^2\theta)} \;=\; \frac{1+\tan^2\theta}{\tan^2\theta(1+\cot^2\theta)}$$

reflexive property of equality

$\sec^2\theta$

$$=\; \frac{?}{\tan^2\theta(1+\cot^2\theta)}$$

identity (13)

$$=\; \frac{\sec^2\theta}{\tan^2\theta\cdot\csc^2\theta}$$

identity (14)

$$=\; \frac{\sec^2\theta}{\dfrac{\sin^2\theta}{\cos^2\theta}\cdot\dfrac{1}{\sin^2\theta}}$$

identities (10), (2)

$$=\; \frac{\dfrac{1}{\cos^2\theta}}{\dfrac{1}{\cos^2\theta}}$$

divide out $\dfrac{\sin^2\theta}{\sin^2\theta}$

$$=\; 1$$

divide

□

There is frequently more than one way to prove a trigonometric identity. We will now solve Example 4A using a different method.

Example 4B □ Reduce $\dfrac{1+\tan^2\theta}{\tan^2\theta(1+\cot^2\theta)}$ to 1.

$$\frac{1+\tan^2\theta}{\tan^2\theta(1+\cot^2\theta)} = \frac{1+\tan^2\theta}{\tan^2\theta(1+\cot^2\theta)} \qquad \text{reflexive property of equality}$$

$$= \frac{\sec^2\theta}{\tan^2\theta + \tan^2\theta \cdot \cot^2\theta} \qquad \text{identity (13), distribute denominator}$$

$$= \frac{\sec^2\theta}{\tan^2\theta + 1} \qquad \text{identity (9)}$$

$$= \frac{\sec^2\theta}{\sec^2\theta} = 1 \qquad \text{identity (13)} \qquad □$$

Which way is better?

It doesn't matter. Follow any steps that occur to you. The important thing is to get involved with the problem. Then ideas will come. Example 4C is still another possibility.

Example 4C □ Reduce $\dfrac{1+\tan^2\theta}{\tan^2\theta(1+\cot^2\theta)}$ to 1.

$$\frac{1+\tan^2\theta}{\tan^2\theta(1+\cot^2\theta)} = \frac{1+\tan^2\theta}{\tan^2\theta(1+\cot^2\theta)} \qquad \text{reflexive property of equality}$$

$$= \frac{1+\tan^2\theta}{\tan^2\theta + \tan^2\theta \cdot \dfrac{1}{\tan^2\theta}} \qquad \text{distribute denominator}$$

$$= \frac{1+\tan^2\theta}{\tan^2\theta + \tan^2\theta \cdot \dfrac{1}{\tan^2\theta}} \qquad \text{identity (6)}$$

$$= \frac{1+\tan^2\theta}{1+\tan^2\theta} = 1 \qquad □$$

Chapter 5 Basic Trigonometric Identities

Example 5 □ Reduce $\sec x - \cos x$ to $\tan x \sin x$.

$$\sec x - \cos x \quad = \quad \sec x - \cos x \qquad \text{reflexive property of equality}$$

$$= \quad \frac{1}{\cos x} - \cos x \qquad \text{identity (4)}$$

$\dfrac{\cos x}{\cos x}$

$$= \quad \frac{1}{\cos x} - \cos x \left(\underline{\quad ? \quad}\right) \qquad \text{multiply by one to get a common denominator}$$

$$= \quad \frac{1 - \cos^2 x}{\cos x} \qquad \text{add}$$

$\sin^2 x$

$$= \quad \frac{?}{\cos x} \qquad \text{identity (12)}$$

$$= \quad \frac{\sin x}{\cos x} \cdot \frac{\sin x}{1} \qquad \text{factor } \sin^2 x$$

$$= \quad \tan x \cdot \sin x \qquad \text{identity (10)} \qquad \square$$

Example 6 □ Reduce $(\sin x + \cos x)^2$ to $1 + 2\sin x \cos x$.

$$(\sin x + \cos x)^2 \quad = \quad (\sin x + \cos x)^2 \qquad \text{reflexive property of equality}$$

$$= \quad \sin^2 x + 2\sin x \cos x + \cos^2 x \qquad \text{square of binomial}$$

$$= \quad \cos^2 x + \sin^2 x + 2\sin x \cos x \qquad \text{commutative property of addition}$$

$$= \quad 1 + 2\sin x \cos x \qquad \text{identity (12)} \qquad \square$$

Problem Set 5.1

Match each expression in problems 1 through 20 with an expression from A through T on the following page to form a fundamental identity. You may use some answers more than once or not at all. (Some problems may have more than one answer.)

1. $\csc u$ _____
2. $\sin u$ _____
3. $\sin^2 u + \cos^2 u$ _____
4. $\cos u \cdot \sec u$ _____
5. $\sec^2 u - 1$ _____
6. $\dfrac{\sin u}{\cos u}$ _____
7. $1 + \tan^2 u$ _____
8. $\csc^2 u - \cot^2 u$ _____
9. $\tan u$ _____
10. $1 - \sin^2 u$ _____
11. $\sin u \cdot \csc u$ _____
12. $\cos u$ _____
13. $\sec^2 u - \tan^2 u$ _____
14. $1 - \cos^2 u$ _____
15. $\tan u \cdot \cot u$ _____
16. $\cot u$ _____
17. $\csc^2 u - 1$ _____
18. $\sec u$ _____
19. $\dfrac{\cos u}{\sin u}$ _____
20. $1 + \cot^2 u$ _____

A.	$\tan^2 u$	B.	$\cot u$	C.	$\dfrac{1}{\sec u}$	D.	1
E.	$\sec^2 u$	F.	$\cot^2 u$	G.	$\dfrac{1}{\cot u}$	H.	$\csc^2 u$
I.	$\sin^2 u$	J.	$\dfrac{1}{\sin u}$	K.	$\dfrac{\sin u}{\cos u}$	L.	$\dfrac{1}{\tan u}$
M.	$\dfrac{\sec u}{\csc u}$	N.	$\dfrac{1}{\cos u}$	O.	$\dfrac{\cot u}{\tan u}$	P.	$\dfrac{1}{\csc u}$
Q.	$\dfrac{\csc u}{\sec u}$	R.	$\cos^2 u$	S.	$\dfrac{\cos u}{\sin u}$	T.	$\tan u$

21. Use the definitions of the trigonometric functions to prove identity 9 in a manner similar to the way the book proves identities 7 and 8.

Reduce the first expression to the second in each of the following.

22. $\sin \theta \sec \theta, \tan \theta$

23. $\cos u \csc u, \cot u$

24. $\tan \theta \csc \theta, \sec \theta$

25. $\cot \theta \sin \theta, \cos \theta$

26. $\cos \theta \tan \theta, \sin \theta$

27. $1 - \cos \theta \sin \theta \tan \theta, \cos^2 \theta$

28. $\cot^2 x \cdot \sec^2 x, \csc^2 x$

29. $\sin x(\csc x - \sin x), \cos^2 x$

30. $\sec x \tan x \cos x \cot x, 1$

31. $\tan^2 x \csc^2 x - 1, \tan^2 x$

32. $\dfrac{\csc^4 x - 1}{1 + \csc^2 x}, \cot^2 x$

33. $\cos^2 \theta \left(\tan^2 \theta + 1\right), 1$

34. $\cos x \cdot \sec x + \tan^2 x, \sec^2 x$

35. $\dfrac{1 - \sin^4 x}{\cos^2 x}, 1 + \sin^2 x$

36. $\csc x - \sin x, \cos x \cot x$

37. $\dfrac{\sin x}{\tan x \sec x}, \cos^2 x$

38. $\cos^2 x - \cos^4 x, \cos^2 x \sin^2 x$

39. $\cot^2 x \tan^2 x + \cot^2 x, \csc^2 x$

40. $\dfrac{\cot^2 x}{\cos^2 x}, \csc^2 x$

41. $\dfrac{\cot x \cdot \sec x}{\sin x}, \csc^2 x$

42. $\cot^2 x \sin^2 x \csc^2 x \tan^2 x, 1$

43. $\tan x(\cos x + \cot x)(1 - \sin x), \cos^2 x$

44. $\dfrac{\sin^2 x + \cos^2 x}{\sec^2 x}, \cos^2 x$

45. $\csc \theta - \sin \theta, \cot \theta \cos \theta$

46. $\cot^2 x \sin^2 x, \cos^2 x$

47. $1 - \sin x \cos x \cot x, \sin^2 x$

48. $\left(1 - \sin^2 \theta\right)\left(\cot^2 \theta + 1\right), \cot^2 \theta$

49. $\tan^2 \theta \cos^2 \theta, \sin^2 \theta$

50. $\sec u \cot u \sin u, 1$

51. $\left(1 + \tan^2 u\right)\sin^2 u, \tan^2 u$

52. $\cos x(\csc x + \sec x), \cot x + 1$

53. $\cos x + \tan x \sin x, \sec x$

54. $\cos^2 \theta + \sin^2 \theta + \cot^2 \theta, \csc^2 \theta$

55. $\dfrac{\sin \theta}{1 - \sin^2 \theta}, \tan \theta \sec \theta$

56. $\dfrac{1 - \cos^4 \theta}{1 + \cos^2 \theta}, \sin^2 \theta$

57. $\sin^2 \theta \left(\cot^2 \theta + 1\right), 1$

58. $\sin x \tan x \sec x + 1, \sec^2 x$

59. $\cot \theta + \tan \theta, \sec \theta \csc \theta$

60. $\cot \theta(\sin \theta + \tan \theta), \cos \theta + 1$

61. $\dfrac{\cot^2 \theta}{\csc^2 \theta}, \cos^2 \theta$

62. $\sin \theta + \dfrac{\cos^2 \theta}{\sin \theta}, \csc \theta$

63. $(\sec^2 \theta - 1)\cos^2 \theta, \sin^2 \theta$

64. $\csc^2 \theta(1 - \sin^2 \theta), \cot^2 \theta$

65. $\dfrac{\sin^2 \theta + \cos^2 \theta}{\sin^2 \theta}, \csc^2 \theta$

66. $\cot^2 x - \cos^2 x, \cot^2 x \cos^2 x$

67. $\sec x - \cos x, \tan x \sin x$

68. $\sec^4 \theta, \sec^2 \theta \tan^2 \theta + \sec^2 \theta$

69. $1 - (\cos \theta - \sin \theta)^2, 2 \sin \theta \cos \theta$

70. $\dfrac{(1 + \sin \theta)(1 - \sin \theta)}{\cos \theta}, \cos \theta$

Chapter 5 Basic Trigonometric Identities

5.2 Opposite Angle Identities

The ability to express functions of negative angles as functions of positive angles helps us to simplify expressions. By examining the symmetry of the circle, we can develop several useful relationships between trigonometric functions of positive and negative angles.

☐ Figure 5.3

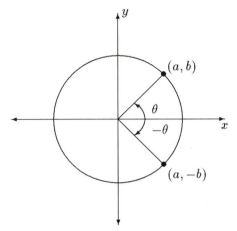

Notice that for any angle $-\theta$, the x-coordinate of a point on the circle is exactly the same as for $+\theta$.

Because $\cos\theta = \dfrac{x}{r}$ we can say

(15) $$\cos(-\theta) = \cos\theta \text{ for all } \theta$$

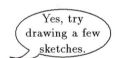

Looking at the same figure, notice that if the y-coordinate for $+\theta$ is b, the y-coordinate for $-\theta$ is $-b$.

☐ Figure 5.4

This negative sign just says the y-coordinate is opposite of b. It doesn't say that y is negative.

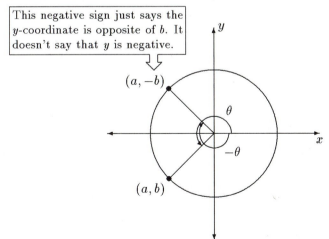

Since $\sin\theta = \dfrac{y}{r}$

(16) $\qquad\qquad\qquad\qquad \sin(-\theta) = -\sin\theta \text{ for all } \theta$

Using the same logic as we used to derive identity 16, we can determine $\tan(-\theta)$ from $\tan\theta$.

Because $\tan\theta = \dfrac{y}{x}$ and the x-coordinate is the same for $+\theta$ or $-\theta$

(17) $\qquad\qquad\qquad\qquad \tan(-\theta) = -\tan\theta \text{ for all } \theta$

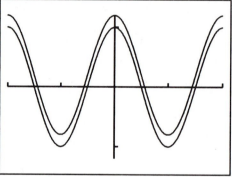

Using Your Graphing Calculator

You can use your graphing calculator to test the validity of identities 15, 16, and 17. Try plotting the graph of $y = \cos x$ and $y = \cos(-x)$ from $X_{\min} = -6.28$ to $X_{\max} = 6.28$. The reason you didn't see the graph of $y = \cos(-x)$ plot is because it is exactly the same as the plot of $y = \cos x$. To convince yourself this is true, change the equation of $y = \cos x$ to $y = \cos x + 0.2$ and replot the graphs.

□ Figure 5.5

Example 7 □ Find $\sin(-50°)$.

Using a calculator to find $\sin(+50°)$

$$
\begin{aligned}
\sin 50° &\approx 0.7660 \\
\text{Therefore } \sin(-50°) &= -\sin(50°) \\
&\approx -0.7660
\end{aligned}
$$

□

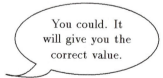

Why not use the negative sign on the calculator and calculate $\sin(-50°)$?

You could. It will give you the correct value.

There are two ways to evaluate the sine of a negative number. Either use the identity to find the sine of a positive number or use your calculator to directly evaluate the sine of the negative angle.

Proving Identities

To prove an identity, apply a series of previously established identities to reduce one side of the identity until it matches the other side.

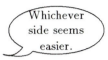

Pointers for Proving Identities

1. Generally, pick the more complicated side of the identity to work on.
2. If you can't make headway on one side, try the other.
3. If one side has an indicated operation, perform it.
4. If one side of the identity contains more than one function, try to write it in terms of a single function.
5. Reducing an entire side to sines and cosines and simplifying usually establishes whether or not a given expression is an identity.

Example 8 □ Prove $\cos(-\theta)\tan(-\theta) = \sin(-\theta)$.

Select the left side to work with.

$$
\begin{array}{rcll}
\cos(-\theta)\tan(-\theta) & = & \cos(-\theta)\tan(-\theta) & \text{reflexive property} \\
& & & \text{of equality} \\[1em]
& = & (\cos\theta)(-\tan\theta) & \text{identities (15), (17)} \\[1em]
& = & \cos\theta\dfrac{(-\sin\theta)}{\cos\theta} & \text{identity (10)} \\[1em]
& = & -\sin\theta & \text{cancel } \cos\theta \\[1em]
\cos(-\theta)\tan(-\theta) & = & \sin(-\theta) & \text{identity (16)}
\end{array}
$$

□

> In proving identities we confine our work to one side because we are not solving an equation but are simplifying an expression.

Group Writing Activity

Why can't you prove an identity by manipulating both sides of the equation, as in the following?

$$
\begin{array}{rcll}
\sin x & = & \cos x & \\
\sin x + 2 & = & \cos x + 2 & \\
3\sin x + 6 & = & 3\cos x + 6 & \\
3\sin x + 6 - 6 & = & 3\cos x + 6 - 6 & \\
\dfrac{3\sin x}{3} & = & \dfrac{3\cos x}{3} & \\
\sin x & = & \cos x & \text{done}
\end{array}
$$

What is wrong with this reasoning?

Example 9 □ Prove $\cos^2\alpha(1 + \tan^2\alpha) = 1$.

We will work this example two ways.

Method I

$$\cos^2\alpha(1 + \tan^2\alpha) = \cos^2\alpha(1 + \tan^2\alpha) \qquad \text{reflexive property of equality}$$

$$= \cos^2\alpha(\sec^2\alpha) \qquad \text{identity (13)}$$

$$= \cos^2\alpha\frac{(1)}{(\cos^2\alpha)} \qquad \text{identity (3)}$$

$$\cos^2\alpha(1 + \tan^2\alpha) = 1 \qquad \text{reduce}$$

Method II

$$\cos^2\alpha(1 + \tan^2\alpha) = \cos^2\alpha(1 + \tan^2\alpha) \qquad \text{reflexive property of equality}$$

$$= \cos^2\alpha + \cos^2\alpha\tan^2\alpha \qquad \text{distributive property}$$

$$= \cos^2\alpha + \cos^2\alpha\frac{\sin^2\alpha}{\cos^2\alpha} \qquad \text{identity (10)}$$

$$= \cos^2\alpha + \sin^2\alpha \qquad \text{reduce}$$

$$\cos^2\alpha(1 + \tan^2\alpha) = 1 \qquad \text{identity (12)}$$

□

Which way is better?

It doesn't matter; try anything that comes to you and see where it leads.

Example 10 □ Prove $1 + \sin^2 x = (1 + \sin x)^2$

Start by squaring the right hand side.

$$1 + \sin^2 x \overset{?}{=} (1 + \sin x)^2$$

$$\overset{?}{=} 1 + 2\sin x + \sin^2 x \qquad \text{squaring}$$

$$\overset{?}{=} 1 + \sin^2 x + 2\sin x \qquad \text{commutative}$$

$$\underline{1 + \sin^2 x} \overset{?}{=} \underline{1 + \sin^2 x} + 2\sin x$$

| These parts match exactly | This part is extra |

Because $2\sin x$ is not always equal to zero the right side cannot be equal to the left side; thus this cannot be an identity.

You can use your graphing calculator to check the validity of an identity by graphing each side of the equation and seeing whether or not the graphs coincide.

Using Your Graphing Calculator

You can confirm that $1 + \sin^2 x \neq (1 + \sin x)^2$ with your graphing calculator. Set

$$
\begin{aligned}
Y_1 &= 1 + \sin^2 x \\
Y_2 &= (1 + \sin x)^2
\end{aligned}
$$

With range

$$
\begin{array}{llll}
X_{\min} &= -6.28 & \qquad Y_{\min} &= -4.2 \\
X_{\max} &= 6.28 & \qquad Y_{\max} &= 4.2
\end{array}
$$

And plot

□ Figure 5.6

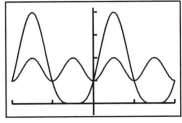

The two graphs do not coincide; therefore, the left side does not equal the right side, so this cannot be an identity.

> If the graphs appear to coincide, must I still prove the identity?

> Yes. In mathematics just because something appears to be true doesn't mean that it is. It must be proven true by mathematical means.

Try graphing the following equations on your graphing calculator.

$$
\begin{array}{ll}
y = (\sin x)^2 + .2 \sin x & \text{from } X_{\min} = -6.28 \\
\text{and } y = (\sin x + .1)^2 & \text{to } X_{\max} = 6.28
\end{array}
$$

The graphs appear to coincide

□ Figure 5.7

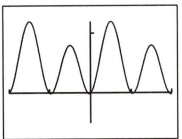

but if you zoom in once or twice, you will see a difference between the 2 curves.

$$
\begin{array}{lll}
\text{In fact} & (\sin x + .1)^2 & \neq \quad \sin^2 x + .2 \sin x \\
\text{because} & \sin^2 x + .2 \sin x + .01 & \neq \quad \sin^2 x + .2 \sin x
\end{array}
$$

Example 11 □ Prove $\dfrac{(1 + \cos x)(1 - \cos x)}{\sin x} = \sin x$.

First check the validity by using your graphing calculator. Seeing that the graphs coincide, proceed with the proof.

Because the left side is more complex, we'll reduce it first. Start by performing the indicated multiplication in the numerator.

$$\frac{(1 + \cos x)(1 - \cos x)}{\sin x} = \frac{(1 + \cos x)(1 - \cos x)}{\sin x} \qquad \text{reflexive}$$

$$= \frac{1 - \cos^2 x}{\sin x} \qquad \text{multiply numerator}$$

$$= \frac{\sin^2 x}{\sin x} \qquad \text{identity (12)}$$

$$= \sin x \qquad \text{reduce}$$

□

Example 12 □ Prove $\cos \theta \csc(-\theta) \tan \theta = -1$.

Notice, to use the graphing calculator we must use identities to rewrite the left hand side, so we will proceed with the proof without graphing.

If more than one function is involved, it frequently helps to write the side you are working on in terms of $\sin \theta$ and $\cos \theta$.

$$\cos \theta \csc(-\theta) \tan \theta = \cos \theta \csc(-\theta) \tan \theta \qquad \text{reflexive}$$

$$= \cos \theta \cdot \frac{1}{\sin(-\theta)} \cdot \frac{\sin \theta}{\cos \theta} \qquad \text{identities (2), (10)}$$

$$= \frac{1}{-\sin \theta} \cdot \frac{\sin \theta}{1} \qquad \text{identity (16) and reduce}$$

$$= -1$$

□

Problem Set 5.2

Express each of the following as functions of sine and/or cosine, then simplify the results. Leave the answer in terms of functions of sine and/or cosine.

1. $\cos u \tan u$

2. $\dfrac{\tan \theta}{\sec \theta}$

3. $\cos^2 \theta \sec^2 \theta \csc^2 \theta$

4. $\sec^2 \theta \cot^2 \theta$

5. $\sin \alpha \sec \alpha$

6. $\left(\csc^2 \theta - 1 \right) \left(\sin^2 \theta \right)$

7. $\sec \theta \cot \theta \sin \theta \cos \theta$

8. $\sin \theta + \dfrac{\cos^2 \theta}{\sin \theta}$

9. $\tan u + \cot u$

10. $\cos^2 \theta + \tan^2 \theta + \sin^2 \theta$

11. $\cos \theta \left(1 + \cot^2 \theta \right)$

12. $\dfrac{\cot x}{\csc x}$

13. $\left(\sec^2 x - 1 \right) \cos^2 x$

14. $\sec x - \dfrac{1}{\sec x}$

15. $\sin x \cot x \cos x$

16. $\csc x \tan x \sin x \cos x$

17. $\cos x + \dfrac{\sin^2 x}{\cos x}$

18. $\left(\csc^2 x - 1 \right) \sin^2 x$

19. $\csc^2 x - \cot^2 x + \tan^2 x$

20. $\sin x \cot x \cos x$

Test the validity of the following identities using a graphing calculator. If the graphs coincide prove the identity. If the equations are not identities, give a value of x that proves that they are not identities.

21. $\dfrac{\cos^2 \theta}{1 - \sin(-\theta)} = 1 - \sin \theta$

22. $\dfrac{1 - \cos^4 \theta}{\sin^2 \theta} = 1 + \cos^2 \theta$

23. $\dfrac{\cos x + \sin x}{\cos x - \sin x} = \dfrac{1 - \tan x}{1 + \tan x}$

24. $\dfrac{\sin x}{1 - \cos x} = \dfrac{\cos x}{1 - \sin x}$

25. $(1 - \cos^2 x)(1 + \tan^2 x) = \tan^2 x$

26. $\cos(-x) = -\cos x$

27. $\dfrac{1}{3} \cos 3x = \cos x$

28. $\dfrac{1 + \sin^2 x}{\cos^2 x} = 1 + 2\tan^2 x$

29. $\dfrac{1 - \sin^4 \theta}{1 + \sin^2 \theta} = \cos^2 \theta$

30. $2 \sin x = \sin 2x$

31. $\dfrac{\sin^2 \theta}{1 - \cos(-\theta)} = 1 + \cos \theta$

32. $\cos(x + \pi) = \cos x$

33. $\dfrac{(1 + \sin \theta)(1 - \sin \theta)}{\cos \theta} = 1 + \cos \theta$

34. $\dfrac{1 - (\cos \theta - \sin \theta)^2}{\cos \theta} = 2 \sin \theta$

35. $\dfrac{1}{\csc x - \sin x} = \dfrac{\tan x}{\cos x}$

36. $2\cos^2 \theta - 1 = 1 - 2\sin^2 \theta$

37. $\dfrac{\cos^3 x + \sin^3 x}{\cos x + \sin x} = 1 - \sin x \cos x$

38. $\dfrac{1}{\sin x \cos x} - \dfrac{\sin x}{\cos x} = \tan x$

39. $1 - 2 \sin^2 \theta + \sin^4 \theta = \cos^4 \theta$

40. $\sin^4 \theta + 2 \cos^2 \theta \sin^2 \theta + \cos^4 \theta = 1$

5.2 Opposite Angle Identities   167

Prove or disprove the following identities. Work on one side only.

41. $\dfrac{\cos(-\theta)}{1-\sin\theta} = \sec\theta + \tan\theta$

42. $\dfrac{\tan^2\theta + 1}{\csc^2\theta} = \tan^2\theta$

43. $\dfrac{\cos(-\theta)}{1-\cos^2\theta} = \cot\theta\csc\theta$

44. $\dfrac{\cot x}{\tan^2 x - \sec^2 x} = \dfrac{1}{\tan(-x)}$

45. $\csc x - \cot x \cos x = \sin x$

46. $\dfrac{\cos(-x) - \sin(-x)}{\sin x} = \cot x + 1$

47. $\dfrac{\cos(-x)}{\csc x + \sin(-x)} = \tan x$

48. $\dfrac{1}{\sin x \cos x} + \dfrac{\sin(-x)}{\cos(-x)} = \cot x$

49. $\dfrac{\csc^2 x - 1}{\cos^2 x} = \csc^2 x$

50. $(\csc x - 1)(\csc x + 1) = \cot^2 x$

51. $(1 - \sin^2 x)(1 + \cot^2 x) = \cot^2 x$

52. $\sec^2 x[1 - \sin(-x)] = \dfrac{1}{1 - \sin x}$

53. $\dfrac{\csc x}{\sin x} + \dfrac{\sec x}{\cos x} = \sec^2 x \csc^2 x$

54. $\dfrac{\sin x + \cos x}{\sec x + \csc x} = \sin x \cos x$

55. $\dfrac{\sin(-\theta)\csc\theta}{\tan(-\theta)} = \cot\theta$

56. $\dfrac{\cos^4 u - \sin^4 u}{\sin^2 u \cos^2 u} = \cot^2 u - \tan^2 u$

57. $\csc^2\theta - (\sin^2\theta + \cot^2\theta) = \cos^2\theta$

58. $\dfrac{1}{\cot\theta - \tan(-\theta)} = \dfrac{\cos\theta}{\csc\theta}$

59. $\dfrac{\sin(-\theta)}{1 - \sin^2\theta} = -\tan\theta\sec\theta$

60. $\dfrac{\cot^2\theta + 1}{\cot^2\theta} = \sec^2\theta$

61. $\dfrac{\tan\theta}{\sin^2\theta + \cos^2\theta} = \dfrac{1}{\cot\theta}$

62. $\dfrac{\tan\theta}{\sec\theta - \cos\theta} = \dfrac{\sec\theta}{\tan\theta}$

63. $\dfrac{\cos\theta}{\sec\theta + \tan\theta} = 1 - \sin\theta$

64. $\dfrac{\cot\theta - \cos\theta\sin\theta}{\csc\theta} = \cos^3\theta$

65. $\sec^2\theta + \csc^2\theta = \sec^2\theta\csc^2\theta$

66. $\dfrac{1 + \csc x}{\cos x + \cot x} = \sec x$

67. $\csc^2 x - \sec^2 x = \cot^2 x - \tan^2 x$

68. $\dfrac{\sec^2 x - \tan^2 x}{\csc^2 x} = \sin^2 x$

69. $\csc^4 x - 2\csc^2 x \cot^2 x + \cot^4 x = 1$

70. $\sec^4\theta - \tan^4\theta = \sec^2\theta + \tan^2\theta$

71. $\cos^2 x \csc^2 x + 1 = \csc^2 x$

72. Draw the graph of $y_1 = \sin^3 x + \dfrac{1 - \cos^4 x}{\sin x}$. From the graph find a function y_2 to make the identity $y_1 = y_2$. Prove this identity.

5.3   Additional Techniques to Prove Identities

In addition to the suggestions in section 5.2 here are some other ways to prove identities.

Pointers for Proving Identities

1. If one side of an identity contains a sum or difference of two functions, find the least common denominator and perform the indicated operation.

2. If one side consists of several terms over a single denominator, try breaking the fraction into a sum of separate fractions. (This is the opposite of suggestion 1.)

3. Factor any terms that can be factored.

4. Try multiplying both numerator and denominator of a fraction by the same expression as you do in simplifying complex fractions.

Example 13 $\square$ Prove $\dfrac{\sin x}{1 - \cos x} - \dfrac{1 + \cos x}{\sin x} = 0$.

We'll try suggestion 1 and find a common denominator.

$$\frac{\sin x}{1 - \cos x} - \frac{1 + \cos x}{\sin x} = \frac{\sin x}{1 - \cos x} - \frac{1 + \cos x}{\sin x}$$

$$\frac{\sin x}{1 - \cos x} \cdot \frac{(\sin x)}{(\sin x)} - \frac{1 + \cos x}{\sin x} \cdot \frac{(1 - \cos x)}{(1 - \cos x)} = \qquad \text{multiplication by 1}$$

$$\frac{\sin^2 x}{\sin x\,(1 - \cos x)} - \frac{1 - \cos^2 x}{\sin x\,(1 - \cos x)} = \qquad \text{perform indicated operations in the numerators}$$

$$\frac{\sin^2 x - \left(\sin^2 x\right)}{\sin x\,(1 - \cos x)} = \qquad \text{add, identity (12)}$$

$$\frac{0}{\sin x\,(1 - \cos x)} =$$

$$0 = \frac{\sin x}{1 - \cos x} - \frac{1 + \cos x}{\sin x}$$

$\square$

Example 14 □ Prove $\dfrac{1 + \tan\beta\sin^2\beta - \tan\beta}{\sin\beta} = \csc\beta - \cos\beta$.

This time we'll start on the left side with suggestion 2.

$$\frac{1 + \tan\beta\sin^2\beta - \tan\beta}{\sin\beta} = \frac{1 + \tan\beta\sin^2\beta - \tan\beta}{\sin\beta}$$

$$\frac{1}{\sin\beta} + \frac{\tan\beta\sin^2\beta}{\sin\beta} - \frac{\tan\beta}{\sin\beta} = \qquad \text{make separate fractions}$$

$$\frac{1}{\sin\beta} + \frac{\sin\beta}{\cos\beta}\cdot\frac{\sin^2\beta}{\sin\beta} - \frac{\sin\beta}{\cos\beta}\cdot\frac{1}{\sin\beta} = \qquad \text{identity (10)}$$

$$\frac{1}{\sin\beta} + \frac{\sin^2\beta}{\cos\beta} - \frac{1}{\cos\beta} = \qquad \text{simplify}$$

Now compare what you have with what you want.

$$\frac{1}{\sin\beta} + \frac{\sin^2\beta}{\cos\beta} - \frac{1}{\cos\beta} \stackrel{?}{=} \csc\beta - \cos\beta$$

This is $\csc\beta$.

This should be $\cos\beta$. Notice the common denominator.

$$\csc\beta + \frac{\sin^2\beta - 1}{\cos\beta} = \qquad \text{identity (2)}$$

What now?

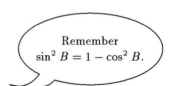

Remember
$\sin^2 B = 1 - \cos^2 B$.

$$\csc\beta + \frac{(1 - \cos^2\beta) - 1}{\cos\beta} = \qquad \text{identity (12)}$$

$$\csc\beta - \frac{\cos^2\beta}{\cos\beta} = \qquad \text{simplify}$$

$$\csc\beta - \cos\beta = \frac{1 + \tan\beta\sin^2\beta - \tan\beta}{\sin\beta}$$

□

Example 15 □ Prove $\sin^2\alpha - \cos^2\alpha = \sin^4\alpha - \cos^4\alpha$.

Start on the right side because it is of higher degree than the left. It is also factorable.

$$
\begin{aligned}
\sin^4\alpha - \cos^4\alpha &= \sin^4\alpha - \cos^4\alpha \\[2mm]
&= \left(\sin^2\alpha - \cos^2\alpha\right)\left(\sin^2\alpha + \cos^2\alpha\right) \quad \text{factor} \\[2mm]
\sin^4\alpha - \cos^4\alpha &= \left(\sin^2\alpha - \cos^2\alpha\right) \cdot 1 \qquad\qquad\quad \text{identity (12)}
\end{aligned}
$$

□

Example   16 □ Prove $\dfrac{1 + \sin x}{1 - \sin x} = (\sec x + \tan x)^2$.

Because $\dfrac{\sin x}{\cos x} = \tan x$, we will start on the left side and introduce a denominator of $\cos x$.

$$
\frac{1 + \sin x}{1 - \sin x} = \frac{1 + \sin x}{1 - \sin x}
$$

$$
\frac{\dfrac{1}{\cos x} + \dfrac{\sin x}{\cos x}}{\dfrac{1}{\cos x} - \dfrac{\sin x}{\cos x}} = \qquad\qquad \text{multiply numerator and denominator by } \dfrac{1}{\cos x}
$$

$$
\frac{\sec x + \tan x}{\sec x - \tan x} = \qquad\qquad \text{identities (4), (10)}
$$

$$
\frac{\sec x + \tan x}{\sec x - \tan x} \cdot \frac{(\sec x + \tan x)}{(\sec x + \tan x)} = \qquad\qquad \text{multiply by 1}
$$

$$
\frac{(\sec x + \tan x)^2}{\sec^2 x - \tan^2 x} = \qquad\qquad \text{multiply denominator}
$$

$$
(\sec x + \tan x)^2 = \frac{1 + \sin x}{1 - \sin x} \qquad\qquad \text{identity (13) and divide by 1}
$$

□

5.3 Additional Techniques to Prove Identities

Algebra Reminder

A Special Factoring Technique

It is possible to use the distributive property to rewrite sums as products. That process is called **factoring**. When you rewrite trigonometric identities, it is sometimes useful to "factor" expressions that do not appear in the sum. Consider the following examples.

Example 17 □ Factor x from $x^2 + 4x$.

$$x^2 + 4x = x(x + 4)$$

Now ask yourself "How do you know that the example is factored correctly?" The answer is

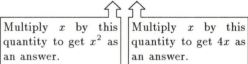

$$x^2 + 4x \quad = \quad x(x + 4)$$

| Multiply x by this quantity to get x^2 as an answer. | Multiply x by this quantity to get $4x$ as an answer. |

□

Example 18 □ Factor x^2 from $x^2 + 4x$.
Use the reasoning above.

$$x^2 + 4x = x^2\left(1 + \frac{4}{x}\right)$$

□

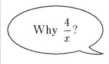

Why $\dfrac{4}{x}$?

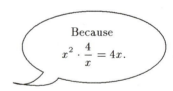

Because

$$x^2 \cdot \frac{4}{x} = 4x.$$

Example 19 □ Factor 4 from $x^2 + 4x$.

$$x^2 + 4x = 4\left(\frac{x^2}{4} + x\right)$$

□

We can use this line of reasoning to factor expressions that do not appear in each term.

Example 20 □ Factor b from $4x^2 + x$.

By what would you multiply b to get $4x^2$? _____

By what would you multiply b to get x? _____

The factored expression is:

$$4x^2 + x = b\left(\frac{4x^2}{b} + \frac{x}{b}\right)$$

□

$\dfrac{4x^2}{b}$

$\dfrac{x}{b}$

Chapter 5 Basic Trigonometric Identities

We can use the concept in the algebra reminder above to rewrite trigonometric expressions.

Example 21 □ Factor $\sin \theta$ from $\cos \theta + \tan \theta$. First rewrite $\cos \theta + \tan \theta$ in terms of $\sin \theta$ and $\cos \theta$.

$$\cos \theta + \tan \theta \quad = \quad \cos \theta + \frac{\sin \theta}{\cos \theta}$$

$$\sin \theta \quad = \quad \sin \theta \left(\frac{\cos \theta}{\sin \theta} + \frac{1}{\cos \theta} \right) \qquad \text{Factoring } \sin \theta$$

□

Why would you want to do that?

This factoring technique is a powerful tool in rewriting expressions to look the way you want them to look. Read below.

Example 22 □ Prove $\sin x + \cot x = \cos x \left(\tan x + \csc x \right)$.

We will work on the left side to demonstrate the power of factoring. Because the right side has a factor of $\cos x$, we will factor $\cos x$ from the left.

$$\sin x + \cot x \quad = \quad \cos x \left(\tan x + \csc x \right)$$

$$\cos x \left(\frac{\sin x}{\cos x} + \frac{1}{\cos x} \cdot \cot x \right) \quad =$$

Now to show that the quantity in the parentheses on the left side is equivalent to the quantity in the parentheses on the right side, we will rewrite $\cot x$:

$$\cos x \left(\frac{\sin x}{\cos x} + \frac{1}{\cos x} \cdot \frac{\cos x}{\sin x} \right) \quad =$$

$$\cos x \left(\tan x + \frac{1}{\sin x} \right) \quad =$$

$$\cos x \left(\tan x + \csc x \right) \quad = \quad \cos x \left(\tan x + \csc x \right)$$

□

Problem Set 5.3

Prove the following identities using the suggestions in this section. Work on one side only.

1. $\dfrac{1}{\sec^2 x} + \dfrac{1}{\csc^2 x} = 1$

2. $\dfrac{\cot x - \tan x}{1 - \tan x} = \cot x + 1$

3. $\dfrac{\cos^2 x}{\csc x - 1} = \dfrac{\sin x + 1}{\csc x}$

4. $\dfrac{\cos x}{1 + \sin x} + \tan x = \sec x$

5. $\csc x + \dfrac{\sin(-x)}{1 + \cos(-x)} = \cot x$

6. $\csc x + \sin(-x) = \cos x \cot x$

7. $\csc^4 x + \cot^4 x = 1 + 2\csc^2 x \cot^2 x$

8. $\dfrac{1 - \cot^2 x}{1 - \tan^2 x} = 1 - \csc^2 x$

9. $\cot x - \tan x = \dfrac{2\cos^2 x - 1}{\sin x \cos x}$

10. $\dfrac{1}{1 - \cos x} - \dfrac{1}{1 + \cos x} = 2 \cot x \csc x$

11. $\dfrac{\cos x}{\cot x} - \dfrac{\sin x}{\tan x} = \sin x - \cos x$

12. $\dfrac{\cot x - \tan x}{\csc^2 x - \sec^2 x} = \dfrac{1}{\csc x \sec x}$

13. $\cot^2 x - \cos^2 x = \cot^2 x \cos^2 x$

14. $\dfrac{\tan x}{\sec x} - \dfrac{\cot x}{\csc x} = \sin x - \cos x$

15. $\dfrac{\sin x \cos x}{1 - \sin x} - \tan x = \sec x - \cos x$

16. $\sec x + \cos x = 2 \sec x - \tan x \sin x$

17. $\dfrac{\sin^2 x + \cot^2 x - 1}{\cos^2 x} = \cot^2 x$

18. $\dfrac{1}{1 - \sin x} + \dfrac{1}{1 + \sin x} = 2\sec^2 x$

19. $\dfrac{2 + \sec x}{\csc x} + 2\sin(-x) = \tan x$

20. $\sec^2 x + \tan^2 x \sec^2 x = \sec^4 x$

21. $\dfrac{1 - \cos(-x)}{\sin x} - \dfrac{\sin(-x)}{1 - \cos x} = 2 \csc x$

22. $(\sec x - \tan x)(\csc x + 1) = \cot x$

23. $\dfrac{\cos x}{1 - \csc x} - \dfrac{\cos x}{1 + \csc x} = -2 \tan x$

24. $\dfrac{\sin x \cos x}{\cos^2 x - \sin^2 x} = \dfrac{\tan x}{1 - \tan^2 x}$

25. $\dfrac{1 + \sin x}{\cos x} - \dfrac{\cos(-x)}{\sin(-x)} = \dfrac{1 + \sin x}{\sin x \cos x}$

26. $\dfrac{\cot x - \tan x}{\cot x + 1} = 1 - \tan x$

27. $\dfrac{1}{\cos^2 x} - \dfrac{1}{\cot^2 x} = 1$

28. $\dfrac{\sin x}{\sec x + 1} + \dfrac{\sin x}{\sec x - 1} = 2 \cot x$

29. $\csc x - \dfrac{\sin x}{1 + \cos x} = \cot x$

30. $\dfrac{3\cos^2 x + 2\cos x - 1}{\sin^2 x} = \dfrac{3\cos x - 1}{1 - \cos x}$

31. $\sec x - \dfrac{\cos x}{1 + \sin x} = \tan x$

32. $\dfrac{1 - \tan^4 x}{\sec^2 x} = 2 - \sec^2 x$

Test the validity of the following identities using a graphing calculator. If the graphs coincide, prove the identity. If the equations are not identities, give a value of x that proves that they are not identities.

33. $\dfrac{3 - 4\sin x + \sin^2 x}{\cos^2 x} = \dfrac{3 - \sin x}{1 + \sin x}$

34. $\dfrac{\sin x}{1 + \cos x} + 1 = \dfrac{\sin x + 1}{1 + \cos x}$

35. $\dfrac{1 + 2\sin x \cos x}{\cos^2 x} = (1 + \tan x)^2$

36. $\tan^2 x - \sin^2 x = \tan^2 x \sin^2 x$

37. $\dfrac{\sin x - 1}{\cos x} + \dfrac{1 - \cos x}{\sin x} = 1$

38. $\dfrac{1}{\sin^2 x} - \dfrac{1}{\tan^2 x} = 1$

39. $\dfrac{1 + \sin x - 2\sin^2 x}{\cos^2 x} = \dfrac{2\sin x + 1}{\sin x + 1}$

40. $\dfrac{\tan x}{\cos x} - \dfrac{\sin x}{\tan x} = \cos x - \sin x$

41. $\dfrac{\sin^2 x - 4\sin x + 3}{\cos^2 x} = \dfrac{3 - \sin x}{1 + \cos x}$

42. $\sin^2 x + \tan^2 x \sin^2 x = 1 + \sin^2 x$

5.1 **1.** A conditional equation is true for some values of the variable, but not necessarily all of them.

2. An identity is an equation that is true for all values of its variable.

3. Reciprocal trigonometric identities:

$$\sin u = \frac{1}{\csc u} \qquad \csc u = \frac{1}{\sin u} \qquad \tan u = \frac{1}{\cot u}$$

$$\cos u = \frac{1}{\sec u} \qquad \sec u = \frac{1}{\cos u} \qquad \cot u = \frac{1}{\tan u}$$

4. Ratio trigonometric identities:

$$\tan u = \frac{\sin u}{\cos u} \qquad \cot u = \frac{\cos u}{\sin u}$$

5. Pythagorean trigonometric identities:

$$\cos^2 u + \sin^2 u = 1$$
$$1 + \tan^2 u = \sec^2 u$$
$$1 + \cot^2 u = \csc^2 u$$

6. Use the trigonometric identities to transform one trigonometric expression into another one.

5.2 Trigonometric identities for opposite angles:

$$
\begin{aligned}
\sin(-u) &= -\sin u \\
\cos(-u) &= \cos u \\
\tan(-u) &= -\tan u
\end{aligned}
$$

Use the trigonometric identities for opposite angles to change trigonometric functions of negative angles into functions of positive angles.

5.3 When proving trigonometric identities:

1. Work on one side only.

2. Use the fundamental trigonometric identities to transform an expression.

3. Use algebraic operations such as:
   a. adding or subtracting fraction,
   b. factoring any member that can be factored,
   c. breaking a fraction into separate fractions over a common denominator.

4. Graphing calculators may first be used to check the validity of an identity before proceeding with the proof.

Chapter 5 Review Test

In problems 1 through 5 match with one of the forms on the right to form a fundamental identity. **(5.1)**

1. $\tan u$ _____

2. $\sec u$ _____

3. $1 + \tan^2 u$ _____

4. $\sin u$ _____

5. $\dfrac{\cos u}{\sin u}$ _____

A. $\csc^2 u$

B. $\dfrac{1}{\sin u}$

C. $\dfrac{\sin u}{\cos u}$

D. $\cot u$

E. $\dfrac{1}{\csc u}$

F. $\sec^2 u$

G. $\dfrac{1}{\cos u}$

H. $\dfrac{1}{\sec u}$

Reduce the first expression to the second in each of the following. **(5.1)**

6. $\sin u \left(\csc u - \sin u \right),\ \cos^2 u$

7. $\csc^2 u - 4,\ \cot^2 u - 3$

8. $\sec u - \cos u,\ \tan u \sin u$

9. $\cos u \left(\sec u + \cos u \csc^2 u \right),\ \csc^2 u$

10. $\dfrac{\cos u}{\csc^2 u - 1},\ \sin u \tan u$

Prove the following identities. Work on one side only. **(5.2, 5.3)**

11. $\dfrac{\csc u}{\cot^2 u} = \tan u \sec u$

12. $\csc u - \sin u = \cot u \cos u$

13. $\dfrac{\sec u}{\tan u + \cot u} = \sin u$

14. $\dfrac{\tan u}{1 - \tan u} + \dfrac{\cot u}{1 - \cot u} + 1 = 0$

15. $\dfrac{\cot^2 u - 1}{\cot^2 u + 1} = \dfrac{\csc^2 u - \sec^2 u}{\csc^2 u + \sec^2 u}$

16. $\dfrac{1 + 3\cos u - 4\cos^2 u}{\sin^2 u} = \dfrac{1 + 4\cos u}{1 + \cos u}$

17. $\dfrac{\sin(-u)}{1 - \cos(-u)} - \dfrac{1 + \cos u}{\sin(-u)} = 0$

18. $\dfrac{\sec^2 u - \tan^2 u + \tan u}{\sec u} = \sin u + \cos u$

19. $\dfrac{2\cos u}{\sin u + \csc u + \cos^2 u \csc u} = \sin^2 u \cot u$

20. $\dfrac{\left(\cot^2 u - 1 \right) \sin^2 u}{\cos u - \sin u} = \cos u + \sin u$

21. $\dfrac{1}{\csc u + \cot u} + \cot u = \csc u$

Test the validity of the following identities using a graphing calculator. If the graphs coincide, prove the identity. If the equations are not identities, give a value of x that proves that they are not identities. **(5.2), (5.3)**

22. $\dfrac{\sin^3 u + \cos^3 u}{\sin u + \cos u} = 1 - \sin u \cos u$

23. $\dfrac{\tan x}{1 - \sin x} + \dfrac{\sin x}{1 - \tan x} + 1 = \cos x + \tan x$

24. $\dfrac{\sin x}{\tan x + \cot x} = \cos x$

25. $\dfrac{\sin x + \cos^3 x - \cos x}{\cos x} = \tan x - \sin^2 x$

Chapter 6

Sum and Difference Identities

Contents

Preview

Notice that $\cos(60°) = 0.5$ and $\cos(30°) \approx 0.86$ but $\cos(60° - 30°)$ is not equal to the cosine of 60° minus the cosine of 30°. This chapter will use the basic identities from the last chapter and the formulas for the distance between two points to develop an identity for $\cos(60° - 30°)$.

From the formula for the cosine of the difference of two angles and the basic identity, this chapter will then develop the complete set of sum and difference identities. The final section will use the sum and difference identities to express several useful identities involving an angle and a multiple of $\frac{1}{2}\pi$.

6.1 Sum and Difference Formulas for Cosine

Derivation of cos $(\alpha - \beta)$

To find a formula for the cosine of the difference of two angles, we will study two angles on a unit circle.

Consider any point on a circle with radius r. It is possible to express the x- and y-coordinates of that point in terms of r and θ.

□ Figure 6.1

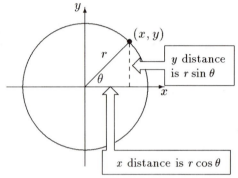

Because $\sin \theta = \dfrac{y}{r}$

$y = r \sin \theta$

A point (x, y) on a circle can be represented as $(r \cos \theta, r \sin \theta)$. Recall the formula for the distance between two points.

□ Figure 6.2

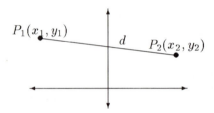

$$d = \sqrt{(x_1 - x_2)^2 + (y_1 - y_2)^2}$$

On a unit circle label points P_1 and P_2 corresponding to angles α and β.

□ Figure 6.3

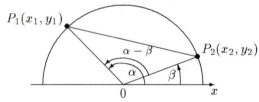

We have identified two angles on a unit circle and the coordinates of the points where the angles intersect the unit circle. Now we will work to express the difference between the angles, $\alpha - \beta$, in terms of those points.

We represent the distance d from P_1 to P_2 as $D(P_1, P_2)$. The distance is given by the distance formula:

$$D(P_1, P_2) = \sqrt{(x_1 - x_2)^2 + (y_1 - y_2)^2}$$

The measure of the central angle between P_1 and P_2 is $P_1 O P_2 = \alpha - \beta$.

Next locate another point, P_3, at angle $(\alpha - \beta)$ with the initial side on the x-axis. Then identify P_4 at a 0 angle on the x-axis.

     Chapter 6 Sum and Difference Identities

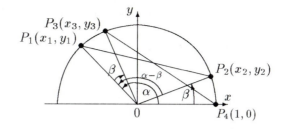

The central angle between P_3 and P_4 is also $\alpha - \beta$. We now have two central angles in the circle with the measure $\alpha - \beta$.

Equal angles intercept equal chords on the circle, therefore the distance from P_3 to P_4 must equal the distance from P_1 to P_2. With equal distances we can use the distance formula to find $\cos(\alpha - \beta)$.

The distance from P_3 to P_4 is

$$D(P_3, P_4) = \sqrt{(x_3 - x_4)^2 + (y_3 - y_4)^2}$$

Because the circle has a radius of 1, point P_4 has coordinates (1,0) and

$$D(P_3, P_4) = \sqrt{(x_3 - 1)^2 + (y_3 - 0)^2}$$

Knowing $D(P_1, P_2) = D(P_3, P_4)$ we can set the expressions for these distances equal to each other. Starting with the expression we developed earlier for the distance between P_1 and P_2, we will simplify each expression separately.

$$[D(P_1, P_2)]^2 = (x_1 - x_2)^2 + (y_1 - y_2)^2$$

This is a unit circle with radius 1, therefore:

$$\begin{aligned} x_1 &= 1 \cdot \cos \alpha & y_1 &= 1 \cdot \sin \alpha \\ x_2 &= 1 \cdot \cos \beta & y_2 &= 1 \cdot \sin \beta \end{aligned}$$

Substituting in the equation for $[D(P_1, P_2)]^2$ above,

$$[D(P_1, P_2)]^2 = (\cos \alpha - \cos \beta)^2 + (\sin \alpha - \sin \beta)^2$$

Expanding each term on the right yields:

$$= \boxed{\cos^2 \alpha} - 2 \cos \alpha \cos \beta + \boxed{\cos^2 \beta} + \boxed{\sin^2 \alpha} - 2 \sin \alpha \sin \beta + \boxed{\sin^2 \beta}$$

Each marked expression adds to one:

$$= 1 + 1 - 2 \, \cos \alpha \cos \beta - 2 \, \sin \alpha \sin \beta$$

Now we have the distance from P_1 to P_2 in terms of trigonometric functions of α and β.

Next we will find the distance from P_3 to P_4 in terms of α and β:

$$[D(P_3, P_4)]^2 = (x_3 - 1)^2 + y_3{}^2$$

Notice $x_3 = 1 \cdot \cos(\alpha - \beta)$, and $y_3 = 1 \cdot \sin(\alpha - \beta)$.

Now we can substitute for x_3 and y_3.

$$\begin{aligned}
&= && [\cos(\alpha - \beta) - 1]^2 + \sin^2(\alpha - \beta) \\
&= && \cos^2(\alpha - \beta) - 2\cos(\alpha - \beta) + 1 + \sin^2(\alpha - \beta) && \text{squaring the} \\
& && && \text{first term} \\
&= && 1 + 1 - 2\cos(\alpha - \beta) && \text{identity (12)}
\end{aligned}$$

Because both distances are equal, we can set the two trigonometric expressions for the distance equal to each other.

$$\begin{aligned}
[D(P_3, P_4)]^2 &= [D(P_1, P_2)]^2 \\
1 + 1 - 2\cos(\alpha - \beta) &= 1 + 1 - 2\cos\alpha\cos\beta - 2\sin\alpha\sin\beta \\
-2\cos(\alpha - \beta) &= -2\cos\alpha\cos\beta - 2\sin\alpha\sin\beta
\end{aligned}$$

Dividing both sides by -2 gives an identity for $\cos(\alpha - \beta)$

(18) $$\boldsymbol{\cos(\alpha - \beta) = \cos\alpha\cos\beta + \sin\alpha\sin\beta}$$

Example 1 □ Find $\cos 15°$ from $\cos 30°$ and $\cos 45°$.

$$\cos 15° = \cos(45° - 30°)$$

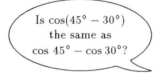

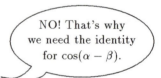

Substituting $45°$ and $30°$ in the identity for $\cos(\alpha - \beta)$ yields:

$$\begin{aligned}
\cos(\alpha - \beta) &= \cos\alpha\cos\beta + \sin\alpha\sin\beta \\
\cos(45° - 30°) &= \cos 45° \cdot \cos 30° + \sin 45° \cdot \sin 30°
\end{aligned}$$

Substituting exact values

$$\begin{aligned}
&= \frac{\sqrt{2}}{2} \cdot \frac{\sqrt{3}}{2} + \frac{\sqrt{2}}{2} \cdot \frac{1}{2} \\
&= \frac{\sqrt{6}}{4} + \frac{\sqrt{2}}{4} \\
&= \frac{\sqrt{6} + \sqrt{2}}{4} \\
&\approx 0.9659
\end{aligned}$$

□

Derivation of cos $(\alpha + \beta)$

To derive an identity for $\cos(\alpha + \beta)$ we substitute $-\beta$ in the identity for $\cos(x - y)$.

$$\cos(x - y) = \cos x \cos y + \sin x \sin y$$

Let $x = \alpha$ and $y = -\beta$

$$\cos[\alpha - (-\beta)] = \cos\alpha\cos(-\beta) + \sin\alpha\sin(-\beta)$$
$$\cos(\alpha + \beta) = \cos\alpha\cos(-\beta) + \sin\alpha\sin(-\beta)$$

Because $\cos(-\beta) = \cos\beta$ and $\sin(-\beta) = -\sin\beta$

(19) $$\mathbf{\cos\ (\alpha + \beta) = \cos\ \alpha\ \cos\ \beta - \sin\ \alpha\ \sin\ \beta}$$

Example 2 □ Simplify $\cos 20° \cdot \cos 30° + \sin 20° \cdot \sin 30°$.

We need to recognize this expression is an application of $\cos(\alpha - \beta)$.

How do we tell it's $\cos(\alpha - \beta)$ and not $\cos(\alpha + \beta)$?

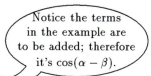
Notice the terms in the example are to be added; therefore it's $\cos(\alpha - \beta)$.

$$\cos(\alpha - \beta) = \cos\alpha\cos\beta + \sin\alpha\sin\beta$$
$$\cos(20° - 30°) = \cos 20° \cdot \cos 30° + \sin 20° \cdot \sin 30°$$
$$\cos(-10°) = \cos 20° \cdot \cos 30° + \sin 20° \cdot \sin 30°$$

How did you know it's $20° - 30°$ and not $30° - 20°$?

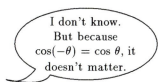
I don't know. But because $\cos(-\theta) = \cos\theta$, it doesn't matter.

$$\cos(-10°) = \cos(10°)$$

□

With the sum and difference identities we can prove several other useful identities:

Example 3 □ Prove $\cos(\pi + x) = -\cos x$.

$$\cos(\pi + x) = \cos \pi \cos x - \sin \pi \sin x$$

Because $\cos \pi = -1$ and $\sin \pi = 0$, we obtain $(-1)\cos x - 0(\sin x)$.

(20) $$\cos (\pi + x) = -\cos x$$

□

Example 4 □ Simplify $\cos \left(\frac{1}{2}\pi - \alpha \right)$.

Using the identity for the $\cos(\alpha - \beta)$

$$
\begin{aligned}
\cos \left(\frac{1}{2}\pi - \alpha \right) &= \cos \left(\frac{1}{2}\pi \right) \cos \alpha + \sin \left(\frac{1}{2}\pi \right) \sin \alpha \\
&= 0 \cos \alpha + 1 \cdot \sin \alpha
\end{aligned}
$$

(21) $$\cos \left(\frac{1}{2}\pi - \alpha \right) = \sin \alpha$$

□

Notice $\alpha + \left(\frac{1}{2}\pi - \alpha \right)$ adds up to $\frac{1}{2}\pi$. Recall that if the sum of two angles is 90° or $\frac{1}{2}\pi$, the angles are called complementary angles.

Another way to express this identity is:

The sine of an angle is equal to the cosine of its complement.

Because of this relationship, the cosine and sine are sometimes called cofunctions.

Substituting $\theta = \frac{1}{2}\pi - \alpha$ in identity (21) above yields the converse of this identity.

(22) $$\cos \theta = \sin \left(\frac{1}{2}\pi - \theta \right)$$

Derivation of sin $(\alpha \pm \beta)$

To derive an identity for $\sin(\alpha + \beta)$, start with the identity proved in Example 4. The sine of an angle is the cosine of its complement.

$$\sin \theta = \cos \left(\frac{1}{2}\pi - \theta \right)$$

Replace θ with $(\alpha + \beta)$:

$$
\begin{aligned}
\sin(\alpha + \beta) &= \cos \left[\frac{1}{2}\pi - (\alpha + \beta) \right] \\
\sin(\alpha + \beta) &= \cos \left[\left(\frac{1}{2}\pi - \alpha \right) - \beta \right] \qquad \text{regrouping}
\end{aligned}
$$

Using the identity

$$\cos(x - y) = \cos x \cos y + \sin x \sin y$$

let $x = \dfrac{1}{2}\pi - \alpha$ and $y = \beta$

$$\sin(\alpha + \beta) = \cos\left(\frac{1}{2}\pi - \alpha\right)\cos\beta + \sin\left(\frac{1}{2}\pi - \alpha\right)\sin\beta$$

Because the sine of an angle is the cosine of its complement:

(23) $$\mathbf{\sin\,(\alpha + \beta) = \sin\,\alpha\,\cos\,\beta + \cos\,\alpha\,\sin\,\beta}$$

If you substitute $-\beta$ for β in the identity above, you can derive

(24) $$\mathbf{\sin\,(\alpha - \beta) = \sin\,\alpha\,\cos\,\beta - \cos\,\alpha\,\sin\,\beta}$$

Example 5 □ Find $\sin 105°$ from the trigonometric functions of $60°$ and $45°$.

$$
\begin{aligned}
\sin(105°) &= \sin(60° + 45°) \\
&= \sin 60° \cos 45° + \cos 60° \sin 45° \\
&= \frac{\sqrt{3}}{2} \cdot \frac{\sqrt{2}}{2} + \frac{1}{2} \cdot \frac{\sqrt{2}}{2} \\
&= \frac{\sqrt{6}}{4} + \frac{\sqrt{2}}{4} \\
&= \frac{\sqrt{6} + \sqrt{2}}{4}
\end{aligned}
$$

□

Example 6 □ Convert $\cos(\theta - 30°)$ to an expression using $\sin\theta + \cos\theta$ only.
We can use the formula for $\cos(\alpha - \beta)$.

$$
\begin{aligned}
\cos(\theta - 30°) &= \cos\theta \cos 30° + \sin\theta \sin 30° \\
&= \cos\theta \cdot \frac{\sqrt{3}}{2} + \sin\theta \cdot \frac{1}{2} \\
&= \frac{\sqrt{3}}{2}\cos\theta + \frac{1}{2}\sin\theta \\
&= \frac{1}{2}\left(\sqrt{3}\cos\theta + \sin\theta\right)
\end{aligned}
$$

□

Example 7 □ Find the exact value of $\cos(\alpha + \beta)$, given $\sin\alpha = \dfrac{4}{5}$ and $\cos\beta = -\dfrac{3}{5}$, α in quadrant I and β in quadrant II. Do not use a calculator.
Start with the identity for $\cos(\alpha + \beta)$:

$$\cos(\alpha + \beta) = \cos\alpha \cos\beta - \sin\alpha \sin\beta$$

We are given $\sin\alpha$ and $\cos\beta$. Find $\cos\alpha$ and $\sin\beta$ by using a circle and its equation.

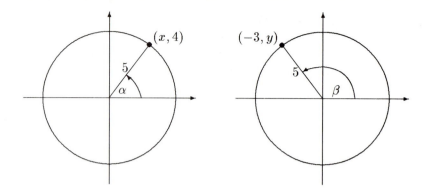

$$\begin{aligned} x^2 + y^2 &= r^2 \\ x^2 + 4^2 &= 5^2 \\ x^2 + 16 &= 25 \\ x^2 &= 9 \\ x &= 3 \end{aligned} \qquad\qquad \begin{aligned} x^2 + y^2 &= r^2 \\ (-3)^2 + y^2 &= 5^2 \\ 9 + y^2 &= 25 \\ y^2 &= 16 \\ y &= 4 \end{aligned}$$

Therefore,

$$\cos \alpha = \frac{3}{5} \text{ and } \sin \beta = \frac{4}{5}$$

Thus

$$\begin{aligned} \cos(\alpha + \beta) &= \cos \alpha \cos \beta - \sin \alpha \sin \beta \\ &= \left(\frac{3}{5}\right)\left(-\frac{3}{5}\right) - \left(\frac{4}{5}\right)\left(\frac{4}{5}\right) \\ &= -\frac{9}{25} - \frac{16}{25} \\ &= -\frac{25}{25} \\ &= -1 \end{aligned}$$

□

Example 8 □ Prove $\dfrac{\cos(x + y)}{\cos x \sin y} = \cot y - \tan x$.

Work on the left side

$$\begin{aligned} \frac{\cos(x + y)}{\cos x \sin y} &= \frac{\cos(x + y)}{\cos x \sin y} \\ \frac{\cos x \cos y - \sin x \sin y}{\cos x \sin y} &= && \text{identity (19)} \\ \frac{\cos x \cos y}{\cos x \sin y} - \frac{\sin x \sin y}{\cos x \sin y} &= && \text{separate fractions} \\ \cot y - \tan x &= \frac{\cos(x + y)}{\cos x \sin y} && \text{identities (10), (11)} \end{aligned}$$

□

Problem Set 6.1

Using the sine and cosine values of 30°, 45°, and 60°, find the following exact values. Check these values using your calculator.

1. $\cos 75°$

2. $\cos 105°$

3. $\cos 15°$ (use $60° - 45°$)

4. $\sin 15°$ (use $60° - 45°$)

5. $\sin 75°$

6. $\sin 120°$ (use $60° + 60°$)

7. $\cos 15°$ (use $45° - 30°$)

8. $\sin 120°$ (use $90° + 30°$)

Use addition or subtraction identities to show that the following are true.

9. $\cos(2\pi + x) = \cos x$

10. $\cos(\pi - x) = -\cos x$

11. $\cos(\pi + x) = -\cos x$

12. $\sin(2\pi + x) = \sin x$

13. $\sin(\pi - x) = \sin x$

14. $\sin(\pi + x) = -\sin x$

Using only sum and difference identities, convert each of the following expressions to expressions using $\sin x$ and $\cos x$.

15. $\cos\left(\dfrac{1}{2}\pi + x\right)$

16. $\cos\left(\dfrac{3}{2}\pi - x\right)$

17. $\cos(2\pi - x)$

18. $\sin\left(\dfrac{1}{2}\pi + x\right)$

19. $\sin\left(\dfrac{1}{2}\pi - x\right)$

20. $\sin(2\pi - x)$

21. $\sin\left(x - \dfrac{1}{2}\pi\right)$

22. $\cos\left(x - \dfrac{1}{2}\pi\right)$

Use the sum or difference formula to simplify the following.

23. $\cos 15° \cos 75° - \sin 15° \sin 75°$

24. $\cos 80° \cos 20° + \sin 80° \sin 20°$

25. $\sin 45° \cos 30° - \cos 45° \sin 30°$

26. $\sin 45° \cos 40° + \cos 45° \sin 40°$

Find the exact function values without a calculator, using the information given and the appropriate identities. A sketch helps.

Functions	Values	Quadrant	
		α	β
27. $\cos(\alpha + \beta)$	$\cos\alpha = \dfrac{3}{5}, \ \tan\beta = \dfrac{4}{3}$	I	III
28. $\cos(\alpha - \beta)$	$\sin\alpha = -\dfrac{4}{5}, \ \cot\beta = -\dfrac{3}{4}$	III	II
29. $\cos(\alpha - \beta)$	$\cos\alpha = -\dfrac{5}{13}, \ \sin\beta = \dfrac{5}{13}$	III	II
30. $\cos(\alpha + \beta)$	$\sin\alpha = -\dfrac{3}{5}, \ \cos\beta = \dfrac{\sqrt{8}}{3}$	III	IV
31. $\sin(\alpha - \beta)$	$\tan\alpha = -\dfrac{12}{5}, \ \sin\beta = \dfrac{4}{5}$	II	I
32. $\sin(\alpha - \beta)$	$\cos\alpha = -\dfrac{1}{3}, \ \tan\beta = \dfrac{1}{2}$	II	III
33. $\sin(\alpha + \beta)$	$\cos\alpha = -\dfrac{4}{5}, \ \cos\beta = \dfrac{12}{13}$	II	IV
34. $\sin(\alpha + \beta)$	$\sin\alpha = -\dfrac{1}{3}, \ \sin\beta = \dfrac{\sqrt{8}}{3}$	III	II

Prove the following identities:

35. $\dfrac{\cos(\alpha - \beta)}{\cos\alpha\sin\beta} = \cot\beta + \tan\alpha$

36. $\dfrac{\cos(x + y)}{\cos x\cos y} = 1 - \tan x\tan y$

37. $\cos(x + y)\sec y = \sin x(\cot x - \tan y)$

38. $\cos y(\cot x - \tan y) = \cos(x + y)\csc x$

39. $\cos(\alpha + \beta) + \cos(\alpha - \beta) = 2\ \cos\alpha\cos\beta$

40. $\dfrac{\cos(x - y) - \cos(x + y)}{\cos x\sin y} = 2\ \tan x$

41. $\dfrac{\cos(\alpha - \beta)}{\cos(\alpha + \beta)} = \dfrac{\cot\alpha + \tan\beta}{\cot\alpha - \tan\beta}$

42. $\cos(x + y)\cdot\cos(x - y) = \cos^2 x - \sin^2 y$

43. $\dfrac{\cos(x - y)}{\sin x\sin y} = \cot x(\cot y + \tan x)$

44. $\dfrac{\cos(\alpha + \beta)}{\sin\alpha\cos\beta} = \cos\alpha(\csc\alpha - \sec\alpha\tan\beta)$

45. $\dfrac{\cos(x + y) - \cos(x - y)}{\sin x\sin y} = -2$

46. $\cos(x + y)\cdot\cot y = \cos^2 y(\cos x\csc y - \sin x\sec y)$

 ## Group Writing Activity

We have a lot of identities to memorize up to this point. One way to help remember formulas is to sometimes use specific values to test what we think is the formula. What specific values could you use to remember the sum and difference formulas? Why do these work for you?

6.2   Some Identities Useful in Calculus

In section 6.1 we developed two identities:

$$(21) \qquad\qquad \cos\left(\frac{1}{2}\pi - u\right) \quad = \quad \sin u$$

$$(22) \qquad\qquad \sin\left(\frac{1}{2}\pi - u\right) \quad = \quad \cos u$$

In words the identities say **"the sine of an angle is the cosine of its complement."**

If angle α and angle β are complementary then $\alpha + \beta = \dfrac{1}{2}\pi$, and so for any $0 \le x \le 1$ we can say

$$\alpha = \text{Sin}^{-1}x \text{ and } \beta = \text{Cos}^{-1}x$$

Then substituting these expressions for α and β in

$$\alpha + \beta = \frac{1}{2}\pi$$

yields

$$\textbf{(25)} \qquad\qquad \mathbf{Sin^{-1}\textit{x} + \ Cos^{-1}\textit{x} = \frac{1}{2}}$$

We can combine this identity with the Pythagorean identities to yield some results that are very useful in calculus.

Example 9 □ Write $\cos\left(\text{Sin}^{-1}x\right)$ as an algebraic expression.

Start with the Pythagorean identity.

$$\cos^2 u + \sin^2 u = 1$$
$$\text{Then } \cos^2 u = 1 - \sin^2 u$$

Let $u = \text{Sin}^{-1}x$

$$\cos^2(\text{Sin}^{-1}x) = 1 - \sin^2(\text{Sin}^{-1}x)$$

Recall $\sin(\text{Sin}^{-1}x) = x$ by the definition of the inverse trigonometric functions.

$$\cos^2(\text{Sin}^{-1}x) = 1 - x^2$$
$$|\cos(\text{Sin}^{-1}x)| = \sqrt{1 - x^2}$$

$\text{Sin}^{-1}x$ has a range between $-\frac{1}{2}\pi$ and $\frac{1}{2}\pi$. With this range as the domain of $\cos u$, $\cos u$ is always positive and we can drop the absolute value signs.

(26) $$\mathbf{\cos\left(Sin^{-1}x\right) = \sqrt{1 - x^2}}$$

□

Example 10 □ Find the exact value of $\cos\left(\text{Sin}^{-1}\frac{1}{2}\right)$.

$$\cos(\text{Sin}^{-1}u) = \sqrt{1 - u^2}$$
$$\cos\left(\text{Sin}^{-1}\frac{1}{2}\right) = \sqrt{1 - \left(\frac{1}{2}\right)^2}$$
$$\cos\left(\text{Sin}^{-1}\frac{1}{2}\right) = \frac{\sqrt{3}}{2}$$

□

You can arrive at the same solution by constructing a right triangle with a hypotenuse of 2 units and a leg of 1 unit.

Example 11 □ Write an algebraic expression for $\csc(\text{Cot}^{-1}x)$.

Start with the identity
$$\csc^2 u = 1 + \cot^2 u$$

Let $u = \text{Cot}^{-1}x$

$$\csc^2(\text{Cot}^{-1}x) = 1 + \cot^2(\text{Cot}^{-1}x)$$
$$\csc^2(\text{Cot}^{-1}x) = 1 + x^2$$

(27) $$\mathbf{\csc\left(Cot^{-1}x\right) = \sqrt{1 + x^2}}$$

We use the positive square root because $\text{Cot}^{-1}x$ has a range between zero and π, and in that range $\csc u$ is positive. □

Problem Set 6.2

Use the method of Example 9 to prove the following identities.

1. $\sec\left(\text{Tan}^{-1}x\right) = \sqrt{1+x^2}$ **Identity (28)**   **2.** $\tan\left(\text{Sec}^{-1}x\right) = \sqrt{x^2-1}$ **Identity (29)**

3. $\sin\left(\text{Cos}^{-1}x\right) = \sqrt{1-x^2}$ **(30)**   **4.** $\cot\left(\text{Csc}^{-1}x\right) = \sqrt{x^2-1}$ **(31)**

Use identities 26 through 31 to find the exact value of each of the following.

5. $\sin\left(\text{Cos}^{-1}\dfrac{1}{2}\right)$ **6.** $\tan\left(\text{Sec}^{-1}3\right)$ **7.** $\csc\left[\text{Cot}^{-1}(-2)\right]$

8. $\cos\left[\text{Sin}^{-1}\left(-\dfrac{\sqrt{3}}{2}\right)\right]$ **9.** $\cot\left(\text{Csc}^{-1}.7\right)$ **10.** $\sec\left(\text{Tan}^{-1}51\right)$

6.3   Tan $(\alpha \pm \beta)$

Section 6.1 developed identities for $\cos(\alpha+\beta)$ and $\sin(\alpha+\beta)$. This section will complete the sum and difference identities by deriving identities for $\tan(\alpha+\beta)$ and $\tan(\alpha-\beta)$.

Tan $(\alpha + \beta)$

Now that we have identities for $\sin(\alpha \pm \beta)$ and $\cos(\alpha \pm \beta)$, we can easily derive an identity for $\tan(\alpha \pm \beta)$.

$$\tan(\alpha+\beta) = \frac{\sin(\alpha+\beta)}{\cos(\alpha+\beta)}$$

$$= \frac{\sin\alpha\cos\beta + \cos\alpha\sin\beta}{\cos\alpha\cos\beta - \sin\alpha\sin\beta}$$

The identity for $\tan(\alpha+\beta)$ should be in terms of $\tan\alpha$ and $\tan\beta$. Examine the equation above. Multiply each term of the numerator and denominator by $\dfrac{1}{\cos\alpha\cos\beta}$

$$\tan(\alpha+\beta) = \frac{\dfrac{\sin\alpha\cos\beta}{1} + \dfrac{\cos\alpha\sin\beta}{1}}{\dfrac{\cos\alpha\cos\beta}{1} - \dfrac{\sin\alpha\sin\beta}{1}} \cdot \left[\dfrac{\dfrac{1}{\cos\alpha\cos\beta}}{\dfrac{1}{\cos\alpha\cos\beta}}\right]$$

$$= \frac{\dfrac{\sin\alpha\cos\beta}{\cos\alpha\cos\beta} + \dfrac{\cos\alpha\sin\beta}{\cos\alpha\cos\beta}}{\dfrac{\cos\alpha\cos\beta}{\cos\alpha\cos\beta} - \dfrac{\sin\alpha\sin\beta}{\cos\alpha\cos\beta}} \quad \boxed{\begin{array}{l}\text{This is}\\ \text{multiplication}\\ \text{by one}\end{array}}$$

$$= \frac{\dfrac{\sin\alpha}{\cos\alpha} + \dfrac{\sin\beta}{\cos\beta}}{1 - \dfrac{\sin\alpha}{\cos\alpha} \cdot \dfrac{\sin\beta}{\cos\beta}}$$

$$(32) \qquad \tan(\alpha + \beta) = \frac{\tan \alpha + \tan \beta}{1 - \tan \alpha \tan \beta}$$

Proof of the following identity is left as an exercise in problem set 6.3.

$$(33) \qquad \tan(\alpha - \beta) = \frac{\tan \alpha - \tan \beta}{1 + \tan \alpha \tan \beta}$$

Example 12 □ Prove $\tan(x + \pi) = \tan x$.

$\tan(x + \pi) = \tan x$
because the period
of $\tan x$ is π.

$$\begin{aligned}
\tan(x + \pi) &= \frac{\tan x + \tan \pi}{1 - \tan x \tan \pi} \\
&= \frac{\tan x + 0}{1 - \tan x(0)} \\
&= \tan x
\end{aligned}$$

□

Example 13 □ Prove $\dfrac{\sin(x + y)}{\sin(x - y)} = \dfrac{\tan x + \tan y}{\tan x - \tan y}$.

$$\begin{aligned}
\frac{\sin(x + y)}{\sin(x - y)} &= \frac{\sin x \cos y + \cos x \sin y}{\sin x \cos y - \cos x \sin y} && \text{identity (23)} \\
&&& \text{identity (24)} \\[2mm]
&= \frac{\dfrac{\sin x \cos y}{1} + \dfrac{\cos x \sin y}{1}}{\dfrac{\sin x \cos y}{1} - \dfrac{\cos x \sin y}{1}} \cdot \left[\frac{\dfrac{1}{\cos x \cos y}}{\dfrac{1}{\cos x \cos y}} \right] \\[2mm]
&= \frac{\dfrac{\sin x \cos y}{\cos x \cos y} + \dfrac{\cos x \sin y}{\cos x \cos y}}{\dfrac{\sin x \cos y}{\cos x \cos y} - \dfrac{\cos x \sin y}{\cos x \cos y}} \\[2mm]
&= \frac{\dfrac{\sin x}{\cos x} + \dfrac{\sin y}{\cos y}}{\dfrac{\sin x}{\cos x} - \dfrac{\sin y}{\cos y}} \\[2mm]
\frac{\sin(x + y)}{\sin(x - y)} &= \frac{\tan x + \tan y}{\tan x - \tan y}
\end{aligned}$$

□

Example 14 □ Prove $\tan(x + y) \cdot \tan(x - y) = \dfrac{\tan^2 x - \tan^2 y}{1 - \tan^2 x \tan^2 y}$.

Since the left side involves sums and differences, reduce it to the right side.

$$\begin{aligned}
\tan(x + y) \cdot \tan(x - y) &= \tan(x + y) \cdot \tan(x - y) \\[2mm]
\frac{\tan x + \tan y}{1 - \tan x \tan y} \cdot \frac{\tan x - \tan y}{1 + \tan x \tan y} &= && \text{identities (32), (33)} \\[2mm]
\frac{\tan^2 x - \tan^2 y}{1 - \tan^2 x \tan^2 y} &= \tan(x + y) \tan(x - y)
\end{aligned}$$

□

Example 15 □ Prove

$$\frac{\sin(x+h) - \sin x}{h} = \sin x \left(\frac{\cos h - 1}{h}\right) + \cos x \left(\frac{\sin h}{h}\right)$$

Start on the left by rewriting:

$$\frac{\sin(x+h) - \sin x}{h} = \frac{\sin(x+h) - \sin x}{h}$$

$$\frac{\sin x \cos h + \cos x \sin h - \sin x}{h} = \qquad \text{expand } \sin(x+h)$$

$$\frac{\sin x \cos h - \sin x + \cos x \sin h}{h} = \qquad \text{rearrange terms}$$

$$\frac{\sin x \cos h - \sin x}{h} + \frac{\cos x \sin h}{h} = \qquad \text{separate fraction}$$

$$\frac{\sin x(\cos h - 1)}{h} + \frac{\cos x(\sin h)}{h} = \qquad \text{factor}$$

$$\sin x \left(\frac{\cos h - 1}{h}\right) + \cos x \left(\frac{\sin h}{h}\right) = \qquad \text{separate factors}$$

□

Problem Set 6.3

Using 30°, 45°, and 60°, find the following exact values. You may check these values using your calculator.

1. $\tan 105°$ **2.** $\tan 120°$ **3.** $\tan 75°$

4. $\tan 15°$ (use $60° - 45°$)   **5.** $\tan 15°$ (use $45° - 30°$)   **6.** $\tan 90°$ (use $45° + 45°$)

In the following problems give exact values.

7. Compute $\tan 165°$ from the functions of 135° and 30°.

8. Compute $\tan 165°$ from the functions of 120° and 45°.

9. Compute $\tan 195°$ from the functions of 240° and 45°.

10. Compute $\tan 165°$ from the functions of 210° and 45°.

Develop a formula for each of the following expressions so that each argument is expressed as a single angle.

11. $\tan \left(\frac{1}{4}\pi + \theta\right)$ **12.** $\tan(2\pi - \theta)$ **13.** $\tan(2\pi + \theta)$ **14.** $\tan \left(\frac{1}{2}\pi + \theta\right)$

15. Use a process similar to the proof of $\tan(\alpha + \beta) = \dfrac{\tan \alpha + \tan \beta}{1 - \tan \alpha \tan \beta}$ to prove

$$\tan(\alpha - \beta) = \frac{\tan \alpha - \tan \beta}{1 + \tan \alpha \tan \beta}$$

16. By substituting $-\beta$ for β in identity (32), prove that identity (33) is true.

            Chapter 6 Sum and Difference Identities

Prove the following identities.

17. $\dfrac{\sin(x+y)}{\sin x \cos y} = 1 + \cot x \tan y$ **18.** $\sin(x-y) \cdot \csc y = \sin x (\cot y - \cot x)$

19. $\dfrac{\tan x - \tan(x-y)}{1 + \tan x \tan(x-y)} = \tan y$ **20.** $\dfrac{\sin(x+y) + \sin(x-y)}{\sin^2 x \cos^2 y} = 2 \csc x \sec y$

21. $\dfrac{\sin(x+y)}{\sin x \sin y} = \cot y + \cot x$ **22.** $\sin(x+y) \cdot \sin(x-y) = \sin^2 x - \sin^2 y$

23. $\tan(x-y) + \tan y = \dfrac{\sin x \sec y}{\cos(x-y)}$ **24.** $\tan(x-y) = \dfrac{\cot y - \cot x}{\cot x \cot y + 1}$

25. $\cot(x+y) = \dfrac{\cot x \cot y - 1}{\cot x + \cot y}$ **26.** $1 - \tan y \tan(x-y) = \dfrac{\cos x \sec y}{\cos(x-y)}$

Group Writing Activity

In problem 14 you are asked to find a formula for $\tan\left(\dfrac{1}{2}\pi + \theta\right)$. What problem do you encounter doing this? How can you take care of this problem?

6.4 Identities Involving Sums and Differences of π or $\dfrac{1}{2}\pi$

The addition/subtraction identities can be used to develop a whole series of identities that involve addition or subtraction of π or $\dfrac{1}{2}\pi$ to or from an angle θ.

Example 16 $\square$ Simplify $\sin(\theta + \pi)$.

Using identity (23):

$$
\begin{aligned}
\sin(\theta + \pi) &= \sin\theta \cos\pi + \cos\theta \sin\pi \\
&= \sin\theta(-1) + \cos\theta(0) \\
&= -\sin\theta
\end{aligned}
$$

$\square$

Example 17 $\square$ Simplify $\cos(\pi - \theta)$.

$$
\begin{aligned}
\cos(\pi - \theta) &= \cos\pi \cos\theta + \sin\pi \sin\theta \qquad \text{identity (18)} \\
&= -1 \; \cos\theta + (0) \; \sin\theta \\
&= -\cos\theta
\end{aligned}
$$

$\square$

The strategy used in Examples 16 and 17 can prove each identity in this entire class. However, by using our knowledge of the symmetry of the circle, we can derive any of these identities very quickly.

To derive an identity for $\cos(\pi + \theta)$:

Step 1

Sketch a unit circle with θ in the first quadrant. From the definition of the cosine function, $\cos\theta = a$.

□ Figure 6.6

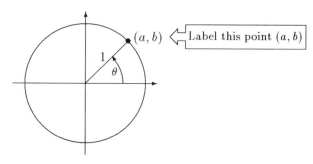

Step 2

Expand your sketch to include $\pi + \theta$.

□ Figure 6.7

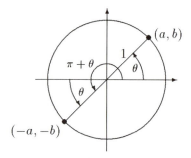

$$\cos(\pi + \theta) \;=\; -a$$
$$\text{and } \cos\theta \;=\; +a$$

Determine the coordinates of the point at $\pi + \theta$ by examining the symmetry of the circle.

Therefore $\cos(\pi + \theta) = -\cos\theta$.

A similar line of reasoning works with sums or differences involving $\frac{1}{2}\pi$.

Example 18 □ Find $\sin\left(\frac{1}{2}\pi - \theta\right)$.

□ Figure 6.8

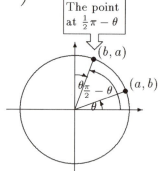

$$\sin\left(\frac{1}{2}\pi - \theta\right) \;=\; a$$
$$\cos(\theta) \;=\; a$$

Therefore,

$$\sin\left(\frac{1}{2}\pi - \theta\right) \;=\; \cos\theta$$

□

Chapter 6 Sum and Difference Identities

All the following identities can be derived by using the addition formulas or directly from the symmetry of the circle. For ease of identification, the identities have been numbered.

$$\textbf{(34)} \quad \sin(\pi + \theta) = -\sin\theta \qquad\qquad \textbf{(36)} \quad \sin(\pi - \theta) = \sin\theta$$

$$\textbf{(20)} \quad \cos(\pi + \theta) = -\cos\theta \qquad\qquad \textbf{(37)} \quad \cos(\pi - \theta) = -\cos\theta$$

$$\textbf{(35)} \quad \tan(\pi + \theta) = \tan\theta \qquad\qquad \textbf{(38)} \quad \tan(\pi - \theta) = -\tan\theta$$

$$\textbf{(39)} \quad \sin\left(\frac{1}{2}\pi + \theta\right) = \cos\theta \qquad\qquad \textbf{(22)} \quad \sin\left(\frac{1}{2}\pi - \theta\right) = \cos\theta$$

$$\textbf{(40)} \quad \cos\left(\frac{1}{2}\pi + \theta\right) = -\sin\theta \qquad\qquad \textbf{(21)} \quad \cos\left(\frac{1}{2}\pi - \theta\right) = \sin\theta$$

$$\textbf{(41)} \quad \tan\left(\frac{1}{2}\pi + \theta\right) = -\cot\theta \qquad\qquad \textbf{(42)} \quad \tan\left(\frac{1}{2}\pi - \theta\right) = \cot\theta$$

Why don't the numbers of the identities above follow in order?

We tried to arrange this group in a logical sequence. Identities 20, 21, 22 were proven earlier. We have numbered all identities in the order we prove them.

Example 19 □ Prove $\dfrac{\tan(\pi + x)\sin\left(\dfrac{1}{2}\pi - x\right)}{\sin(\pi - x)} = 1.$

$$\frac{\tan(\pi + x)\sin\left(\dfrac{1}{2}\pi - x\right)}{\sin(\pi - x)} = \frac{\tan(\pi + x)\sin\left(\dfrac{1}{2}\pi - x\right)}{\sin(\pi - x)}$$

$$\frac{\tan x(\cos x)}{\sin x} = \qquad\qquad \text{identities (22), (35), (36)}$$

$$\frac{\sin x}{\cos x}\cdot\frac{\cos x}{\sin x} =$$

$$1 = \frac{\tan(\pi + x)\sin\left(\dfrac{1}{2}\pi - x\right)}{\sin(\pi - x)}$$

□

For convenient reference the identities are listed below.

6.4 The Fundamental Trigonometric Identities

(numbers indicate the order in which the identities were derived)

Reciprocal Identities

(1) $\sin u = \dfrac{1}{\csc u}$

(2) $\csc u = \dfrac{1}{\sin u}$

(3) $\cos u = \dfrac{1}{\sec u}$

(4) $\sec u = \dfrac{1}{\cos u}$

(5) $\tan u = \dfrac{1}{\cot u}$

(6) $\cot u = \dfrac{1}{\tan u}$

(7) $\sin u \cdot \csc u = 1$

(8) $\cos u \cdot \sec u = 1$

(9) $\tan u \cdot \cot u = 1$

Ratio Identities

(10) $\tan u = \dfrac{\sin u}{\cos u}$

(11) $\cot u = \dfrac{\cos u}{\sin u}$

Pythagorean Identities

| Identity | Alternate Forms |

(12) $\cos^2 u + \sin^2 u = 1$

$\sin^2 u = 1 - \cos^2 u$
$\cos^2 u = 1 - \sin^2 u$

(13) $1 + \tan^2 u = \sec^2 u$

$1 = \sec^2 u - \tan^2 u$
$\tan^2 u = \sec^2 u - 1$

(14) $\cot^2 u + 1 = \csc^2 u$

$1 = \csc^2 u - \cot^2 u$
$\cot^2 u = \csc^2 u - 1$

Opposite Angle Identities

(15) $\cos(-u) = \cos u$ (16) $\sin(-u) = -\sin u$

(17) $\tan(-u) = -\tan u$

Sum and Difference Identities

(18) $\cos(\alpha - \beta) = \cos \alpha \cos \beta + \sin \alpha \sin \beta$ (19) $\cos(\alpha + \beta) = \cos \alpha \cos \beta - \sin \alpha \sin \beta$

(23) $\sin(\alpha + \beta) = \sin \alpha \cos \beta + \cos \alpha \sin \beta$ (24) $\sin(\alpha - \beta) = \sin \alpha \cos \beta - \cos \alpha \sin \beta$

(32) $\tan(\alpha + \beta) = \dfrac{\tan \alpha + \tan \beta}{1 - \tan \alpha \tan \beta}$ (33) $\tan(\alpha - \beta) = \dfrac{\tan \alpha - \tan \beta}{1 + \tan \alpha \tan \beta}$

Identities Involving Inverses

(25) $\text{Sin}^{-1} u + \text{Cos}^{-1} u = \dfrac{1}{2}\pi$ (26) $\cos\left(\text{Sin}^{-1} u\right) = \sqrt{1 - u^2}$

(27) $\csc\left(\text{Cot}^{-1} u\right) = \sqrt{1 + u^2}$ (28) $\sec\left(\text{Tan}^{-1} u\right) = \sqrt{1 + u^2}$

(29) $\tan\left(\text{Sec}^{-1} u\right) = \sqrt{u^2 - 1}$ (30) $\sin\left(\text{Cos}^{-1} u\right) = \sqrt{1 - u^2}$

(31) $\cot\left(\text{Csc}^{-1} u\right) = \sqrt{u^2 - 1}$

Cofunction Identities

(21) $\sin u = \cos\left(\dfrac{1}{2}\pi - u\right)$ (22) $\cos u = \sin\left(\dfrac{1}{2}\pi - u\right)$

(42) $\cot u = \tan\left(\dfrac{1}{2}\pi - u\right)$

Other Identities Involving Multiples of $\dfrac{1}{2}\pi$

(34) $\sin(\pi + u) = -\sin u$ (36) $\sin(\pi - u) = \sin u$

(20) $\cos(\pi + u) = -\cos u$ (37) $\cos(\pi - u) = -\cos u$

(35) $\tan(\pi + u) = \tan u$ (38) $\tan(\pi - u) = -\tan u$

(39) $\sin\left(\dfrac{1}{2}\pi + u\right) = \cos u$ (41) $\tan\left(\dfrac{1}{2}\pi + u\right) = -\cot u$

(40) $\cos\left(\dfrac{1}{2}\pi + u\right) = -\sin u$

Problem Set 6.4

In problems 1-20, match each expression on the left with an expression on the right to form a fundamental identity. The values on the right may be used once, more than once, or not at all.

1.	$1 + \tan^2\theta$ _____		**A.**	$\cos\theta$
2.	$\cot\theta$ _____		**B.**	1
3.	$\cos(-\theta)$ _____		**C.**	$\dfrac{1}{\cos\theta}$
4.	$\sin\left(\dfrac{1}{2}\pi - \theta\right)$ _____		**D.**	$\cos\alpha\cos\beta - \sin\alpha\sin\beta$
5.	$\sin\theta$ _____		**E.**	$\sin\theta$
6.	$\cos^2\theta + \sin^2\theta$ _____		**F.**	$\dfrac{\sin\theta}{\cos\theta}$
7.	$\dfrac{\sin\theta}{\cos\theta}$ _____		**G.**	$\csc^2\theta$
8.	$\sin\theta \cdot \csc\theta$ _____		**H.**	$\sin\alpha\cos\beta - \cos\alpha\sin\beta$
9.	$\sec\theta$ _____		**I.**	$-\tan\theta$
10.	$\cos(\alpha + \beta)$ _____		**J.**	$\dfrac{\tan\alpha - \tan\beta}{1 + \tan\alpha\tan\beta}$
11.	$\sin(-\theta)$ _____		**K.**	$\sec^2\theta$
12.	$\csc\theta$ _____		**L.**	$\dfrac{1}{\tan\theta}$
13.	$\sin(\alpha - \beta)$ _____		**M.**	$-\cos\theta$
14.	$\tan(\pi - \theta)$ _____		**N.**	$-\sin\theta$
15.	$\tan(-\theta)$ _____		**O.**	$-\cot\theta$
16.	$\cos(\alpha - \beta)$ _____		**P.**	$\dfrac{1}{\csc\theta}$
17.	$\cos\left(\dfrac{1}{2}\pi + \theta\right)$ _____		**Q.**	$\sin\alpha\cos\beta + \cos\alpha\sin\beta$
18.	$1 + \cot^2\theta$ _____		**R.**	$\cos\alpha\cos\beta + \sin\alpha\sin\beta$
19.	$\cos(\pi - \theta)$ _____		**S.**	$\dfrac{1}{\sin\theta}$
20.	$\tan(\alpha - \beta)$ _____		**T.**	$\tan\theta$

Each of the functions below involves a phase shift. First use your graphing calculator to plot a graph of the function. Then by observing the graph you have plotted, write another function without a phase shift that you believe will produce the same graph.

21. $y = \sin\left(\dfrac{1}{2}\pi + x\right)$ **22.** $y = \tan(\pi - x)$

23. $y = \cos\left(\dfrac{1}{2}\pi + x\right)$ **24.** $y = \tan(2\pi + x)$

Prove or disprove the following identities.

25. $\cos(2\pi - \theta)\tan(\pi + \theta) = \sin\theta$ **26.** $\dfrac{\sin(\pi + \theta)}{\tan(\pi - \theta)} = \cos\theta$

27. $\dfrac{\cos\left(\frac{1}{2}\pi - \theta\right)\cdot\tan\left(\frac{1}{2}\pi + \theta\right)}{\cos(\pi - \theta)} = 1$

28. $\dfrac{\sin\left(\frac{1}{2}\pi + \theta\right)}{\tan\left(\frac{1}{2}\pi - \theta\right)} = \sin\theta$

29. $\dfrac{\tan\left(\frac{1}{2}\pi - \theta\right)\cos\left(\frac{1}{2}\pi + \theta\right)}{\tan\left(\frac{3}{2}\pi - \theta\right)} = -\sin\theta$

30. $\dfrac{\tan(\pi - \theta)}{\sin\left(\frac{1}{2}\pi + \theta\right)\tan(2\pi - \theta)} = \sec\theta$

31. $\dfrac{\cos\left(\frac{3}{2}\pi + \theta\right)\tan\left(\frac{1}{2}\pi + \theta\right)}{\cos(\pi + \theta)} = 1$

32. $\dfrac{\sin\left(\frac{3}{2}\pi - \theta\right)\tan\left(\frac{3}{2}\pi - \theta\right)}{\tan\left(\frac{1}{2}\pi + \theta\right)} = \cos\theta$

33. $\dfrac{\cos(2\pi - \theta)\sin\left(\frac{3}{2}\pi + \theta\right)}{\tan\left(\frac{1}{2}\pi + \theta\right)\cos\left(\frac{3}{2}\pi - \theta\right)} = -\cos\theta$

34. $\dfrac{\cos\left(\frac{1}{2}\pi + \theta\right)\tan\left(\frac{3}{2}\pi + \theta\right)}{\sin(2\pi - \theta)\sin\left(\frac{1}{2}\pi + \theta\right)} = -\csc\theta$

Group Writing Activity

1. Why is the tangent of the angle between a line and the x-axis equal to the slope of the line? Will this relationship always be true?
2. If two lines intersect at right angles, what is the relationship between the tangents of the angles that each line makes with the x-axis?

Chapter 6   **Key Ideas**

6.1   **1.** The cosine of the sum of two angles is
$$\cos(x + y) = \cos x \cos y - \sin x \sin y.$$

        2. The cosine of the difference of two angles is
$$\cos(x - y) = \cos x \cos y + \sin x \sin y.$$

        3. Use the cosine of the sum or difference of two numbers (angles) to transform trigonometric expressions or to prove identities.

        4. The sine of the sum of two angles is
$$\sin(x + y) = \sin x \cos y + \cos x \sin y.$$

        5. The sine of the difference of two angles is
$$\sin(x - y) = \sin x \cos y - \cos x \sin y.$$

6.2   **1.** Because the sine of an angle is the cosine of its complement
$$\text{Sin}^{-1} x + \text{Cos}^{-1} x = \frac{1}{2}\pi$$

        2. Use the inverse identity combined with the Pythagorean identities to derive identities useful in calculus.

6.3 **1.** The tangent of the sum of two angles is
$$\tan(x + y) = \frac{\tan x + \tan y}{1 - \tan x \tan y}.$$

      2. The tangent of the difference of two angles is
$$\tan(x - y) = \frac{\tan x - \tan y}{1 + \tan x \tan y}.$$

      3. Use the sine, cosine, and tangent of the sum or difference of two numbers (angles) to transform trigonometric expression or to prove identities.

6.4 **1.** Use the sum and difference identities to find the cosine, sine, or tangent functions of $\frac{1}{2}\pi + \theta$, $\pi + \theta$, $\frac{3}{2}\pi + \theta$, or any other combinations of multiples of π or $\frac{1}{2}\pi$.

      2. Use the sum and difference identities to prove trigonometric identities.

Chapter 6 Review Test

Using trigonometric function values for $30°$, $45°$, and $60°$, find the exact value for the following. **(6.1, 6.3)**

1. $\tan 15°$ 2. $\cos 105°$ 3. $\sin 105°$ 4. $\sin 120°$

Change each of the following expressions to expressions using $\sin x$, $\cos x$, and $\tan x$ by using only the sum and difference identities, that is, $\sin(\alpha \pm \beta)$, $\cos(\alpha \pm \beta)$, and $\tan(\alpha \pm \beta)$. **(6.1, 6.3)**

5. $\sin\left(\dfrac{3}{2}\pi - x\right)$ 6. $\sin(2\pi + x)$ 7. $\cos(2\pi + x)$

8. $\tan(45° + x)$ 9. $\cos\left(x + \dfrac{1}{3}\pi\right)$ 10. $\sin(x - 30°)$

Without using a calculator, find the exact value using the information given and the appropriate identity. **(6.1, 6.3)**

11. $\cos(\alpha - \beta)$; $\cos\alpha = \dfrac{1}{3}$, $\sin\beta = -\dfrac{3}{5}$, α in quadrant IV and β in quadrant III.

12. $\sin(\alpha + \beta)$; $\sin\alpha = -\dfrac{\sqrt{8}}{3}$, $\tan\beta = -\dfrac{4}{3}$, α in quadrant IV and β in quadrant II.

13. $\tan(\alpha - \beta)$; $\cos\alpha = -\dfrac{12}{13}$, $\sin\beta = \dfrac{3}{\sqrt{34}}$, α in quadrant III and β in quadrant II.

Use identities 26 through 31 to find the exact value of each of the following. **(6.2)**

14. $\cos\left[\operatorname{Sin}^{-1}\left(-\dfrac{1}{2}\right)\right]$ 15. $\csc\left(\operatorname{Cot}^{-1}1\right)$

Prove the following identities. **(6.1, 6.3, 6.4)**

16. $\dfrac{\cos(x - y)}{\sin x \sin y} = \cot x \cot y + 1$ 17. $\sin(x - y) \cdot \sec y = \cos x(\tan x - \tan y)$

18. $\dfrac{\cos(x + y)\cos(x - y)}{\cos^2 x \sin^2 y} = \cot^2 y - \tan^2 x$ 19. $\dfrac{\sin(x - y)}{\sin(x + y)} = \dfrac{\cot y - \cot x}{\cot y + \cot x}$

20. $\dfrac{\sin(x - y)}{\cos(x + y)} = \dfrac{\tan x - \tan y}{1 - \tan x \tan y}$ 21. $\dfrac{\cot y}{\tan(x - y)} - 1 = \dfrac{\cos x \csc y}{\sin(x - y)}$

22. $\dfrac{\sin(\pi - \theta)\tan(2\pi + \theta)}{\cos\left(\dfrac{1}{2}\pi + \theta\right)\sin\left(\dfrac{3}{2}\pi - \theta\right)} = \tan\theta\sec\theta$

Chapters 1–6 Cumulative Review

Name the smallest non-negative angle that is coterminal with the following angles. **(1.1)**

1. $560°$

2. $-\dfrac{17}{6}\pi$

3. Find the degree measure of $\dfrac{7}{3}\pi$ radians. **(1.1)**

4. Find the radian measure of $540°$. **(1.1)**

5. The minute hand of a clock is 6.4 inches long. How far does the tip move in 2 hours? Give your answer correct to two decimal places. **(1.2)**

6. What is the angular velocity in radians per second of a car wheel with an outside diameter of 24 inches, if the car is traveling 60 miles per hour? **(1.3)**

7. Find the values of the six trigonometric functions of an angle θ in standard position for the following coordinates of points on the terminal side of the angle. **(2.1)**

$$(-2, 3)$$

Find the values of the six trigonometric functions of the following angles. **(2.2)**

8. $120°$

9. $\dfrac{7}{6}\pi$ radians

Use your calculator to find the following in radians correct to one decimal place. **(2.3)**

10. $\arctan(1.823)$

11. $\arccos(0.1439)$

Solve the following right triangles. **(2.4)**

12. $a = 3.4, c = 5.1$

13. $\alpha = 38.4°, b = 0.493$

14. To measure the height of a cliff a surveyor measured the angle of elevation at a certain distance from the base of the cliff to be $48.4°$. At a point 90 feet farther from the base of the cliff, he measured the angle of elevation to be $37.6°$. What is the height of the cliff? **(2.5)**

15. Two lighthouses, 12.2 miles apart on a north-south line, receive a distress signal from a ship at sea. If the ship is 8.6 miles due east of the north lighthouse, find the ship's distance and bearing from the other lighthouse. **(2.5)**

16. Find the values of the six trigonometric functions of s at the point on the unit circle with the coordinates $\left(\dfrac{4}{5}, -\dfrac{3}{5}\right)$. **(2.6)**

17. Use the relation $\sin^2 s + \cos^2 s = 1$ to solve the following: **(2.6)**

$$\text{Given that } \sin\frac{4}{3}\pi = -\frac{\sqrt{3}}{2}, \text{find } \tan\frac{4}{3}\pi$$

For each of the following functions find the endpoints, quarter point, midpoint, and three-quarter point of the generic box. Then sketch one period. **(3.1)**

18. $y = 3 \sin \dfrac{1}{2}x$ **(3.1)**

19. $y = \dfrac{5}{2} \cos \left(2x + \dfrac{1}{3}\pi \right)$ **(3.2)**

State the period in radians for each of the following. **(3.4)**

20. $\tan 4x$

21. $\cot \dfrac{1}{3}x$

Sketch the graphs of the following functions. Sketch one period, then duplicate one period on the right side of the box. Label the critical points. **(3.4)**

22. $y = \cot \dfrac{1}{2}x$

23. $y = \dfrac{1}{2} \tan \left(x - \dfrac{1}{4}\pi \right)$

Use your graphing calculator to plot the following curves. Use the given information to set the range of your graphing calculator so the graph fits the screen nicely, and the critical points occur at the tick marks. Record the range values from your calculator.

24. $y = 2 \sin 3x$ $\qquad$ Graph 2 periods starting at $-\pi$ **(3.1)**

25. $y = \dfrac{1}{2} \cos \dfrac{1}{3}x$ $\qquad$ Graph 2 periods ending at 9π **(3.1)**

26. $y = \dfrac{3}{4} \sin(3x + \pi)$ $\qquad$ Graph 3 periods ending at $\dfrac{5}{3}\pi$ **(3.2)**

27. $y = -2 \cos \left(\dfrac{1}{2}x - \dfrac{1}{12}\pi \right)$ $\qquad$ Graph 2 periods starting at $-\dfrac{1}{2}\pi$ **(3.2)**

28. $y = 2 \sec \left(2x + \dfrac{1}{3}\pi \right)$ $\qquad$ Graph 2 periods starting at $-\dfrac{7}{4}\pi$ **(3.5)**

29. $y = \cot \left(\dfrac{1}{2}x - \dfrac{1}{12}\pi \right)$ $\qquad$ Graph 3 periods ending at $\dfrac{5}{2}\pi$ **(3.4)**

Use the method of addition of ordinates to sketch the following graphs. **(3.3)**

30. $y = -2 \sin x + 2$

31. $y = \dfrac{1}{2} \cos x + \sin 2x$

Using the same trigonometric function as y_1, give the equation of the function y_2 that meets the given criteria. **(3.2)**

32. $y_1 = \dfrac{1}{2} \cos \left(3x - \dfrac{1}{4}\pi \right)$ $\quad$ y_2 is 3 times as high as y_1 and the graph is shifted to the right $\dfrac{1}{2}\pi$.

Write the equations of two functions that will produce the graph below. **(3.6)**

33.

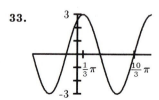

Determine the domain and range of the following equations. Determine if they are one-to-one. (4.1)

34. $y = x^2 + 4$

35. $x = 2y^2 - 1$

For the following equations:
a) Write the inverse. Solve for y in terms of x.
b) Determine if the original equation is a function.
c) Determine if the inverse is a function. (4.1)

36. $y = 2x + 5$

37. $x^2 + 2y^2 = 3$

Give the exact value of the following inverse functions in degrees and radians. (4.2)

38. $\text{Sin}^{-1}\left(\dfrac{1}{2}\right)$

39. $\text{Tan}^{-1}(-1)$

Use your calculator to find the value of the following a) correct to the nearest hundredth of a degree, and b) in radians, correct to three decimal places. (4.2)

40. $\text{Cos}^{-1}(0.3689)$

41. $\text{Tan}^{-1}(-2.862)$

Use your calculator to evaluate the following angles correct to the nearest tenth of a degree and numbers correct to the nearest thousandths. (4.2)

42. $\text{Cos}^{-1}(\sin 36.9°)$

43. $\csc\left[\text{Tan}^{-1}(2.56)\right]$

Without the use of a calculator, evaluate the following. (4.2)

44. $\cot\left[\text{Csc}^{-1}\dfrac{13}{5}\right]$

45. $\sec\left[\text{Tan}^{-1}\dfrac{4}{3}\right]$

Reduce the first expression to the second in each of the following. (5.1)

46. $\cos\theta(\sec\theta - \cos\theta)$, $\sin^2\theta$

47. $\csc\theta - \sin\theta$, $\cot\theta\cos\theta$

Prove the following identities: work on one side only. (5.2), (5.3)

48. $\dfrac{\sec x}{\tan^2 x} = \cot x \csc x$

49. $\dfrac{\tan^2\theta - 1}{\tan^2\theta + 1} = \dfrac{\sec^2\theta - \csc^2\theta}{\sec^2\theta + \csc^2\theta}$

50. $\dfrac{\csc^2\theta - \cot^2\theta + \cot\theta}{\csc\theta} = \cos\theta + \sin\theta$

Test the validity of the following identities using a graphing calculator. If the graphs coincide, prove the identity. If the equations are not identities, give a value of x that proves that they are not identities. (5.2), (5.3)

51. $\dfrac{-2\sin x \cos x}{1 - \sin x + \cos x} = 1 + \sin x + \cos x$

52. $\dfrac{\tan^2\theta - 2\tan\theta}{2\tan\theta - 4} = \tan\theta$

53. Using the values of the trigonometric functions for $30°$, $45°$, and $60°$, find the exact value of $\sin 105°$. (6.1)

Change the following expressions to expressions using $\sin x$, $\cos x$, and $\tan x$ by using only the sum and difference identities, that is, $\sin(\alpha \pm \beta)$, $\cos(\alpha \pm \beta)$ and $\tan(\alpha \pm \beta)$. (6.1), (6.3)

54. $\cos\left(\dfrac{1}{2}\pi - \theta\right)$

55. $\tan(225° + \theta)$

56. Use identities 26 through 31 in section 6.2 to find the exact value of $\sin\left[\text{Cos}^{-1}\left(-\dfrac{\sqrt{3}}{2}\right)\right]$.

Chapters 1–6 Cumulative Review

57. Using the appropriate identity, without using a calculator, find the exact value of $\cos(\alpha - \beta)$ if $\sin \alpha = -\dfrac{\sqrt{8}}{3}$ with α in quadrant III, and $\tan \beta = \dfrac{3}{4}$ with β in quadrant I. **(6.1)**

Prove the following identities. **(6.3)**, **(6.4)**

58. $\sin(x + y) \cos y - \cos(x + y) \sin y = \sin x$

59. $\dfrac{\cos\left(\dfrac{3}{2}\pi - \theta\right) \sin\left(\dfrac{1}{2}\pi + \theta\right)}{\sin(\pi + \theta) \tan(\pi - \theta)} = -\cos\theta \cot\theta$

Chapter 7

Additional Identities

Contents

Preview

The last chapter developed identities for the trigonometric functions of the sum of two angles. If the two angles are equal, these identities yield the double-angle identities. In the physical world, many measurements are literally "done with mirrors" using instruments like sextants and transits. These double-angle and half-angle identities are not limited to surveying and navigation. They are applied in electronics, optics (the study of light), acoustics (the study of sound), and nuclear physics, among other areas.

7.1   Double-Angle Identities

Identities for twice an angle follow directly from the identities for the sum of two angles.

Sin 2θ

If $\alpha = \theta$ and $\beta = \theta$, then $\sin(\alpha + \beta) = \sin 2\theta$

$$
\begin{aligned}
\sin(\alpha + \beta) &= \sin\alpha\cos\beta + \cos\alpha\sin\beta & \text{identity (23)}\\
\sin 2\theta &= \sin\theta\cos\theta + \sin\theta\cos\theta & \text{substituting } \theta \text{ for } \alpha \text{ and } \beta\\
\textbf{(43)} \qquad \mathbf{\sin 2\theta} &= \mathbf{2\sin\theta\cos\theta}
\end{aligned}
$$

Cos 2θ

Using the same logic, we can find an identity for $\cos 2\theta$.

$$
\begin{aligned}
\cos(\alpha + \beta) &= \cos\alpha\cos\beta - \sin\alpha\sin\beta & \text{identity (19)}\\
\cos 2\theta &= \cos\theta\cos\theta - \sin\theta\sin\theta & \text{substituting } \theta \text{ for } \alpha \text{ and } \beta\\
\textbf{(44a)} \qquad \mathbf{\cos 2\theta} &= \mathbf{\cos^2\theta - \sin^2\theta}
\end{aligned}
$$

Two other forms of the identity for $\cos 2\theta$ are frequently used.

$$
\begin{aligned}
\text{Recall that} \qquad \sin^2\theta &= 1 - \cos^2\theta\\
\text{Then} \qquad \cos 2\theta &= \cos^2\theta - \sin^2\theta & \text{identity (44a)}\\
\text{becomes} \qquad &= \cos^2\theta - \left(1 - \cos^2\theta\right) & \text{substituting identity (12)}\\
&= \cos^2\theta - 1 + \cos^2\theta\\
\textbf{(44b)} \qquad \mathbf{\cos 2\theta} &= \mathbf{2\cos^2\theta - 1}\\
\text{Or using} \qquad \cos^2\theta &= 1 - \sin^2\theta & \text{identity (12)}\\
\text{the identity for } \cos 2\theta \qquad \cos 2\theta &= \cos^2\theta - \sin^2\theta & \text{identity (44a)}\\
\text{becomes} \qquad &= 1 - \sin^2\theta - \sin^2\theta\\
\textbf{(44c)} \qquad \mathbf{\cos 2\theta} &= \mathbf{1 - 2\sin^2\theta}
\end{aligned}
$$

Example 1 □ Find $\cos 120°$ from $\sin 60°$ and $\cos 60°$.

$$
\cos 120° = \cos(2 \cdot 60°)
$$

Using identity (44a)

$$
\begin{aligned}
\cos 120° &= \cos^2 60° - \sin^2 60°\\
&= \left(\frac{1}{2}\right)^2 - \left(\frac{\sqrt{3}}{2}\right)^2\\
&= \frac{1}{4} - \frac{3}{4}\\
&= -\frac{1}{2}
\end{aligned}
$$

This isn't the normal way to find $\cos 120°$. We're just illustrating that the identity works.

□

Example 2 □ Find $\cos 120°$ from $\cos 60°$.

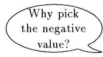

Both identities give the same result.

$$
\begin{aligned}
\cos 120° &= \cos(2 \cdot 60°) \\
&= 2\cos^2 60° - 1 \\
&= 2\left(\frac{1}{2}\right)^2 - 1 \\
&= 2\left(\frac{1}{4}\right) - 1 \\
&= -\frac{1}{2}
\end{aligned}
$$

□

Example 3 □ Given that $\frac{3}{2}\pi < \theta < 2\pi$, $\cos\theta = \frac{3}{8}$, find $\sin 2\theta$.

$\sin 2\theta = 2\sin\theta\cos\theta$, so we can compute $\sin 2\theta$ if we can find $\sin\theta$.

Knowing $\cos\theta$, we can find $\sin\theta$ from the identity:

$$
\begin{aligned}
\sin^2\theta + \cos^2\theta &= 1 \\
\sin^2\theta + \left(\frac{3}{8}\right)^2 &= 1 \\
\sin^2\theta + \frac{9}{64} &= \frac{64}{64} \\
\sin^2\theta &= \frac{55}{64} \\
\sin\theta &= -\frac{\sqrt{55}}{8}
\end{aligned}
$$

Why pick the negative value?

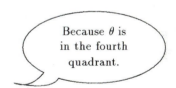

Because θ is in the fourth quadrant.

Now we can find $\sin 2\theta$:

$$
\begin{aligned}
\sin 2\theta &= 2\sin\theta\cos\theta \\
&= (2)\left(-\frac{\sqrt{55}}{8}\right)\left(\frac{3}{8}\right) \\
\sin 2\theta &= -\frac{3\sqrt{55}}{32}
\end{aligned}
$$

□

This value, like most values of the trigonometric functions, is an irrational number. We have expressed it exactly so you can see how the identity is applied. Use your calculator when you want to obtain an approximation.

Example 4 □ Express $\cos 110°$ in terms of functions of $55°$.

We can apply any of the double-angle identities for $\cos 2\theta$.

    a. $\cos 2\theta = \cos^2\theta - \sin^2\theta$
        Therefore: $\cos 110° = \cos^2 55° - \sin^2 55°$

    b. $\cos 2\theta = 2\cos^2\theta - 1$
        Therefore: $\cos 110° = 2\cos^2 55° - 1$

    c. $\cos 2\theta = 1 - 2\sin^2\theta$
        Therefore: $\cos 110° = 1 - 2\sin^2 55°$

□

Example 5 □ Use a double-angle formula to rewrite $1 - 2\sin^2 18°$.

$\cos 2\alpha = 1 - 2\sin^2\alpha$, so if we let $\alpha = 18°$, we can write:

$$\cos 2(18°) = 1 - 2\sin^2(18°)$$
$$\cos 36° = 1 - 2\sin^2 18°$$

□

Example 6 □ Prove $\dfrac{2}{\sin 2\theta} = \tan\theta + \cot\theta$.

When identities have a double angle, it's usually effective to convert the double angle to an expression with a single angle.

$$\frac{2}{\sin 2\theta} = \tan\theta + \cot\theta$$

$$\frac{2}{2\sin\theta\cos\theta} = \qquad\qquad\text{identity (43)}$$

$$\frac{1}{\sin\theta\cos\theta} =$$

> That doesn't look like the right side.

> Not yet, but recall that $\sin^2\theta + \cos^2\theta = 1$.

$$\frac{\sin^2\theta + \cos^2\theta}{\sin\theta\cos\theta} =$$

Now separate the fraction

$$\frac{\sin^2\theta}{\sin\theta\cos\theta} + \frac{\cos^2\theta}{\sin\theta\cos\theta} =$$

$$\frac{\sin\theta}{\cos\theta} + \frac{\cos\theta}{\sin\theta} =$$

$$\tan\theta + \cot\theta = \tan\theta + \cot\theta$$

> How did you know to separate the fraction?

> One of the things books try to show is useful *techniques* that you can add to your bag of *tools*.

□

Example 7 □ Write an expression for $\cos 3\theta$.

Think of $\cos 3\theta$ as $\cos(2\theta + \theta)$.

$$\cos(2\theta + \theta) = \cos 2\theta \cos \theta - \sin 2\theta \sin \theta \qquad \text{identity (19)}$$

Now substituting for $\cos 2\theta$ and $\sin 2\theta$,

$$
\begin{aligned}
&= (\cos^2\theta - \sin^2\theta)\cos\theta - (2\sin\theta\cos\theta)\sin\theta \\
&= \cos^3\theta - \sin^2\theta\cos\theta - 2\sin^2\theta\cos\theta \\
&= \cos^3\theta - 3\sin^2\theta\cos\theta
\end{aligned}
$$

□

Example 8 □ Prove $\sin 2\alpha = (\tan\alpha)(1 + \cos 2\alpha)$.

Both sides have double angles. We'll work on the right side since it appears more complex.

$$
\begin{aligned}
(\tan\alpha)(1 + \cos 2\alpha) &= (\tan\alpha)(1 + \cos 2\alpha) \\
&= \frac{\sin\alpha}{\cos\alpha}(1 + \cos^2\alpha - \sin^2\alpha) \\
&= \frac{\sin\alpha}{\cos\alpha} + \sin\alpha\cos\alpha - \frac{\sin^3\alpha}{\cos\alpha}
\end{aligned}
$$

This is going nowhere.

Sometimes you'll make a bad start. Try $\cos 2\alpha = 2\cos^2\alpha - 1$ on the right side.

$$
\begin{aligned}
\frac{\sin\alpha}{\cos\alpha}(1 + 2\cos^2\alpha - 1) &= \frac{\sin\alpha}{\cos\alpha}(1 + 2\cos^2\alpha - 1) \\
&= \frac{\sin\alpha}{\cos\alpha}(2\cos^2\alpha) \qquad \text{Because } 1 - 1 = 0 \\
&= 2\sin\alpha\cos\alpha \\
&= \sin 2\alpha
\end{aligned}
$$

□

Tan 2θ

An identity for $\tan 2\theta$ follows directly from the identity for $\tan(\alpha + \beta)$.

$$\tan(\alpha + \beta) = \frac{\tan\alpha + \tan\beta}{1 - \tan\alpha\tan\beta} \qquad \text{identity (25)}$$

Let $\alpha = \theta$ and $\beta = \theta$

$$\tan 2\theta = \frac{\tan\theta + \tan\theta}{1 - \tan\theta\tan\theta}$$

$$(45) \qquad \boldsymbol{\tan 2\theta = \frac{2\tan\theta}{1 - \tan^2\theta}}$$

Problem Set 7.1

Using the double-angle formula, find the exact value of the following [for example, $90° = 2(45°)$].

1. $\cos 90°$ **2.** $\sin 120°$ **3.** $\tan 60°$ **4.** $\cos 300°$

5. $\sin 450°$ **6.** $\tan 120°$ **7.** $\cos 240°$ **8.** $\sin 240°$

Use a double-angle formula to rewrite the following.

9. $\cos^2 15° - \sin^2 15°$ **10.** $2\sin 12° \cos 12°$ **11.** $\dfrac{2\tan \frac{1}{6}\pi}{1 - \tan^2 \frac{1}{6}\pi}$

12. $2\cos^2 \frac{1}{6}\pi - 1$ **13.** $4\sin \dfrac{A}{2} \cos \dfrac{A}{2}$ **14.** $\cos^2 6A - \sin^2 6A$

Given $\sin \theta = \dfrac{3}{5}$ with $90° \le \theta \le 180°$, find the exact value of:

15. $\sin 2\theta$ **16.** $\cos 2\theta$ **17.** $\tan 2\theta$

Given $\cos = -\dfrac{5}{13}$ with $180° \le \theta \le 270°$, find the exact value of:

18. $\sin 2\theta$ **19.** $\cos 2\theta$ **20.** $\tan 2\theta$

Test the validity of the following identities using a graphing calculator. If the graphs coincide, prove the identity. If the equations are not identities, give a value of x that proves that they are not identities.

21. $\sin 3x = 3\sin x - 4\sin^3 x$ **22.** $\cos 3x = 4\cos^2 x - 3\cos x$

23. $\sin 4x = 4\sin x(1 - 2\sin^2 x)\cos 2x$ **24.** $\cos 4x = 8\cos^4 x - 8\cos^2 x + 1$

Prove the following identities.

25. $(\sin x + \cos x)^2 = 1 + \sin 2x$ **26.** $\cos 2x \sec^2 x = 1 - \tan^2 x$

27. $\csc x - 2\sin x = \dfrac{\cos 2x}{\sin x}$ **28.** $\cot \theta - \tan \theta = \dfrac{2\cos 2\theta}{\sin 2\theta}$

29. $\cos^4 \theta - \sin^4 \theta = \cos 2\theta$ **30.** $\sin 2x = \dfrac{2\cot x}{1 + \cot^2 x}$

31. $\dfrac{\sin 2\theta}{\cos \theta} + \dfrac{\cos 2\theta}{\sin \theta} = \csc \theta$ **32.** $\cot x + \tan 2x = \dfrac{\cot x}{\cos 2x}$

33. $\dfrac{\cot \theta - \tan \theta}{\cot \theta + \tan \theta} = \cos 2\theta$ **34.** $\dfrac{1 - \cos 2\theta}{1 + \cos 2\theta} = \tan^2 \theta$

35. $\dfrac{1 + \tan x}{1 - \tan x} = \dfrac{1 + \sin 2x}{\cos 2x}$ **36.** $1 - \cos 2x = \tan x \sin 2x$

37. $\dfrac{(1 + \cot x)^2}{1 + \cot^2 x} = 1 + \sin 2x$ **38.** $1 - \sin 2x \tan x = \cos 2x$

7.2 Half-Angle Identities

Sin $\dfrac{\theta}{2}$

The half-angle identities are double-angle identities stated in a different form. Consider:

$$\cos 2\alpha = 1 - 2\sin^2\alpha \qquad \text{identity (44c)}$$

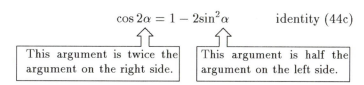

This argument is twice the argument on the right side.

This argument is half the argument on the left side.

If we let $2\alpha = \theta$, then $\alpha = \dfrac{\theta}{2}$

Substituting these values yields

$$\cos\theta = 1 - 2\sin^2\frac{\theta}{2}$$

Solving for $\sin\dfrac{\theta}{2}$,

$$2\sin^2\frac{\theta}{2} = 1 - \cos\theta$$

$$\sin^2\frac{\theta}{2} = \frac{(1 - \cos\theta)}{2}$$

$$\textbf{(46)} \qquad \boldsymbol{\sin\frac{\theta}{2} = \pm\sqrt{\frac{1 - \cos\theta}{2}}}$$

The $\pm$ sign on this identity needs some interpretation. The proper value is determined by the quadrant of $\dfrac{\theta}{2}$.

Cos $\dfrac{\theta}{2}$

To develop an identity for $\cos\dfrac{\theta}{2}$ start with

$$\cos 2\alpha = 2\cos^2\alpha - 1 \qquad \text{identity (44b)}$$

Again let $\theta = 2\alpha$, then

$$\cos\theta = 2\cos^2\frac{\theta}{2} - 1$$

Solving for $\cos\dfrac{\theta}{2}$,

$$2\cos^2\frac{\theta}{2} = \cos\theta + 1$$

$$\textbf{(47)} \qquad \boldsymbol{\cos\frac{\theta}{2} = \pm\sqrt{\frac{\cos\theta + 1}{2}}}$$

You will need to know where $\frac{\theta}{2}$ lies before you can assign the $+$ or $-$ sign to this identity.

Group Writing Activity

Why is the quadrant that $\dfrac{\theta}{2}$ is in important in determining the sign of the formula? How would you use that information to complete a problem involving half-angle identities?

Using Your Graphing Calculator

Recall that the principal square root of x is positive. Therefore, your calculator will always assign a positive sign to a square root.

Use your graphing calculator to plot graphs of both sides of identity (47). Set

$$X_{\min} = -6.28 \qquad X_{\max} = 6.28$$

$$
\begin{aligned}
Y_1 &= \cos(x \div 2) \\
Y_2 &= \boxed{\sqrt{}}((\cos x + 1) \div 2))
\end{aligned}
$$

You should get the graph below:

□ Figure 7.1

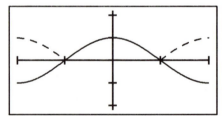

Between $-\pi$ and π there appears to be only one graph. This is because identity (47) is valid for this domain. Using the principal value of the square root in the ranges $-2\pi < x < -\pi$ and $\pi < x < 2\pi$, there are two graphs because in this portion of the domain the identity is

$$\cos \frac{\theta}{2} = -\sqrt{\frac{\cos x + 1}{2}}$$

Example 9 □ Find cos 120° from the cosine of 240°.

$$\cos\frac{240°}{2} = \pm\sqrt{\frac{\cos 240° + 1}{2}}$$

$$\cos 120° = \pm\sqrt{\frac{-\frac{1}{2} + 1}{2}}$$

$$\cos 120° = \pm\sqrt{\frac{\frac{1}{2}}{2}}$$

$$\cos 120° = \pm\sqrt{\frac{1}{4}}$$

$$\cos 120° = -\frac{1}{2}$$

Pick the negative value because 120° is in the second quadrant.

□

Tan $\dfrac{\theta}{2}$

We can use $\tan\alpha = \dfrac{\sin\alpha}{\cos\alpha}$ to derive an identity for $\tan\dfrac{\theta}{2}$:

$$\tan\frac{\theta}{2} = \frac{\sin\dfrac{\theta}{2}}{\cos\dfrac{\theta}{2}}$$

$$= \frac{\pm\sqrt{\dfrac{1 - \cos\theta}{2}}}{\pm\sqrt{\dfrac{1 + \cos\theta}{2}}}$$

$$= \pm\sqrt{\frac{1 - \cos\theta}{2} \cdot \frac{2}{1 + \cos\theta}}$$

(48a) $$\tan\frac{\theta}{2} = \pm\sqrt{\frac{1 - \cos\theta}{1 + \cos\theta}}$$

The proper sign of the identity is determined by the quadrant of $\dfrac{\theta}{2}$.

Rationalizing the denominator of the previous identity yields another useful identity for $\tan\dfrac{\theta}{2}$ where you don't have to worry about the sign of the radical.

$$\tan\frac{\theta}{2} = \frac{\sqrt{1 - \cos\theta}}{\sqrt{1 + \cos\theta}}\left[\frac{\sqrt{1 - \cos\theta}}{\sqrt{1 - \cos\theta}}\right]$$

$$= \frac{\sqrt{(1 - \cos\theta)^2}}{\sqrt{1 - \cos^2\theta}}$$

(48b) $$\tan\frac{\theta}{2} = \frac{1 - \cos\theta}{\sin\theta}$$

If you test the value of θ in all four quadrants, you'll see that the positive radical in each case produces the proper sign for $\tan\dfrac{\theta}{2}$.

Using Your Graphing Calculator

If you plot graphs of $Y_1 = \tan \dfrac{x}{2}$ and $Y_2 = \dfrac{1 - \cos x}{\sin x}$ in the domain $-2\pi \geq x \geq 2\pi$, you'll see they both produce the same graph. This is not the case with identity (48a) where you must be careful of the sign of $\dfrac{x}{2}$.

Example 10 □ Show $\tan \dfrac{\theta}{2} = \dfrac{\sin \theta}{1 + \cos \theta}$.

Start with identity (48b):

$$\tan \frac{\theta}{2} = \frac{1 - \cos \theta}{\sin \theta}$$

Multiply numerator and denominator by $1 + \cos \theta$.

$$
\begin{aligned}
\tan \frac{\theta}{2} &= \frac{1 - \cos \theta}{\sin \theta} \left[\frac{1 + \cos \theta}{1 + \cos \theta} \right] \\[2mm]
&= \frac{1 - \cos^2 \theta}{\sin \theta (1 + \cos \theta)} \\[2mm]
&= \frac{\sin^2 \theta}{\sin \theta (1 + \cos \theta)} \\[2mm]
(48c) \quad \tan \frac{\theta}{2} &= \frac{\sin \theta}{1 + \cos \theta}
\end{aligned}
$$

□

Example 11 □ Prove $\sin \theta = \dfrac{2 \tan \dfrac{\theta}{2}}{1 + \tan^2 \dfrac{\theta}{2}}$.

Working on the right side,

$$
\begin{aligned}
\sin \theta &= \frac{2 \tan \dfrac{\theta}{2}}{1 + \tan^2 \dfrac{\theta}{2}} \\[4mm]
&= \frac{2 \cdot \dfrac{\sin \theta}{1 + \cos \theta}}{1 + \left[\dfrac{\sin \theta}{1 + \cos \theta} \right]^2} \qquad \text{identity (48c)} \\[4mm]
&= \frac{\dfrac{2 \sin \theta}{1 + \cos \theta}}{1 + \dfrac{\sin^2 \theta}{(1 + \cos \theta)^2}}
\end{aligned}
$$

Next get a common denominator and add in the denominator.

$$= \frac{\dfrac{2\sin\theta}{1+\cos\theta}}{\dfrac{(1+\cos\theta)^2 + \sin^2\theta}{(1+\cos\theta)^2}}$$

$$= \frac{\dfrac{2\sin\theta}{1+\cos\theta}}{\dfrac{1 + 2\cos\theta + \cos^2\theta + \sin^2\theta}{(1+\cos\theta)^2}}$$

$$= \frac{2\sin\theta}{1+\cos\theta} \cdot \frac{(1+\cos\theta)^2}{1+2\cos\theta+1} \qquad \text{invert and multiply}$$

$$= \frac{2\sin\theta(1+\cos\theta)}{2(1+\cos\theta)}$$

$$= \sin\theta$$

$\square$

Problem Set 7.2

In problems 1–20 match each expression on the left with an expression on the right to form a fundamental identity. Some answers may be used more than once or not at all.

1.	$1 + \tan^2\theta$ _____	**A.**	$\dfrac{\sin\theta}{1+\cos\theta}$
2.	$\sin(-\theta)$ _____	**B.**	$-\cos\theta$
3.	$\sec\theta$ _____	**C.**	$\sin\theta$
4.	$\tan\dfrac{\theta}{2}$ _____	**D.**	$\cos(\alpha-\beta) - \cos(\alpha+\beta)$
5.	$\sin(\alpha-\beta)$ _____	**E.**	1
6.	$\cos(\pi+\theta)$ _____	**F.**	$-\sin\theta$
7.	$2\cos^2\theta - 1$ _____	**G.**	$\tan 2\theta$
8.	$2\sin\theta\cos\theta$ _____	**H.**	$\cot^2\theta$
9.	$\cos\theta\cdot\sec\theta$ _____	**I.**	$\dfrac{\cos\theta}{\sin\theta}$
10.	$\pm\sqrt{\dfrac{1+\cos\theta}{2}}$ _____	**J.**	$\dfrac{1}{\cos\theta}$
11.	$\tan(-\theta)$ _____	**K.**	$-\tan\theta$
12.	$\csc^2\theta - 1$ _____	**L.**	$\cos\dfrac{\theta}{2}$
13.	$\sin(\pi-\theta)$ _____	**M.**	$\sec^2\theta$
14.	$\dfrac{2\tan\theta}{1-\tan^2\theta}$ _____	**N.**	$\sin\dfrac{\theta}{2}$
15.	$\cos\left(\dfrac{1}{2}\pi - \theta\right)$ _____	**O.**	$\cos 2\theta$
16.	$\cos(\alpha-\beta)$ _____	**P.**	$\sin\alpha\cos\beta - \cos\alpha\sin\beta$
17.	$\tan\theta\cdot\cot\theta$ _____	**Q.**	$\sin 2\theta$
18.	$\tan(\pi-\theta)$ _____	**R.**	$\dfrac{1}{\tan\theta}$
19.	$\cos^2\theta - \sin^2\theta$ _____	**S.**	$\cos\alpha\cos\beta + \sin\alpha\sin\beta$
20.	$\pm\sqrt{\dfrac{1-\cos\theta}{2}}$ _____	**T.**	$-\cot\theta$

Using the half-angle formula, find the exact values of the following $\left[\text{for example, } 15° = \frac{1}{2}(30°)\right].$

21. $\cos 15°$ **22.** $\sin 22\frac{1}{2}°$ **23.** $\tan 30°$ **24.** $\cos 120°$

25. $\sin 75°$ **26.** $\tan 67\frac{1}{2}°$ **27.** $\tan 15°$ **28.** $\sin 67\frac{1}{2}°$

Use a half-angle formula to rewrite the following.

29. $\sqrt{\dfrac{1 - \cos 30°}{2}}$ **30.** $\sqrt{\dfrac{1 - \cos 150°}{1 + \cos 150°}}$ **31.** $\sqrt{\dfrac{1 + \cos 135°}{2}}$

32. $\sqrt{\dfrac{1 - \cos 170°}{2}}$ **33.** $\dfrac{1 - \cos 210°}{\sin 210°}$ **34.** $\sqrt{\dfrac{1 + \cos 130°}{2}}$

Given $\sin \theta = \dfrac{4}{5}$ with $90° \le \theta \le 180°$, find the exact value of:

35. $\sin \dfrac{\theta}{2}$ **36.** $\cos \dfrac{\theta}{2}$ **37.** $\tan \dfrac{\theta}{2}$

Given $\cos \theta = -\dfrac{12}{13}$, with $180° \le \theta \le 270°$, find the exact value of:

38. $\sin \dfrac{\theta}{2}$ **39.** $\cos \dfrac{\theta}{2}$ **40.** $\tan \dfrac{\theta}{2}$

Test the validity of the following identities using a graphing calculator. If the graphs coincide, prove the identity. If the equations are not identities, give a value of x that proves that they are not identities.

41. $\dfrac{2\sin \dfrac{x}{2} - \sin x}{2 - 2\cos \dfrac{x}{2}} = \sin \dfrac{x}{2}$ **42.** $\sin 2x + 2\sin x = 4\sin x \cos^2 \dfrac{x}{2}$

43. $\dfrac{1 + \sin x}{\cos x} = \dfrac{1 + \tan \dfrac{x}{2}}{1 - \tan \dfrac{x}{2}}$ **44.** $\sin^2 \dfrac{x}{2} = \dfrac{\tan x - \sin x}{2\tan x}$

Prove the following identities.

45. $\sin^2 \dfrac{\theta}{2} = \dfrac{\csc \theta - \cot \theta}{2\csc \theta}$ **46.** $\csc x - \cot x = \tan \dfrac{x}{2}$

47. $2\cos^2 \dfrac{x}{2} = \dfrac{\sin^2 x}{1 - \cos x}$ **48.** $\left(\cos \dfrac{\theta}{2} - \sin \dfrac{\theta}{2}\right)^2 = 1 - \sin \theta$

49. $\sin \theta = \tan \dfrac{\theta}{2}(1 + \cos \theta)$ **50.** $\dfrac{\cos x - \cos 2x}{\sin x + \sin 2x} = \tan \dfrac{x}{2}$

51. $\tan \dfrac{\theta}{2} = \csc \theta - \cot \theta$ **52.** $\sin^4 \theta = \dfrac{1}{4} - \dfrac{\cos 2\theta}{2} + \dfrac{\cos^2 2\theta}{4}$

53. $\dfrac{\tan x + \sin x}{2\tan x} = \cos^2 \dfrac{x}{2}$ **54.** $\dfrac{\sin 2x}{2\sin x} = \cos^2 \dfrac{x}{2} - \sin^2 \dfrac{x}{2}$

     Chapter 7 Additional Identities

7.3 Identities to Rewrite Sums and Products

Product-to-Sum Identities

Product-to-sum identities are used less frequently but can be invaluable. Like the distributive property, they allow us to convert a product to a sum or vice versa.

Start with the identities for $\sin(\alpha + \beta)$ and $\sin(\alpha - \beta)$:

$$\sin \alpha \cos \beta + \cos \alpha \sin \beta = \sin(\alpha + \beta)$$
$$\sin \alpha \cos \beta - \cos \alpha \sin \beta = \sin(\alpha - \beta)$$

Adding these identities yields

(49) $\quad 2 \sin \alpha \cos \beta = \sin(\alpha + \beta) + \sin(\alpha - \beta)$

A similar process with $\cos(\alpha + \beta)$ yields

(50) $\quad 2 \cos \alpha \cos \beta = \cos(\alpha + \beta) + \cos(\alpha - \beta)$

also

(51) $\quad 2 \sin \alpha \sin \beta = \cos(\alpha - \beta) - \cos(\alpha + \beta)$
(52) $\quad 2 \cos \alpha \sin \beta = \sin(\alpha + \beta) - \sin(\alpha - \beta)$

Identities (49)–(52) allow us to convert a product of sines and/or cosines into our choice of a sum or difference of sines or cosines.

Sum-to-Product Identities

The following identities let us express sums or differences of sines or cosines as products.

(53) $\quad \cos \alpha + \cos \beta = 2 \cos \left(\dfrac{\alpha + \beta}{2} \right) \cos \left(\dfrac{\alpha - \beta}{2} \right)$

(54) $\quad \cos \alpha - \cos \beta = -2 \sin \left(\dfrac{\alpha + \beta}{2} \right) \sin \left(\dfrac{\alpha - \beta}{2} \right)$

(55) $\quad \sin \alpha + \sin \beta = 2 \sin \left(\dfrac{\alpha + \beta}{2} \right) \cos \left(\dfrac{\alpha - \beta}{2} \right)$

(56) $\quad \sin \alpha - \sin \beta = 2 \cos \left(\dfrac{\alpha + \beta}{2} \right) \sin \left(\dfrac{\alpha - \beta}{2} \right)$

Do I have to memorize these?

No. Just know that they exist in case you need them later.

Pointers to Help Prove Identities

To prove trigonometric identities we must integrate three distinct skills:

1. Remember many identities.

2. Perform algebraic manipulation using trigonometric functions.

3. Use flexible problem-solving techniques.

This book has placed a heavy emphasis on manipulation of identities because the authors believe that this skill will help students to succeed in later math and science courses.

Below are some suggestions on how to prove trigonometric identities.

1. It's usually more productive to reduce the more complicated side of an identity to the simpler side.

2. Perform any indicated operations such as addition, subtraction, multiplication, or division.

3. Try factoring any common factors.

4. Look for expressions that match one of these forms.

$$\left.\begin{array}{r} a^2 - b^2 \\ a^3 + b^3 \\ a^3 - b^3 \\ ax^2 + bx + c \end{array}\right\} \quad \text{Factor these.}$$

5. Sometimes a fraction with a sum in the numerator can be written as a sum of two separate fractions. Each of these may be reducible.

6. Multiplying both the numerator and denominator of a fraction by the same expression may simplify it.

7. Convert all expressions with double angles or half angles to expressions of a single angle.

8. If all else fails, express everything in terms of sines and cosines, and simplify the result.

9. Don't be afraid to give up an unproductive approach for a different attack.

10. Don't lose heart! Eventually you will prove identities well, but to do so will take practice.

The principal identities are listed below.

7.3 The Fundamental Trigonometric Identities

Reciprocal Identities

$(1) \quad \sin u = \dfrac{1}{\csc u}$ $\qquad\qquad$ $(2) \quad \csc u = \dfrac{1}{\sin u}$

$(3) \quad \cos u = \dfrac{1}{\sec u}$ $\qquad\qquad$ $(4) \quad \sec u = \dfrac{1}{\cos u}$

$(5) \quad \tan u = \dfrac{1}{\cot u}$ $\qquad\qquad$ $(6) \quad \cot u = \dfrac{1}{\tan u}$

$(7) \quad \sin u \cdot \csc u = 1$ $\qquad\qquad$ $(8) \quad \cos u \cdot \sec u = 1$

$(9) \quad \tan u \cdot \cot u = 1$

Ratio Identities

$(10) \quad \tan u = \dfrac{\sin u}{\cos u}$ $\qquad\qquad$ $(11) \quad \cot u = \dfrac{\cos u}{\sin u}$

Pythagorean Identities

Identity	Alternate Forms
$(12) \quad \cos^2 u + \sin^2 u = 1$	$\sin^2 u = 1 - \cos^2 u$ $\cos^2 u = 1 - \sin^2 u$
$(13) \quad 1 + \tan^2 u = \sec^2 u$	$1 = \sec^2 u - \tan^2 u$ $\tan^2 u = \sec^2 u - 1$
$(14) \quad \cot^2 u + 1 = \csc^2 u$	$1 = \csc^2 u - \cot^2 u$ $\cot^2 u = \csc^2 u - 1$

Opposite Angle Identities

$(15) \quad \cos(-u) = \cos u$ $\qquad$ $(16) \quad \sin(-u) = -\sin u$

$(17) \quad \tan(-u) = -\tan u$

Sum and Difference Identities

$(18) \quad \cos(\alpha - \beta) = \cos\alpha\cos\beta + \sin\alpha\sin\beta$ $\qquad$ $(19) \quad \cos(\alpha + \beta) = \cos\alpha\cos\beta - \sin\alpha\sin\beta$

$(23) \quad \sin(\alpha + \beta) = \sin\alpha\cos\beta + \cos\alpha\sin\beta$ $\qquad$ $(24) \quad \sin(\alpha - \beta) = \sin\alpha\cos\beta - \cos\alpha\sin\beta$

$(32) \quad \tan(\alpha + \beta) = \dfrac{\tan\alpha + \tan\beta}{1 - \tan\alpha\tan\beta}$ $\qquad$ $(33) \quad \tan(\alpha - \beta) = \dfrac{\tan\alpha - \tan\beta}{1 + \tan\alpha\tan\beta}$

Identities Involving Inverses

$(25) \quad \operatorname{Sin}^{-1} u + \operatorname{Cos}^{-1} u = \dfrac{1}{2}\pi$ $\qquad$ $(26) \quad \cos\left(\operatorname{Sin}^{-1} u\right) = \sqrt{1 - u^2}$

$(27) \quad \csc\left(\operatorname{Cot}^{-1} u\right) = \sqrt{1 + u^2}$ $\qquad$ $(28) \quad \sec\left(\operatorname{Tan}^{-1} u\right) = \sqrt{1 + u^2}$

$(29) \quad \tan\left(\operatorname{Sec}^{-1} u\right) = \sqrt{u^2 - 1}$ $\qquad$ $(30) \quad \sin\left(\operatorname{Cos}^{-1} u\right) = \sqrt{1 - u^2}$

$(31) \quad \cot\left(\operatorname{Csc}^{-1} u\right) = \sqrt{u^2 - 1}$

Cofunction Identities

$(21) \quad \sin u = \cos\left(\dfrac{1}{2}\pi - u\right)$ $\qquad$ $(22) \quad \cos u = \sin\left(\dfrac{1}{2}\pi - u\right)$

$(42) \quad \cot u = \tan\left(\dfrac{1}{2}\pi - u\right)$

Other Identities Involving Multiples of $\dfrac{1}{2}\pi$

$(34) \quad \sin(\pi + u) = -\sin u$ $\qquad$ $(36) \quad \sin(\pi - u) = \sin u$

$(20) \quad \cos(\pi + u) = -\cos u$ $\qquad$ $(37) \quad \cos(\pi - u) = -\cos u$

$(35) \quad \tan(\pi + u) = \tan u$ $\qquad$ $(38) \quad \tan(\pi - u) = -\tan u$

$(39) \quad \sin\left(\dfrac{1}{2}\pi + u\right) = \cos u$ $\qquad$ $(41) \quad \tan\left(\dfrac{1}{2}\pi + u\right) = -\cot u$

$(40) \quad \cos\left(\dfrac{1}{2}\pi + u\right) = -\sin u$

7.3 Trigonometric Identities (continued)

Double-Angle Identities   **Half-Angle Identities**

(43) $\sin 2u = 2\sin u \cos u$ (46) $\sin \dfrac{u}{2} = \pm\sqrt{\dfrac{1-\cos u}{2}}$

(44a) $\cos 2u = \cos^2 u - \sin^2 u$ (47) $\cos \dfrac{u}{2} = \pm\sqrt{\dfrac{\cos u + 1}{2}}$

(44b) $\cos 2u = 2\cos^2 u - 1$ (48a) $\tan \dfrac{u}{2} = \pm\sqrt{\dfrac{1-\cos u}{1+\cos u}}$

(44c) $\cos 2u = 1 - 2\sin^2 u$ (48b) $\tan \dfrac{u}{2} = \dfrac{1-\cos u}{\sin u}$

(45) $\tan 2u = \dfrac{2\tan u}{1-\tan^2 u}$ (48c) $\tan \dfrac{u}{2} = \dfrac{\sin u}{1+\cos u}$

Identities to Rewrite Sums and Products

(49) $2\sin \alpha \cos \beta = \sin(\alpha + \beta) + \sin(\alpha - \beta)$

(50) $2\cos \alpha \cos \beta = \cos(\alpha + \beta) + \cos(\alpha - \beta)$

(51) $2\sin \alpha \sin \beta = \cos(\alpha - \beta) - \cos(\alpha + \beta)$

(52) $2\cos \alpha \sin \beta = \sin(\alpha + \beta) - \sin(\alpha - \beta)$

(53) $\cos \alpha + \cos \beta = 2\cos\left(\dfrac{\alpha + \beta}{2}\right)\cos\left(\dfrac{\alpha - \beta}{2}\right)$

(54) $\cos \alpha - \cos \beta = -2\sin\left(\dfrac{\alpha + \beta}{2}\right)\sin\left(\dfrac{\alpha - \beta}{2}\right)$

(55) $\sin \alpha + \sin \beta = 2\sin\left(\dfrac{\alpha + \beta}{2}\right)\cos\left(\dfrac{\alpha - \beta}{2}\right)$

(56) $\sin \alpha - \sin \beta = 2\cos\left(\dfrac{\alpha + \beta}{2}\right)\sin\left(\dfrac{\alpha - \beta}{2}\right)$

Example 12 □ Prove $\dfrac{\tan x \sin x}{1 - \cos x} = \dfrac{1 + \cos x}{\cos x}$.

Since we know that $1 - \cos^2 x = \sin^2 x$, we'll start by multiplying numerator and denominator on the left side by $1 + \cos x$.

$$\frac{\tan x \sin x}{1 - \cos x} = \frac{\tan x \sin x}{1 - \cos x}$$

$$\frac{\tan x \sin x}{1 - \cos x} \cdot \left(\frac{1 + \cos x}{1 + \cos x}\right) = \qquad \text{multiply by one}$$

$$\frac{\tan x \sin x + \tan x \sin x \cos x}{1 - \cos^2 x} = \qquad \text{multiply numerator and denominator out}$$

$$\frac{\dfrac{\sin x}{\cos x}\sin x + \dfrac{\sin x}{\cancel{\cos x}}\sin x\,\cancel{\cos x}}{\sin^2 x} = \qquad \text{rewrite } \tan x \text{ in numerator, replace } 1 - \cos^2 x \text{ in denominator}$$

$$\frac{\sin^2 x\left(\dfrac{1}{\cos x} + 1\right)}{\sin^2 x} = \qquad \text{factor numerator}$$

$$\frac{\cancel{\sin^2 x}}{1}\left(\frac{1}{\cos x} + 1\right) \cdot \frac{1}{\cancel{\sin^2 x}} = \qquad \text{to divide, invert and multiply}$$

$$\frac{1 + \cos x}{\cos x} = \frac{\tan x \sin x}{1 - \cos x} \qquad \text{find denominator and add}$$

□

Example 13 □ Prove $\dfrac{1 + \cos\alpha(1 + \sin\alpha) - \sin\alpha}{\cos^2\alpha} = \dfrac{1}{1 + \sin\alpha} + \dfrac{\cos\alpha}{1 - \sin\alpha}.$

Since the left side is more complex, start there.

$$\dfrac{\dfrac{1 + \cos\alpha(1 + \sin\alpha) - \sin\alpha}{\cos^2\alpha}}{} =$$

$$\dfrac{\dfrac{1 + \cos\alpha + \cos\alpha\sin\alpha - \sin\alpha}{\cos^2\alpha}}{} =$$

This doesn't appear to lead anywhere. Try an insight.

- **A.** $\cos^2\alpha = 1 - \sin^2\alpha$.
- **B.** Right side has denominators that are factors of $1 - \sin^2\alpha$.
- **C.** Try rewriting the original identity with the denominator of the left side written as $1 - \sin^2\alpha$.

$$\dfrac{1 + \cos\alpha(1 + \sin\alpha) - \sin\alpha}{1 - \sin^2\alpha}$$

Next write the left side as a sum of three separate fractions over the factored denominator:

$$\dfrac{1}{(1 - \sin\alpha)(1 + \sin\alpha)} + \dfrac{\cos\alpha\;\cancel{(1 + \sin\alpha)}}{(1 - \sin\alpha)\;\cancel{(1 + \sin\alpha)}} - \dfrac{\sin\alpha}{(1 - \sin\alpha)(1 + \sin\alpha)} =$$

$$\dfrac{1}{(1 - \sin\alpha)(1 + \sin\alpha)} + \dfrac{\cos\alpha}{(1 - \sin\alpha)} - \dfrac{\sin\alpha}{(1 - \sin\alpha)(1 + \sin\alpha)} =$$

Notice that the first and third term have a common denominator. Add them.

$$\dfrac{\cancel{1 - \sin\alpha}}{\cancel{(1 - \sin\alpha)}(1 + \sin\alpha)} + \dfrac{\cos\alpha}{(1 - \sin\alpha)} =$$

$$\dfrac{1}{1 + \sin\alpha} + \dfrac{\cos\alpha}{1 - \sin\alpha} = \dfrac{1}{1 + \sin\alpha} + \dfrac{\cos\alpha}{1 - \sin\alpha}$$

□

Problem Set 7.3

Test the validity of the following identities using a graphing calculator. If the graphs coincide, prove the identity. If the equations are not identities, give a value of x that proves that they are not identities.

1. $\dfrac{1 - \cos 2x}{\sin 2x} = \tan x$

2. $1 + \sin 2x = \cos^2\dfrac{x}{2}\left(1 + \tan\dfrac{x}{2}\right)^2$

3. $\dfrac{\sin^3 x + \cos^3 x}{\sin x + \cos x} = \dfrac{\sin 2x}{2}$

4. $\dfrac{\tan x}{\sin x - 2\tan x} = \dfrac{1}{\cos x - 2}$

5. $\sin 2x - \cos 2x \tan x = \tan x$

6. $\tan^2 x + \cos x = 1 - \cos 2x \tan^2 x$

7. $\dfrac{\sin x \sin\dfrac{x}{2}}{2\cos\dfrac{x}{2}} + \cos^2\dfrac{x}{2} = 1$

8. $\dfrac{1 + 3\cos x}{1 + \cos x} = \dfrac{1 + 2\cos 2x - 3\cos^2 x}{\sin^2 x}$

Prove the following identities.

9. $\dfrac{\sin x - \cos x}{\sec x - \csc x} = \dfrac{\cos x}{\csc x}$

10. $\dfrac{\tan x - 1}{1 - \cot x} = \dfrac{\sec x}{\csc x}$

11. $(1 - \tan \theta)(1 - \cot \theta) = 2 - \dfrac{2}{\sin 2\theta}$

12. $(\cos \theta + \sin \theta)(\cot \theta + \tan \theta) = \csc \theta + \sec \theta$

13. $\cos 2\theta = \dfrac{\csc^2 \theta - 2}{\csc^2 \theta}$

14. $1 + \sin \theta = \cos^2 \dfrac{\theta}{2}\left(1 + \tan \dfrac{\theta}{2}\right)^2$

15. $\dfrac{2}{\sin 2x} = \tan x + \cot x$

16. $\dfrac{1 + 3\cos x}{1 + \cos x} = \dfrac{1 + 2\cos x - 3\cos^2 x}{\sin^2 x}$

17. $\dfrac{1 - \csc x\sec^3 x + \sec x \csc x}{\csc x} = \sin x - \tan^2 x \sec x$

18. $\dfrac{\cos x(1 + \sin x) + 1 - \sin x}{\cos^2 x} = \dfrac{\cos x}{1 - \sin x} + \dfrac{1}{1 + \sin x}$

19. $\dfrac{\tan^3 x + \sin x \sec x - \sin x \cos x}{\sec x - \cos x} = \sin x + \tan x \sec x$

20. $\dfrac{\csc x + \sec x}{\csc x - \sec x} = \dfrac{1 + \sin 2x}{\cos 2x}$

21. $\sin x - \cos x \tan \dfrac{x}{2} = \tan \dfrac{x}{2}$

22. $\dfrac{\cos x \tan x + \sin x}{\tan x} = 2 \cos x$

23. $1 + \cos 2\theta = \sin 2\theta \cot \theta$

24. $\sin^4 x - \cos^4 x = 1 - 2\cos^2 x$

25. $\sin 2x = \tan x(1 + \cos 2x)$

26. $\tan 2\theta = \dfrac{2 \sin \theta}{2 \cos \theta - \sec \theta}$

27. $\tan \theta - \cot \theta = \sec \theta \csc \theta - 2 \cot \theta$

28. $\sin 2\theta = \dfrac{2 \tan \theta}{1 + \tan^2 \theta}$

29. $\tan \alpha - \tan \beta = \dfrac{\sec \beta \sin(\alpha - \beta)}{\cos \alpha}$

30. $\dfrac{\csc^4 x - 1}{\cot^2 x} = 2 + \cot^2 x$

31. $\tan^4 x - 1 = \sec^4 x - 2\sec^2 x$

32. $\sin(x + y)\sin y + \cos(x + y)\cos y = \cos x$

7.1 **1.** Trigonometric functions of double angles are used to transform trigonometric expressions and to prove identities.

2. Sine of a double angle: $\sin 2\theta = 2\sin\theta\cos\theta$

3. Cosine of a double angle: $\cos 2\theta = \cos^2\theta - \sin^2\theta$

4. Tangent of a double angle: $\tan 2\theta = \dfrac{2\tan\theta}{1 - \tan^2\theta}$

7.2 **1.** Sine of a half angle: $\sin\dfrac{\theta}{2} = \pm\sqrt{\dfrac{1 - \cos\theta}{2}}$

2. Cosine of a half angle: $\cos\dfrac{\theta}{2} = \pm\sqrt{\dfrac{1 + \cos\theta}{2}}$

3. Tangent of a half angle: $\quad \tan\dfrac{\theta}{2} \quad = \quad \pm\sqrt{\dfrac{1 - \cos\theta}{1 + \cos\theta}}$

$\qquad\qquad\qquad$ or $\qquad\qquad \tan\dfrac{\theta}{2} \quad = \quad \dfrac{1 - \cos\theta}{\sin\theta}$

$\qquad\qquad\qquad$ or $\qquad\qquad \tan\dfrac{\theta}{2} \quad = \quad \dfrac{\sin\theta}{1 + \cos\theta}$

7.3 Sum and product identities are used in higher mathematics. They also allow us to convert from sums to products and vice versa.

Chapter 7 Review Test

Using the double-angle formulas, find the exact values of the following [for example, $60° = 2(30°)$]. **(7.1)**

1. $\sin 240°$ **2.** $\cos 180°$ **3.** $\tan 300°$

Using the half-angle formulas, find the exact value of the following $\left[\text{ for example, } 30° = \frac{1}{2}60°\right]$. **(7.2)**

4. $\sin 15°$ **5.** $\cos 75°$ **6.** $\tan 22\frac{1}{2}^°$

Given $\cos x = -\dfrac{3}{5}, 90° \le x \le 180°$, find the exact value for:

7. $\sin 2x$ **8.** $\cos 2x$ **9.** $\tan 2x$ **(7.1)**

10. $\sin \dfrac{x}{2}$ **11.** $\cos \dfrac{x}{2}$ **12.** $\tan \dfrac{x}{2}$ **(7.2)**

Test the validity of the following identities using a graphing calculator. If the graphs coincide, prove the identity. If the equations are not identities, give a value of x that proves that they are not identities. **(7.1, 7.2, 7.3)**

13. $\dfrac{1 - \tan^2 x}{1 + \tan^2 x} = 1 - \sin 2x$ **14.** $\tan^2 \dfrac{x}{2} + \cos 2x = 1 - \cos x \tan^2 \dfrac{x}{2}$

15. $\sin \dfrac{x}{2} \cos \dfrac{x}{2} = \dfrac{\sin x}{2}$ **16.** $2(1 + \cos x) = \dfrac{1 - \cos 2x}{1 - \cos x}$

Prove the following identities. **(7.1, 7.2, 7.3)**

17. $\dfrac{\tan x}{\sin x} + \dfrac{\csc x}{\tan x} = \sec x \csc^2 x$ **18.** $\dfrac{\tan x + \sec^3 x - \sec x}{\sec x} = \tan^2 x + \sin x$

19. $\dfrac{\tan x}{\csc x - \cot x} - \dfrac{\sin x}{\csc x + \cot x} = \sec x + \cos x$ **20.** $\dfrac{\sin 2x}{\sin x} - \dfrac{\cos 2x}{\cos x} = \sec x$

21. $\tan 2x = \dfrac{2}{\cot x - \tan x}$ **22.** $(\sec x - \tan x)^2 = \dfrac{1 - \sin x}{1 + \sin x}$

23. $\dfrac{1}{\sec x + \tan x} + \tan x = \sec x$ **24.** $\dfrac{1 + \tan x \sin^2 x - \tan x}{\sin x} = \csc x - \cos x$

25. $\sec x = \dfrac{\tan^2 \dfrac{x}{2}}{\sin^2 \dfrac{x}{2}\left(1 - \tan^2 \dfrac{x}{2}\right)}$ **26.** $\sin 2x + 2\sin x = 4 \sin x \cos^2 \dfrac{x}{2}$

Chapter 8

Trigonometric Equations

Contents

Preview

Trigonometry is frequently applied in the form of an equation that is to be solved for a specific value. It can answer questions such as:

> What is the angle of a ski slope? How far does a crankshaft rotate? How many degrees is a ray of light refracted as it passes through a prism? What is the best angle of climb for an aircraft? What is the displacement of an airplane's wing as it flexes in flight? When is voltage at its maximum?

To answer questions like these, we must solve trigonometric equations.

In this chapter, we will use our skills from algebra and our skills with trigonometric identities developed over the last three chapters to solve trigonometric equations.

8.1   Solving Basic Trigonometric Equations

Chapter 7 dealt with identities. Identities are true for all permissible replacements of the variable. They are invaluable in simplifying complex expressions. Trigonometric identities are often used to simplify an equation. This application normally provides a set of solutions which solve the equation.

Example 1 □ Find all values of x where $\sin x = \cos x$.

$$\sin x = \cos x$$
$$\frac{\sin x}{\cos x} = 1 \qquad \text{divide both sides by } \cos x$$
$$\tan x = 1 \qquad \text{identity (10)}$$

If you enter Tan^{-1} on your calculator, you will get $\frac{1}{4}\pi$ or $45°$. However, this is only part of the solution set. Tan u is a periodic function. It has a period of π or $180°$. With each repetition of the function, the solutions are repeated. To show the complete solution set, we write

$$x = \frac{1}{4}\pi + k\pi \qquad \text{(where } k \text{ is any integer)}$$

In engineering applications, the same solution is frequently expressed in degree measure as

$$x = 45° + k \cdot 180°$$

□

NOTE: In mathematics and engineering, I, J, K, L, M, and N are frequently used to represent integers. Since we are following this convention, we will not always write "where k is any integer."

Example 2 □ What is the position of the crankshaft when the piston is halfway up the cylinder in an engine?

□ Figure 8.1

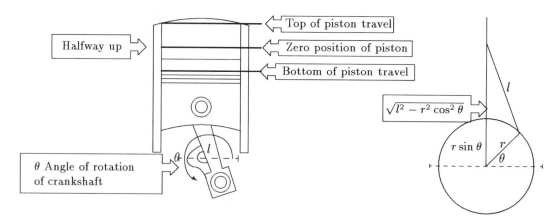

At $\theta = 0°$ the piston is in the zero position in the middle of the cylinder. As θ increases to $90°$, the piston rises.

The equation that gives the height of the crankshaft as a function of the angle of rotation is $y = \sin \theta$.

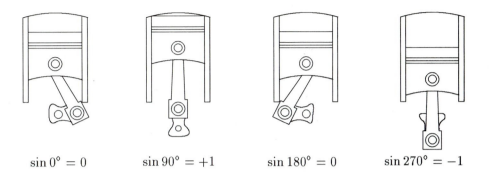

$$\sin 0° = 0 \qquad \sin 90° = +1 \qquad \sin 180° = 0 \qquad \sin 270° = -1$$

The vertical displacement of the piston is also a function of $\sin \theta$. An engineer could determine when the piston will be halfway between the zero position and the top of the cylinder by solving the equation:

$$y_p = r \sin \theta + \sqrt{l^2 - r^2 \cos^2 \theta}$$

To simplify the discussion we will look at the crankshaft only and ask "when is it halfway up as given by the equation $y = \sin \theta$?" At the halfway-up point of the crankshaft rotation

$$\sin \theta = \frac{1}{2}$$

My calculator says $\sin^{-1}\left(\frac{1}{2}\right) = 30°$.

True, but that's only part of the answer.

At $\theta = 30°$, the piston is heading up.

Notice that at $\theta = 150°, \sin 150° = \frac{1}{2}$. This is the value of θ when the piston is halfway up the cylinder and headed down.

There are two distinct sets of answers to this example.

$$\theta = 30° + k360° \qquad \text{(where } k \text{ is any integer)}$$

and

$$\theta = 150° + k360° \qquad \text{(where } k \text{ is any integer)}$$

How do I find the second set of solutions?

Draw a circle and look for values of θ with a reference angle of $30°$ where $\sin \theta$ is positive.

8.1 Solving Basic Trigonometric Equations   225

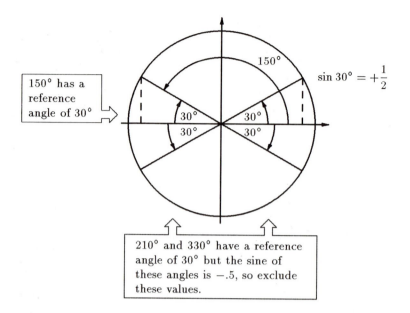

150° has a
reference
angle of 30° ⇨

$\sin 30° = +\dfrac{1}{2}$

210° and 330° have a reference
angle of 30° but the sine of
these angles is −.5, so exclude
these values.

The complete solution set for this example has an infinite number of values because the crankshaft is in the halfway position twice for every turn of the crankshaft. However, these values are all duplicates of the fundamental values. Therefore, we usually define a set of fundamental values for the solutions of trigonometric equations to be values from 0 to, but not including, 2π. □

Example 3 □ Solve for the exact values of x so that $\sqrt{2}\cos x = 1$.

$$\sqrt{2}\cos x \;=\; 1$$
$$\cos x \;=\; \frac{1}{\sqrt{2}} \qquad \text{divide both sides}$$
$$x \;=\; 45° \qquad \boxed{\text{This is one fundamental value}}$$

To find the complete solution, investigate the other angles with a reference angle of 45°. They are:

135°   reject because $\cos 135°$ is negative

225°   reject because $\cos 225°$ is negative

315°   accept because $\cos 315° = \dfrac{1}{\sqrt{2}}$

□ Figure 8.4

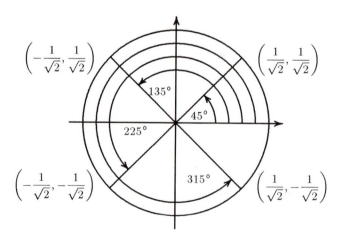

$\left(-\dfrac{1}{\sqrt{2}}, \dfrac{1}{\sqrt{2}}\right)$ $\left(\dfrac{1}{\sqrt{2}}, \dfrac{1}{\sqrt{2}}\right)$

135°   45°   225°   315°

$\left(-\dfrac{1}{\sqrt{2}}, -\dfrac{1}{\sqrt{2}}\right)$ $\left(\dfrac{1}{\sqrt{2}}, -\dfrac{1}{\sqrt{2}}\right)$

The solution set is

$$x = 45° + k \cdot 360°$$

and

$$x = 315° + k \cdot 360°$$

□

Example 4 □ Solve for all the values of θ that satisfy $4\cos^2\theta = 3$.

$$
\begin{aligned}
4\cos^2\theta &= 3 \\
\cos^2\theta &= \frac{3}{4} \\
\cos\theta &= \pm\frac{\sqrt{3}}{2}
\end{aligned}
$$

The square root can be positive or negative. There are two angles with a cosine of $+\dfrac{\sqrt{3}}{2}$ and there are two angles with a cosine of $-\dfrac{\sqrt{3}}{2}$; therefore there are four angles less than $360°$ where $\cos\theta = \pm\dfrac{\sqrt{3}}{2}$.

□ Figure 8.5

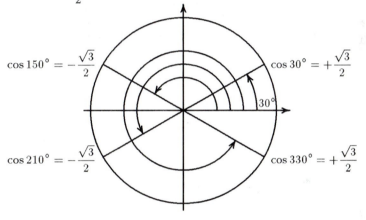

$$\cos 150° = -\frac{\sqrt{3}}{2} \qquad \cos 30° = +\frac{\sqrt{3}}{2}$$

$$30°$$

$$\cos 210° = -\frac{\sqrt{3}}{2} \qquad \cos 330° = +\frac{\sqrt{3}}{2}$$

We could express this answer as four chains of solutions. However, we can write the answer more compactly if we notice that the values in the 1st and 3rd quadrants are $180°$ apart.

$$\theta = 30° + k \cdot 360° \text{ and } \theta = 210° + k \cdot 360°$$

can be written as

$$\theta = 30° + k \cdot 180°$$

Similarly, the values in the 2nd and 4th quadrants can be written

$$\theta = 150° + k \cdot 180°$$

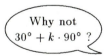

Why not
$30° + k \cdot 90°$?

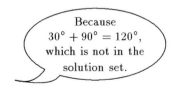

Because
$30° + 90° = 120°$,
which is not in the
solution set.

□

Frequently, in calculus it is more convenient to work in radians instead of degrees. Hence, trigonometric equations usually are solved with the angle expressed in radians.

Example 5 □ Solve $2 \sin x = \sqrt{3}$ for all exact values of x in radian measure.

$$
\begin{aligned}
2 \sin x &= \sqrt{3} \\
\sin x &= \frac{\sqrt{3}}{2} \\
x &= \frac{1}{3}\pi \qquad \text{from our knowledge of standard angles}
\end{aligned}
$$

There is also a symmetrical value of x in the second quadrant where $\sin x = \dfrac{\sqrt{3}}{2}$. It is $x = \dfrac{2}{3}\pi$.

Is that all there is?

No. Each multiple of 2π added to or subtracted from a solution is also a solution.

The complete solution set is

$$
x = \frac{1}{3}\pi + 2k\pi
$$

and

$$
x = \frac{2}{3}\pi + 2k\pi
$$

□

What if the answer isn't a standard angle?

Remember, we only introduced the standard angles so we could use convenient examples. Most of the time you will need a calculator to yield an approximation.

Using Your Graphing Calculator

You can visualize the multiple solutions to a trigonometric equation by using your graphing calculator to plot each side of the equation as a separate function.

To visualize the solutions to

$$2 \sin x = \sqrt{3}$$

set your range to

$$X_{\min} = -6.28$$
$$X_{\max} = 7.3$$
$$X_{\text{scl}} = .5236 \left(\frac{1}{6} \pi \right)$$

$$Y_{\min} = -2$$
$$Y_{\max} = 2$$
$$Y_{\text{scl}} = 0.5$$

Set $Y_1 = 2 \sin x$ and $Y_2 = \sqrt{3}$ and graph

 Figure 8.6

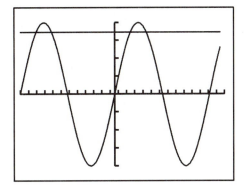

The points where the line $y = \sqrt{3}$ intersects the curve $y = 2 \sin x$ are solutions. Note there is one set of solutions where the sine curve is increasing in value and another set where it is decreasing in value. That is why it took two equations to represent both chains of answers in the previous example.

Trace along the sine curve to find the points of intersection. You can use the zoom feature of your calculator to locate the points of intersection to any degree of accuracy you need. The x-coordinates of the leftmost two points of intersection accurate to four decimal places are: -5.2360 and -4.1888. These are approximations of $-\frac{5}{3}\pi$ and $-\frac{4}{3}\pi$.

Each time you zoom in, the range is reset automatically, which means that you must redefine the range as you search for each successive solution of the equation. Of course, you can never find all the solutions because there are an infinite number of solutions. The method of the previous example is needed to define all solutions.

Example 6 □ Solve $\csc^2 x = 10$ for all real solutions in radian measure.

Because the equation uses $\csc x$ we must convert it to a trigonometric function on a calculator which is $\sin x$.

$$\csc^2 x \;=\; 10$$
$$\sin^2 x \;=\; \frac{1}{10} \qquad\qquad \text{identity (1)}$$
$$\sin^2 x \;=\; 0.1 \qquad\qquad \text{write as a decimal}$$
$$\sin x \;=\; \pm 0.3162 \qquad\qquad \text{use calculator to take}$$
$$\text{square root}$$

□ Figure 8.7

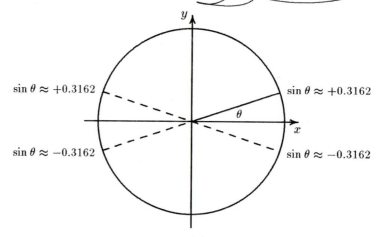

The reference value of θ from the INV SIN key of the calculator is

$$\sin^{-1}(0.3162) \;\approx\; 0.3217$$
$$\theta \;\approx\; 0.3217$$

Adding multiples of 2π yields a chain of solutions

$$x \approx 0.3217 + 2k\pi$$

A second chain of solutions comes from the fundamental value in the second quadrant

$$\pi - \theta \;\approx\; 3.1416 - 0.3217$$
$$x \;\approx\; 2.8199$$

Therefore, we can add to our set of solutions

$$x \approx 2.8199 + 2k\pi$$

In the third quadrant, the fundamental value is

$$\pi + \theta \;\approx\; 3.1416 + 0.3217$$
$$x \;\approx\; 3.4633$$

This generates the third set of solutions

$$x \approx 3.4633 + 2k\pi$$

In the fourth quadrant, the fundamental value is

$$2\pi - \theta \approx 6.2832 - 0.3217$$
$$x \approx 5.9615$$

This yields a fourth solution set

$$x \approx 5.9615 + 2k\pi$$

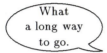

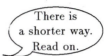

Notice the values in the third quadrant are π away from the values in the first quadrant. The values in the fourth quadrant are π away from the values in the second quadrant. Therefore, we can combine these four chains of answers into only two.

Quadrant	Solution Set	Combined Solution
I	$0.3217 + 2k\pi$	
III	$3.4633 + 2k\pi$	$0.3217 + k\pi$
II	$2.8199 + 2k\pi$	
IV	$5.9615 + 2k\pi$	$2.8199 + k\pi$

$\square$

Problem Set 8.1

Without the use of a calculator, solve the following equations for exact non-negative values less than $360°$.

1. $2\sin\theta = 1$ **2.** $\tan\theta - \sqrt{3} = 0$

3. $3\csc^2\theta - 4 = 0$ **4.** $2\cos\theta = -1$

Without the use of a calculator, solve the following equations for exact non-negative values less than 2π radians.

5. $\sin x = \dfrac{\sqrt{3}}{2}$ **6.** $\sqrt{3}\cot x = -1$

7. $\csc^2 x - 2 = 0$ **8.** $3\tan^2 x - 1 = 0$

Using a calculator, solve the following equations for non-negative values less than $360°$. Give answers correct to two decimal places.

9. $\sqrt{3}\cos\theta - 1 = 0$ **10.** $5\sin^2\theta - 3 = 0$

11. $\sqrt{2}\tan\theta - 1 = 0$ **12.** $\csc\theta + 4 = 0$

Using a calculator, solve the following equations for non-negative values less than 2π radians. Give answers correct to four decimal places.

13. $5 \cot x + 2 = 0$

14. $2 \csc^2 x - 7 = 0$

15. $3 \tan^2 x - 2 = 0$

16. $3 \csc x - 5 = 0$

Without the use of a calculator, solve the following for (a) all values and (b) the fundamental values, that is, non-negative values less than $360°$.

17. $\sqrt{3} \sec \theta + 2 = 0$

18. $-\csc \theta + 2 = 0$

19. $3 \sec^2 \theta - 4 = 0$

20. $2 \sin \theta + \sqrt{3} = 0$

21. $4 \cos^2 \theta = 3$

22. $\sqrt{3} \cot \theta - 3 = 0$

Without the use of a calculator, solve the following for (a) all values and (b) the fundamental values, that is, non-negative values less than 2π radians.

23. $\sqrt{3} \csc x - 2 = 0$

24. $\tan x + \sqrt{3} = 0$

25. $3 \sec^2 \theta - 4 = 0$

26. $2 \cos x - 1 = 0$

27. $\cot x + \sqrt{3} = 0$

28. $4 \sin^2 x - 3 = 0$

Using your calculator, solve the following for (a) all values and (b) the fundamental values, that is, non-negative values less than $360°$. Give answers correct to two decimal places.

29. $2 \tan \theta - \sqrt{3} = 0$

30. $\sqrt{3} \sec \theta + 4 = 0$

31. $6 \cos^2 \theta - 5 = 0$

32. $7 \sin^2 \theta - 6 = 0$

Using your calculator, solve the following for (a) all values and (b) the fundamental values, that is, non-negative values less than 2π. Give answers correct to four decimal places.

33. $4 \cos x + 3 = 0$

34. $5 \csc^2 x - 7 = 0$

35. $4 \tan^2 x - 7 = 0$

36. $3 \sin x + 2 = 0$

8.2 Solving Trigonometric Equations Involving Factoring

Many trigonometric equations can be reduced to a form similar to a quadratic equation. As you will recall, if we can write an equation as $a \cdot b = 0$, then the solution is the solution of $a = 0$ or $b = 0$.

Example 7 □ Solve $\sin x \cdot \cos x = 0$.

$$\sin x \cdot \cos x = 0$$

Since this is a product equal to zero, either

$$\sin x = 0 \qquad \text{or} \qquad \cos x = 0$$
$$x = 0 + k\pi \qquad \qquad x = \frac{1}{2}\pi + k\pi$$

These two chains of answers can be written in the combination form $x = \frac{1}{2}k\pi$.   □

Why did you give the answer in radians?

There is no standard. This book will represent angles in radian measure with letters like x, y, z, and angles in degree measure with Greek letters like α, β, λ, θ.

Example 8 □ Solve $2\sin x \cos x = -\cos x$.

First set this equation equal to zero.

$$2\sin x \cos x = -\cos x$$
$$2\sin x \cos x + \cos x = 0 \qquad \text{add } \cos x$$
$$\cos x(2\sin x + 1) = 0 \qquad \text{factor}$$

Setting each factor equal to zero,

$$\cos x = 0 \qquad \text{or} \qquad 2\sin x + 1 = 0$$
$$x = \frac{1}{2}\pi + k\pi \qquad\qquad \sin x = -\frac{1}{2}$$
$$x = \frac{7}{6}\pi + 2k\pi$$
$$\text{or} \qquad x = \frac{11}{6}\pi + 2k\pi$$

The general solution is

$$x = \frac{1}{2}\pi + k\pi \quad \text{or} \quad x = \frac{7}{6}\pi + 2k\pi \quad \text{or} \quad x = \frac{11}{6}\pi + 2k\pi$$

The fundamental solutions are

$$x = \frac{1}{2}\pi, \frac{3}{2}\pi, \frac{7}{6}\pi, \frac{11}{6}\pi$$

□

Group Writing Activity

To solve trigonometric equations involving more than one function, why can't we eliminate a function by either dividing or multiplying both sides of the equation by it?

Example 9 □ Solve $\sin^2\theta + 2\sin\theta - 3 = 0$ for values of $0° \le \theta < 360°$.

This is a quadratic-type equation; therefore, it can be factored.

$$\sin^2\theta + 2\sin\theta - 3 = 0$$
$$(\sin\theta + 3)(\sin\theta - 1) = 0$$

Setting each factor equal to zero,

$$\sin\theta + 3 = 0 \qquad \text{or} \qquad \sin\theta - 1 = 0$$
$$\sin\theta = -3 \qquad\qquad \sin\theta = 1$$
$$\theta = 90°$$

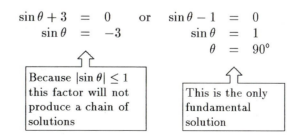

Because $|\sin\theta| \le 1$ this factor will not produce a chain of solutions

This is the only fundamental solution

□

Algebra Reminder
Quadratic Equations

Recall for equations of the form $ax^2 + bx + c = 0$ the solutions are:

$$x = \frac{-b \pm \sqrt{b^2 - 4ac}}{2a}$$

Example 10 □ Solve $2\tan^2\theta + 3\tan\theta = 1$ for the fundamental solutions and the general solution.

First set the equation equal to zero:

$$2\tan^2\theta + 3\tan\theta = 1$$
$$2\tan^2\theta + 3\tan\theta - 1 = 0$$

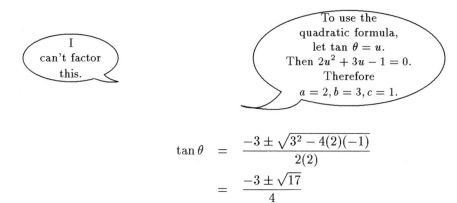

I can't factor this.

To use the quadratic formula, let $\tan\theta = u$. Then $2u^2 + 3u - 1 = 0$. Therefore $a = 2, b = 3, c = 1$.

$$\tan\theta = \frac{-3 \pm \sqrt{3^2 - 4(2)(-1)}}{2(2)}$$
$$= \frac{-3 \pm \sqrt{17}}{4}$$

Now use your calculator to find decimal approximations:

$$\tan\theta \approx \frac{-3 \pm 4.123}{4}$$
$$\tan\theta \approx -1.7808 \qquad \text{or} \qquad \tan\theta \approx 0.28077$$
$$\theta \approx -60.68° \qquad\qquad\qquad \theta \approx 15.68°$$

The fundamental solutions are values of $0° \leq \theta < 360°$. In this case they are:

First Chain

$$\theta \approx -60.68° + 180°$$
$$\approx 119.32°$$

Because 119.32 is less than 180°, we can still add 180° to get another fundamental solution in the first chain. It is

$$\theta \approx 119.32° + 180°$$
$$\approx 299.32°$$

Second Chain

$$\theta \approx 15.68°$$

and

$$\theta \approx 15.68° + 180°$$
$$\approx 195.68°$$

Chapter 8 Trigonometric Equations

The fundamental solutions are

$$\theta \approx 15.68°, 119.32°, 195.68°, 299.32°$$

The general solution is

$$\theta \approx 15.68° + k \cdot 180° \quad \text{and} \quad \theta \approx 119.32° + k \cdot 180°$$

$\square$

Problem Set 8.2

Without the use of a calculator, solve the following equations for non-negative values less than 360°.

1. $2 \sin \theta \tan \theta = \tan \theta$ **2.** $2 \cos^2 \theta + \cos \theta - 1 = 0$ **3.** $\sqrt{3} \cot^2 \theta + \cot \theta = 0$

4. $2 \cos \theta \cot \theta = \sqrt{3} \cot \theta$ **5.** $2 \sin^2 \theta + \sin \theta = 1$ **6.** $\tan^2 \theta + \sqrt{3} \tan \theta = 0$

Without the use of a calculator, solve the following equations for non-negative values less than 2π radians.

7. $\sqrt{3} \cot^2 x + 2 \cot x - \sqrt{3} = 0$ **8.** $\csc^2 x - 2 \csc x = 0$

9. $4 \sin^3 x - 3 \sin x = 0$ **10.** $\sqrt{3} \tan^2 x - 4 \tan x + \sqrt{3} = 0$

11. $2 \cos^2 x - \sqrt{3} \cos x = 0$ **12.** $3 \sec^3 x - 4 \sec x = 0$

Using a calculator, solve the following equations for non-negative values less than 360°. Round off answers correct to two decimal places.

13. $3 \sin^2 \theta + 5 \sin \theta - 2 = 0$ **14.** $6 \tan^2 \theta + 5 \tan \theta - 4 = 0$

15. $2 \sec^2 \theta + 5 \sec \theta - 3 = 0$ **16.** $4 \cos^2 \theta + 7 \cos \theta - 2 = 0$

Using a calculator, solve the following equations for non-negative values less than 2π radians. Round off answers correct to four decimal places.

17. $6 \sec^2 x - 17 \sec x + 5 = 0$ **18.** $6 \tan^2 x + 5 \tan x - 6 = 0$

19. $6 \csc^2 x + \csc x - 12 = 0$ **20.** $4 \cos^2 x + 3 \cos x - 1 = 0$

Using a calculator when necessary, solve the following equations for (a) all values and (b) the fundamental values, that is, non-negative values less than 360°. (Give answers correct to hundredths of degree.)

21. $2 \sin^2 \theta - 5 \sin \theta - 3 = 0$ **22.** $3 \cos^2 \theta + 4 \cos \theta - 4 = 0$

23. $6 \sin \theta \cos \theta - 3 \sin \theta + 2 \cos \theta - 1 = 0$ **24.** $\tan^2 \theta - 2 \tan \theta - 4 = 0$

25. $\cos \theta \csc \theta - 2 \cos \theta - \csc \theta + 2 = 0$ **26.** $3 \cos^2 \theta - 1 = 0$

27. $12 \cos^2 \theta - 7 \cos \theta - 12 = 0$ **28.** $4 \cos \theta \csc \theta - \csc \theta + 12 \cos \theta - 3 = 0$

Using a calculator when necessary, solve the following equations for (a) all values and (b) the fundamental values, that is, non-negative values less than 2π radians. (Give answers correct to four decimal places.)

29. $\sec x \cot x + \sqrt{5} \sec x - 4 \cot x - 4\sqrt{5} = 0$ 30. $\cos x \tan x - 2 \cos x + \tan x - 2 = 0$

31. $4 \sin x \sec x - 12 \sin x + \sec x - 3 = 0$ 32. $2 \sin^2 x + 3 \sin x - 4 = 0$

33. $\cos^2 x - 4 \cos x - 3 = 0$ 34. $\sec^2 x - 2 \sec x - 4 = 0$

35. $2 \cos^2 x - 3 \cos x - 3 = 0$ 36. $2 \sin^2 x + \sin x - 4 = 0$

8.3 Solving Trigonometric Equations Where the Argument is a Function

Some trigonometric equations involve trigonometric functions where the argument of the function is itself a function. We need to distinguish three elements in a trigonometric equation.

1. The trigonometric function(s) involved
2. The argument of the functions
3. The variable

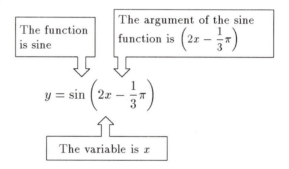

To solve a trigonometric equation means to find the set of values that the variable x may assume that makes the equation true.

Example 11 □ Solve $2 \cos(3\theta) = 1$.

$$2 \cos(3\theta) = 1$$

Write the equation in the form cos ('argument') = a.

$$\cos(3\theta) = \frac{1}{2}$$

We know that

$$\cos(60° + k \cdot 360°) = \frac{1}{2} \qquad \text{first chain}$$

and

$$\cos(300° + k \cdot 360°) = \frac{1}{2} \qquad \text{second chain}$$

So there are two distinct chains of solutions.

Chapter 8 Trigonometric Equations

Now set the argument equal to the values it must assume and solve for the variable.

First Chain

$$3\theta = 60° + k \cdot 360°$$
$$\theta = 20° + k \cdot 120°$$

This chain yields fundamental solutions of $\theta = 20°,\ 140°,\ 260°$.

Second Chain

$$3\theta = 300° + k \cdot 360°$$
$$\theta = 100° + k \cdot 120°$$

This chain yields solutions of $\theta = 100°,\ 220°,\ 340°$.
There are six fundamental solutions:

$$\theta = 20°,\ 100°,\ 140°,\ 220°,\ 260°,\ 340°$$

□

Example 12 □ Solve $\sin\left(2x - \dfrac{1}{3}\pi\right) = 1$ for the fundamental solutions.
This equation can be visualized as

$$\sin(\text{'argument'}) = 1$$

Since we know that

$$\sin\left(\frac{1}{2}\pi + 2k\pi\right) = 1$$

we can conclude

$$\text{'argument'} = \frac{1}{2}\pi + 2k\pi$$

Substituting the given expression for 'argument'

$$2x - \frac{1}{3}\pi = \frac{1}{2}\pi + 2k\pi$$

Now solving for x,

$$2x - \frac{1}{3}\pi\left(\frac{2}{2}\right) = \frac{1}{2}\pi\left(\frac{3}{3}\right) + 2k\pi \qquad \text{common denominator}$$

$$2x - \frac{2}{6}\pi = \frac{3}{6}\pi + 2k\pi$$

$$2x = \frac{5}{6}\pi + 2k\pi \qquad \text{add } \frac{2}{6}\pi$$

$$x = \frac{5}{12}\pi + k\pi \qquad \text{divide both sides by 2}$$

This expression is the general solution. Now substituting $k = 0, 1, 2 \ldots$ we can find the fundamental solutions.

$$k = 0 \qquad x = \frac{5}{12}\pi$$

$$k = 1 \qquad x = \frac{5}{12}\pi + \pi$$

$$= \frac{17}{12}\pi$$

$$k = 2 \qquad x = \frac{5}{12}\pi + 2\pi$$

Because the value of x for $k = 2$ is greater than 2π, we do not consider it a fundamental solution. The fundamental solutions are:

$$x = \frac{5}{12}\pi, \ \frac{17}{12}\pi$$

□

Example 13 □ Solve $2\cos x \sin 2x - \cos x = 0$ for all values of x and then give the fundamental solutions.

$$2\cos x \sin 2x - \cos x = 0$$
$$\cos x(2\sin 2x - 1) = 0 \qquad \text{factor } \cos x$$

$$\cos x = 0 \qquad \text{or} \quad \sin 2x = \frac{1}{2}$$

$$x = \frac{1}{2}\pi + k\pi \qquad 2x = \frac{1}{6}\pi + 2k\pi \quad \text{or} \quad 2x = \frac{5}{6}\pi + 2k\pi$$

$$x = \frac{1}{12}\pi + k\pi \qquad x = \frac{5}{12}\pi + k\pi$$

The general solutions are:

$$x = \frac{1}{2}\pi + k\pi \quad \text{or} \quad x = \frac{1}{12}\pi + k\pi \quad \text{or} \quad x = \frac{5}{12}\pi + k\pi$$

Above are the general solutions. To find the fundamental solutions, substitute values for k until you get values outside the range $0 \le x < 2\pi$.

First Chain

$$k = 0 \quad x = \frac{1}{2}\pi + 0 \cdot \pi = \frac{1}{2}\pi$$
$$k = 1 \quad x = \frac{1}{2}\pi + 1 \cdot \pi = \frac{3}{2}\pi$$
$$k = 2 \quad x = \frac{1}{2}\pi + 2 \cdot \pi \quad \Longleftarrow \boxed{\text{This value is not a fundamental solution}}$$

Making similar substitutions in the other chains of solutions yields:

Second Chain

$$x = \frac{1}{12}\pi, \qquad \frac{13}{12}\pi$$

Third Chain

$$x = \frac{5}{12}\pi, \qquad \frac{17}{12}\pi$$

All the fundamental solutions are:

$$x = \frac{1}{12}\pi, \ \frac{5}{12}\pi, \ \frac{1}{2}\pi, \ \frac{13}{12}\pi, \ \frac{17}{12}\pi, \ \frac{3}{2}\pi$$

□

Chapter 8 Trigonometric Equations

Problem Set 8.3

Solve the following equations for (a) all values and (b) the fundamental values (non-negative values less than $360°$). Do not use a calculator.

1. $2\sin 2\theta = \sqrt{3}$

2. $2\cos 3\theta - 1 = 0$

3. $3\tan 3\theta - \sqrt{3} = 0$

4. $\sin(2\theta - 30°) = -\dfrac{1}{2}$

5. $\sqrt{2}\sec 3\theta = 2$

6. $\csc 2\theta + \sqrt{2} = 0$

7. $\tan(3\theta + 30°) = \sqrt{3}$

8. $\sin(4\theta - 120°) = -\dfrac{\sqrt{3}}{2}$

Solve the following equations for (a) all values and (b) the fundamental values (non-negative values less than 2π radians). Do not use a calculator. Express your answer as a rational function of π rather than as a decimal.

9. $\sin x \sec 2x - 2\sin x = 0$

10. $\sqrt{3}\cos x \cot 2x + \cos x = 0$

11. $\cos 4x \tan 3x - \cos 4x + \tan 3x - 1 = 0$

12. $2\sin\left(4x - \dfrac{1}{3}\pi\right) + 1 = 0$

13. $\cos x \csc 2x - 2\cos x = 0$

14. $\sin\left(2x + \dfrac{1}{2}\pi\right) = -\dfrac{1}{2}$

15. $2\sin 3x \tan 2x - \tan 2x + 2\sin 3x - 1 = 0$

16. $2\sqrt{3}\cos 2x \cot 3x - 3\cot 3x + 2\cos 2x - \sqrt{3} = 0$

In the following equations, use your calculator when necessary to find (a) all values and (b) the fundamental solutions (non-negative values less than 2π radians). Round off answers to four decimal places.

17. $2\cot 3x + \sqrt{5} = 0$

18. $3\sec^2 3x - 8 = 0$

19. $5\sin^2 2x - \sqrt{3} = 0$

20. $3\sin^2\left(3x - \dfrac{1}{3}\pi\right) - 2 = 0$

21. $2\sec^2\left(4x - \dfrac{1}{6}\pi\right) - 5 = 0$

22. $4\csc^2\left(2x + \dfrac{1}{4}\pi\right) - 7 = 0$

23. $3\tan 2x - \sqrt{7} = 0$

24. $5\cos^2\left(3x - \dfrac{1}{3}\pi\right) - 2 = 0$

25. $3\csc^2\left(2x + \dfrac{1}{3}\pi\right) - 5 = 0$

26. $2\tan^2\left(4x + \dfrac{1}{2}\pi\right) - 5 = 0$

27. $5\cot^2\left(3x - \dfrac{1}{2}\pi\right) - 4 = 0$

28. $5\sin^2\left(2x - \dfrac{1}{3}\pi\right) - 3 = 0$

Group Writing Activity

In a trigonometric equation in which the argument is a function, can you derive a method for determining the number of solutions that will occur between 0 and 2π without solving the equation? What is your explanation of why this works or does not work?

8.4 Using Identities to Solve Trigonometric Equations

Many trigonometric equations require manipulation using trigonometric identities before they can be solved. A good place to start is by expressing the equation in terms of a single function.

Example 14 $\square$ Find the general solutions of $\cos x - 2\sec x = 1$.

$$\cos x - 2\sec x = 1$$
$$\cos x - \frac{2}{\cos x} = 1 \qquad \text{substituting for } \sec x$$

Next, clear the fractions by multiplying both sides by $\cos x$:

$$\cos^2 x - 2 = \cos x$$

Multiplying by a variable may introduce extraneous roots.

Write as a quadratic:

$$\cos^2 x - \cos x - 2 = 0$$
$$(\cos x - 2)(\cos x + 1) = 0 \qquad \text{factor}$$

Investigate each chain of solutions:

First Chain

$$\cos x - 2 = 0$$
$$\cos x = 2$$

This is impossible; therefore, there are no solutions in this chain.

Second Chain

$$\cos x + 1 = 0$$
$$\cos x = -1$$
$$x = \pi + 2k\pi$$

Check in the original equations to be sure this isn't an extraneous root.

$$\cos \pi - 2\sec \pi \stackrel{?}{=} 1$$
$$-1 - 2(-1) = 1 \qquad \text{This checks.}$$

$\square$

Example 15 $\square$ Find the fundamental solutions of $2\sec^2 x - \tan x = 3$. Express your answers in radian measure.

This equation involves both $\tan x$ and $\sec x$. Fortunately there is an identity that allows us to write $\sec^2 x$ as $\tan^2 x + 1$.

$$2\sec^2 x - \tan x = 3$$
$$2(\tan^2 x + 1) - \tan x = 3 \qquad \text{substituting}$$
$$2\tan^2 x + 2 - \tan x = 3$$
$$2\tan^2 x - \tan x - 1 = 0 \qquad \text{write as a quadratic}$$
$$(2\tan x + 1)(\tan x - 1) = 0 \qquad \text{factor}$$

Chapter 8 Trigonometric Equations

First Chain

$$2\tan x + 1 = 0$$
$$2\tan x = -1$$
$$\tan x = -\frac{1}{2}$$
$$\tan^{-1}(-0.5) \approx -0.4636 + k\pi$$

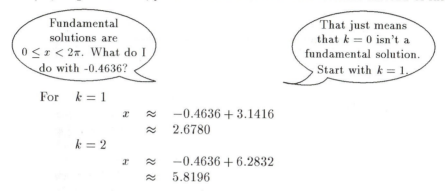

(If you got $-26.56°$, your calculator was set on DEGREES instead of RADIANS.)

For $k = 1$
$$x \approx -0.4636 + 3.1416$$
$$\approx 2.6780$$

$k = 2$
$$x \approx -0.4636 + 6.2832$$
$$\approx 5.8196$$

Second Chain

$$\tan x - 1 = 0$$
$$\tan x = 1$$
$$x = \frac{1}{4}\pi + k\pi$$

For $k = 0$
$$x = \frac{1}{4}\pi$$

$k = 1$
$$x = \frac{1}{4}\pi + \pi$$
$$= \frac{5}{4}\pi$$

The fundamental solutions are: $x \approx 0.7854, 2.6780, 3.9270, 5.8196$

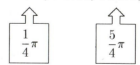

Example 16 □ Find the general solutions of $\cos 2x + \cos x + 1 = 0$.

Notice that this equation involves $\cos x$ and $\cos 2x$. First write everything in terms of $\cos x$:

$$\cos 2x + \cos x + 1 = 0$$
$$(2\cos^2 x - 1) + \cos x + 1 = 0 \qquad \text{substitute for } \cos 2x$$
$$2\cos^2 x + \cos x = 0$$

Now factor
$$\cos x(2\cos x + 1) = 0$$

Set each factor to zero to develop two chains of solutions.

8.4 Using Identities to Solve Trigonometric Equations

First Chain

$$\begin{aligned} \cos x &= 0 \\ x &= \frac{1}{2}\pi + k\pi \end{aligned}$$

Second Chain

$$\begin{aligned} 2\cos x + 1 &= 0 \\ \cos x &= -\frac{1}{2} \\ x &= \frac{2}{3}\pi + 2k\pi \end{aligned}$$

or

$$x = \frac{4}{3}\pi + 2k\pi$$

□

Pointers to Remember About Trigonometric Equations

1. Quadratic-type equations should be factored.

After factoring, set each factor equal to zero, and find the values of the argument by inspection when possible; otherwise use your calculator.

2. Equations containing more than one function:

a. Should be factored, if possible, into factors containing one function only.

b. If unable to factor or separate functions, use the fundamental identities to write all functions in terms of the same function. Then factor and solve for the value of the argument. Finally, follow each chain of solutions to find the values of the variable that satisfy the equation.

3. Equations involving multiples of the argument.

After preliminary steps, such as factoring, solve for the argument as a multiple number, as in

$$\tan 3x = 1$$

Solve the equation for $3x$.

$$3x = \frac{1}{4}\pi + 2k\pi \qquad \text{or} \qquad 3x = \frac{5}{4}\pi + 2k\pi$$

Then write the general solution by solving for x.

$$x = \frac{1}{12}\pi + \frac{2}{3}k\pi \qquad \text{or} \qquad x = \frac{5}{12}\pi + \frac{2}{3}k\pi$$

To find the fundamental solutions, use values of k that generate values of x between 0 and 2π.

Be careful not to lose or gain roots if you multiply or divide both sides of the equation by an expression containing the variable. Be sure to check your solution in the original equation.

Chapter 8 Trigonometric Equations

Using Your Graphing Calculator to Estimate Solutions of a Trigonometric Equation

Example 17 □ Estimate all the solutions in the interval $0 \leq x < 2\pi$ of $\sin x = \cos x$.

□ Figure 8.8

Graph $Y_1 = \sin x$ Set Range
and $Y_2 = \cos x$ $X_{\min} = 0$ $Y_{\min} = -1.2$
                          $X_{\max} = 6.28$ $Y_{\max} = 1.2$

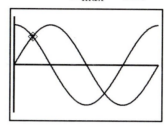

$x = .8017021$ $y = .6954847$

Push $\boxed{\text{trace}}$ and $\boxed{\triangleright}$ until you arrive at the point of intersection of the two graphs. The first x value is approximately 0.8. To find a more accurate value you could use the zoom function, but for our purposes this is reasonably close to $\frac{1}{4}\pi$. The second point of intersection occurs approximately at $x = 3.9$, which is reasonably close to $\frac{3}{4}\pi$.

It is possible to find multiple solutions to any designed degree of accuracy to trigonometric equations using your graphing calculator.   □

Why don't you just solve the equation $\sin x = \cos x$ by dividing both sides by $\cos x$?

The new equation will be $\tan x = 1$ and the solutions to this equivalent equation will be the same.

Try adding the graphs of $Y_3 = \tan x$ and $Y_4 = 1$ to the viewing screen. You should get the picture below.

□ Figure 8.9

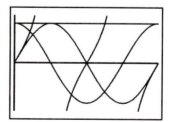

How come the graphs don't intersect at the same points?

Since these are different graphs they intersect at different y values. See the following Algebra Reminder.

Algebra Reminder

When solving an equation, we are looking for a value of x that makes the equation a true statement. When we reduce each equation to a simpler equivalent form, the same x value makes the equivalent equation a true statement. In other words, the same value of x is also a solution for the new equation. We continue this until we get the equivalent equation whose solution is obvious:

$$x = \text{ some value,}$$

which has the same solution as the original equation.

Example 18 □ $\sin x = \cos x$

$\sin x = \cos x$ ⟵ original equation

$\dfrac{\sin x}{\cos x} = 1$ ⟵ equivalent equation

$\tan x = 1$ ⟵ equation with obvious solution

The solution to the last equation is $x = \dfrac{1}{4}\pi + k\pi$. Therefore, the solution to the first equation is the same. Notice on the viewing screen you can see where the original pictures intersect. You can also see where the functions that represent each side of the equivalent equations intersect. Both share the same x coordinates or solution set, but the y values of the points of intersection are different for each equivalent equation. □

Problem Set 8.4

Use an appropriate trigonometric identity to reduce the following equations to a form solvable by one of the methods studied. Without using a calculator, find the exact value. Find (a) all values and (b) the fundamental values (non-negative values less than 2π radians).

1. $\sqrt{3}\tan x \cot x + \sqrt{3}\tan x - 1 - \cot x = 0$

2. $\tan x = \cot^2 x$

3. $\tan x = \sin x$

4. $2\sin^2 x = -3\cos x$

5. $\sin \dfrac{1}{2}x = \cos x$

6. $\cos 2x + 2\cos^2 \dfrac{1}{2}x = 1$

7. $\sin 2x = \cos x$

8. $\cos 2x = -\cos x$

9. $2\sin^2 x + 3\cos x - 3 = 0$

10. $4\tan^2 x - 3\sec^2 x = 0$

11. $\cos 2x - 3\sin x - 2 = 0$

12. $2\sin x - \csc x + \cot x = 0$

13. $2\cos^2 x + 3\sin x - 3 = 0$

14. $\cos^2 x + 3\sin x - 3 = 0$

Use an appropriate trigonometric identity to reduce the following equations to a form solvable by one of the methods studied. Using a calculator, find the values correct to four decimal places. Find (a) all values and (b) the fundamental values (non-negative values less than 2π radians).

15.	$\cos 2x = 3\sin x$	**16.**	$2 - \cos x = \cos 2x$
17.	$\sin 2x + 3\tan x \sin 2x = 0$	**18.**	$\cos 3x + 3\sin 3x \cos 3x = 0$
19.	$2\cos 4x + 3\cos 2x + 1 = 0$	**20.**	$\sin^2 3x + \sin 3x = \cos^2 3x + 1$
21.	$\tan x = 3\sin 2x$	**22.**	$\cot^2 x \csc^2 x = 16 + \cot^2 x$
23.	$\cot x = \tan^2 x$	**24.**	$\cos 2x = 4\sin x$
25.	$\tan^2 x - 6\sec x - 1 = 0$	**26.**	$2\sin^2\dfrac{x}{2} - 2 = \cos^2 x$

Using your graphing calculator, estimate all the solutions in the interval $0 \le x < 2\pi$ for the following equations. Find the solutions by non-graphical means correct to four decimal places. Use the graphical solution to test the reasonability of the solution found by non-graphical means.

27.	$4\sin^2 x - 3 = 0$	**28.**	$3\tan^2 x = 2$
29.	$3\tan^2 x + 5\tan x - 2 = 0$	**30.**	$\sin^2 x + \cos x \sin x = 0$
31.	$\cos x \tan x - 2\cos x + \tan x = 2$	**32.**	$4\cos 3x = 1$
33.	$\sqrt{3}\tan\left(\dfrac{1}{2}x - \dfrac{1}{4}\pi\right) = -1$	**34.**	$5\sin^2\left(2x + \dfrac{1}{3}\pi\right) = 2$
35.	$\cos 2x = \sin x$	**36.**	$2\tan x = \sin 2x$

8.5   Applications

This section is a problem set consisting of brief explanations of how trigonometric equations are used to solve practical problems. Each brief explanation is then followed with a few problems that let you apply your skill at solving trigonometric equations.

Problem Set 8.5

In this section, express your answer to the degree of accuracy you can reasonably expect.

Captain Zoom makes his living being shot out of a cannon at state fairs. The horizontal distance he flies is given by the formula for the range of a projectile:

$$R = \frac{V^2 \sin 2\theta}{g}$$

where R = range (horizontal distance)

V = velocity out of the cannon

θ = angle of elevation of the cannon

g = acceleration of gravity (32 ft/sec^2)

1. Captain Zoom's cannon fires him through the air at 66 feet per second. At what angle should the captain set the cannon in order to land in a net 135 feet away?

2. Captain Zoom wants to fly through the air for the maximum time before he strikes the net. Therefore, he picks the higher angle of elevation for each shot. If Captain Zoom wants to fly 100 feet, what are his choices for cannon angles if he uses enough charge to launch himself at 70 feet per second?

3. In traffic investigations, officers use the same formula as Captain Zoom. A motorcyclist traveling at 60 miles per hour (88 feet per second) crashed into a concrete divider. At what angle did the investigating officer determine the cyclist left the motorcycle in order to land on the highway 150 feet from the point of impact?

4. At what angle should an artillery shell with an initial velocity of 600 miles per hour be fired to hit $\frac{1}{2}$ mile horizontally from the firing site? (Note: There are two answers.)

5. At what angle θ should an object be projected to go the maximum distance with given initial velocity? (Hint: Ask yourself, "What is the effect of sin 2θ in the equation for the range of a projectile?")

6. An offer of large bucks was made to Captain Zoom of problem 1 if he could fly 300 feet through the air and land on a Super Soft Sleeper Mattress. Because Captain Zoom passed trigonometry he knows that the answer to problem 5 is 45°. With what velocity will the captain have to leave the cannon in order to fly 300 feet? (Neglect wind resistance.)

We roll heavy objects up ramps rather than lift them because it takes less force to move the object even though we have to move the object through a longer distance and do the same amount of work. If it takes 1 pound of force to roll a 10-pound barrel up a ramp, the ramp will have an Ideal Mechanical Advantage of 10.

The formula IMA = csc θ can be used to determine the Ideal Mechanical Advantage of a simple machine such as an inclined plane. θ is the angle between the inclined plane and the horizontal.

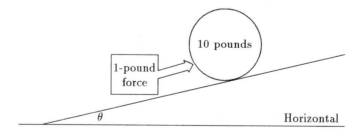

7. To roll a dolly with a 480-pound refrigerator up a ramp using 100 pounds of force would require an Ideal Mechanical Advantage of 4.8. What is the angle between the inclined plane and the horizontal to produce a 4.8 IMA?

8. At an office building ramps were installed to accommodate handicapped workers. It was determined that the Ideal Mechanical Advantage of the ramp should be 9.6. At what angle should the ramp be inclined with the horizontal to produce the correct Ideal Mechanical Advantage?

Polaroid glasses are commonly used to protect a person's eyes on a bright sunny day. Since light is an electromagnetic wave, a polarizing sheet can be used to filter light. A polarizing sheet has the characteristic that it allows waves in one direction only to pass through. If a polarizing sheet is placed upon a second polarizing sheet, light transmission may be restricted by rotating one or both of the sheets. Polaroid glasses use two such polarizing sheets to filter sunlight.

The formula $I = I_m \cos^2 \theta$ is used to compare unpolarized light transmission with polarized light transmission.

I is the intensity of polarized light.

I_m is the intensity of transmitted light.

θ is the angle between the polarizing direction of the polarized sheets when rotated with reference to each other.

$\cos^2 \theta$ gives the fraction of the total light that passes through the sheets.

9. Through what angle must either polarized sheet be turned to filter out three-fourths of the transmitted light? Restrict your answer to values $0° \leq \theta \leq 180°$.

10. A person purchasing a pair of polaroid glasses may be concerned about the amount of protection available. What angle θ is necessary for a 20 percent drop in the light transmission?

The angle between two intersecting lines is given by the formula $\tan \beta = \dfrac{m_2 - m_1}{1 + m_1 m_2}$, where m_1 is the slope of the first line and m_2 is the slope of the second line.

11. Find the angle formed by the intersection of the lines $y = -\dfrac{1}{2}x + 4$ and $y = 2x - 5$.

12. Find the angle formed by the intersection of the lines $y = \dfrac{3}{4}x - 3$ and $y = -\dfrac{5}{8}x + 4$.

The formula $E = \dfrac{I}{s^2} \cos \theta$ is used to determine the illumination on a surface when light rays from some source shine obliquely on the surface.

E is the illumination in foot-candles.
I is the intensity of the light in candlepower.
s is the distance in feet from the source of light.
θ is the angle between the incident light ray and a line perpendicular to the surface upon which the light is shining.

13. Find the angle θ if $E = 0.17$ foot-candles, I is 40 candlepower, and s is 10 feet.

14. Find the angle θ if $E = 0.25$ foot-candles, $I = 80$ candlepower, and s is 16 feet.

The reason we have seasons is that the amount of light (and the corresponding heat) that is absorbed by the earth's surface depends upon the angle at which the rays strike the surface. The earth's axis of rotation is tilted 23° from its orbital plane. This means that the angle of incidence of the sun's rays is constantly changing throughout the year.

At the winter solstice (shortest day of the year) the intensity of light striking the earth is $I = I_{max} \cos(\text{latitude} + 23°)$, where I_{max} is the maximum intensity of light at the equator. This means that at Saint Joseph, Missouri, which is at 40° north latitude, the light that strikes the earth on the shortest day of the year is:

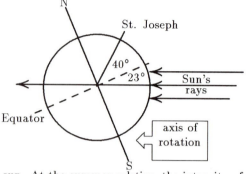

$$\begin{aligned} I &= I_{max} \cos(40° + 23°) \\ &= I_{max} \cos 63° \\ &\approx 0.45\, I_{max} \end{aligned}$$

In the summer the tilt of the earth is toward the sun. At the summer solstice, the intensity of light striking Saint Joseph, Missouri is:

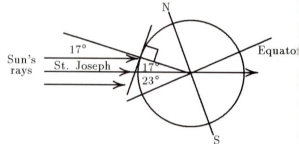

$$\begin{aligned} I &= I_{max} \cos(40° - 23°) \\ &= I_{max} \cos 17° \\ &\approx 0.96 I_{max} \end{aligned}$$

This means that the light and heat striking Saint Joseph, Missouri, vary from 45 to 96 percent of the light available at the hottest time of the year at the equator.

15. Find the latitude on the earth's surface where only 10 percent of the sun's energy is available on the shortest day of the year.

16. Find the latitude on the earth's surface where 75 percent of the sun's energy is available on the longest day of the year.

17. On the shortest day of the year in the northern hemisphere, what percent of the sun's energy strikes the equator?

18. On the longest day of the year in the northern hemisphere, what percentage of the sun's energy strikes the equator?

The amount of ultrasonic energy a remote control for a television actually delivers to the sensor is a function of the distance the beam is from the sensor and the angle of impact the beam makes with the sensor. At any given distance, the fraction of available energy to strike the sensor is given by:

$$E = E_{max} \cos \theta$$ where θ is the angle between the beam and the perpendicular to the sensor.

19. A sensor will activate with as little as 25 percent of the maximum energy a remote control unit can deliver at 10 feet. Through what range of angles will the remote activate the set at 10 feet?

When light waves pass from air into another substance there is a change in the speed of light causing the light beam to bend or be refracted. How much a medium will bend a ray of light is given by Snell's law, which says the index of refraction is $n = \dfrac{\sin \alpha}{\sin \beta}$, where α is the angle of incidence and β is the angle of refraction.

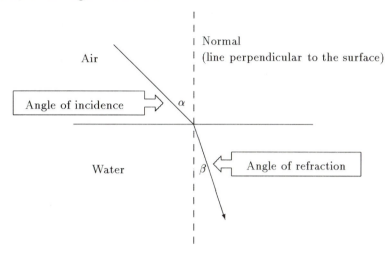

20. The index of refraction for glass is 1.50 and the angle of incidence is 32°. Find the angle of refraction.

21. If the index of refraction for water is 1.33 and the angle of incidence is 48°, find the angle of refraction.

The index of refraction of a substance can be found by rotating it as it is struck by a ray of light. At some point in the rotation, there is a minimum angle of deviation of the light ray through the substance. The formula for finding the index of refraction of light passing through a prism from the minimum angle of deviation is given by:

$$n = \frac{\sin \left[\dfrac{1}{2}(A + \beta) \right]}{\sin \left(\dfrac{1}{2} A \right)}$$

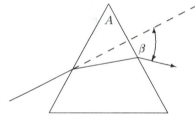

where n is the index of refraction, β is the angle of minimum deviation, and A is the angle at the top of the prism.

22. Find the angle of minimum deviation of light passing through a prism if the index of refraction is 1.4 and the refracting angle of the prism is 60°.

23. Light passes through a prism with an index of refraction of 1.8. If the refracting angle of the prism is 50°, find the angle of minimum deviation.

Most highway curves are banked for safety reasons. Usually speed limit signs are posted to indicate the safe speed to negotiate the curve. Large trucks occasionally turn over if the safe speed limit sign is ignored.

The formula used to determine the angle at which the roadway must be banked for a given speed is given by $\tan \theta = \dfrac{V^2}{Rg}$, where θ is the banking angle, V is the velocity or speed of the vehicle, R is the radius of the curve, and g is the acceleration of gravity, which is 32 feet per second squared. This formula also applies to the bank angle of an airplane in a turn.

24. An engineer wishes to design a freeway ramp so traffic can maintain a speed of 40 miles per hour on the ramp. At the point where the radius of curvature of the ramp is 340 feet, what is the appropriate bank angle for the roadway? (Notice: Convert 40 miles per hour to feet per second.)

25. At an airport an effort was made to reduce the noise level of jet liners landing and taking off by keeping them close to the airport. It was determined that jets coming in to land should remain within a 2-mile radius in their turning approach to the runway. If a jet is approaching the airport at 200 miles per hour, what is the bank angle required to make a turn with a 2-mile radius?

To afford predictability to the air traffic control system, airplanes are expected to make standard rate turns. A standard rate turn is a turn that takes 2 minutes to turn 360°, or a full circle.

Because an airplane traveling 120 miles per hour will travel 4 miles in 2 minutes, the airplane will fly in a circle with a circumference of 4 miles while making a standard rate turn at 120 miles per hour.

26. After finding the radius of a circle with a circumference of 4 miles, use the bank angle formula to find the bank angle of an airplane flying at 120 miles per hour in a standard rate turn.

27. Calculate the radius of turn and bank angle for an aircraft flying at 240 miles per hour in a standard rate turn.

Chapter 8 Trigonometric Equations

Using Your Graphing Calculator

Consider a piston and a crankshaft similar to the one in Example 2 of this chapter with a piston rod 3 inches long on a crankshaft with a radius of 1 inch. We can use a graphing calculator to find the angle of the crankshaft when the piston is halfway between the top and the bottom of its travel.

Even though the piston is at the top of the cylinder at $\theta = 90°$ and at the bottom at $\theta = 270°$ it is not at the center at $\theta = 180°$. To find when it is at the center, we analyze the problem this way.

The halfway point in the cylinder is halfway between the top of the stroke and the bottom of the stroke. Using the formula for the height of the piston:

$$h = r \sin \theta + \sqrt{l^2 - r^2 \cos^2 \theta}$$

and using $\qquad r = 1$ and $l = 3$

we have $\qquad h = \sin \theta + \sqrt{9 - \cos^2 \theta}$

when $\quad \theta = 90° \qquad h \quad = \quad \sin 90° + \sqrt{9 - (\cos 90°)^2}$
$\qquad\qquad\qquad\qquad\quad = \quad 4$

when $\quad \theta = 270° \qquad h \quad = \quad \sin 270° + \sqrt{9 - (\cos 270°)^2}$
$\qquad\qquad\qquad\qquad\quad = \quad 2$

Therefore, the halfway point occurs at $h = 3$. On your graphing calculator set:

$y_1 = \sin \theta + \sqrt{9 - (\cos \theta)^2}$ $\qquad$ Range
$y_2 = 3$ $\qquad\qquad\qquad\qquad\qquad$ $X_{min} = 0 \qquad\qquad Y_{min} = 0$
$\qquad\qquad\qquad\qquad\qquad\qquad\quad$ $X_{max} = 6.28 \qquad Y_{max} = 5$

Execute the graph:

□ Figure 8.10

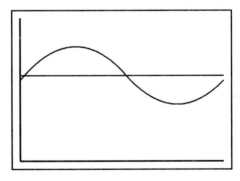

This is a plot with the x-axis in radians. When you trace and zoom in on the 2 points of intersection you get

$$x = .16 \text{ radians} \qquad \text{and} \qquad x = 2.98 \text{ radians}$$

Converting radians to degrees we find the angle of the crankshaft that places the piston at the halfway point is approximately
$\qquad\qquad$ 9.17° on the way up
and $\qquad$ 170.74° on the way down.

Use your graphing calculator to solve the following problems.

28. Approximate the angle of the crankshaft that places the piston at the point halfway between the bottom of its travel and the top of its travel in the cylinder when the radius of the crankshaft is 2 inches and the length of the piston rod is 5 inches.

29. For the above engine approximate the angle of the crankshaft that places the piston at a point one fourth of the way from the top.

Chapter 8	**Key Ideas**

8.1   **1.** Conditional trigonometric equations are of the type $2\sin\theta = 1$ or $\sec^2\theta = 5$.

2. The general solution of a trigonometric equation includes all real numbers in degrees or radians in the interval $0° \le \theta < 360°$ or $0° \le \theta < 2\pi$.

3. The fundamental solution of a trigonometric equation includes all solutions in degrees or radians in the interval $0° \le \theta < 360°$ or $0 \le \theta < 2\pi$.

8.2   **1.** Quadratic-type equations should be factored if possible. Then solve each factor for the variable.

2. Use the quadratic formula to solve quadratic-type equations that are not factorable. Use resulting values to find the values of θ or x in terms of degrees or radians. Usually a calculator is needed to find the values of θ or x.

3. Non-quadratic equations are to be factored. Then solve each factor for the variable.

8.3   **1.** Some equations involve trigonometric functions where the argument of the function is itself a function of the angle θ or x. They require an additional step. Solve the argument for the variable to determine the solution set.

2. Equations containing more than one function are factored to separate the functions so that each factor contains only one function.

8.4   Some equations require the use of trigonometric identities before factoring can be done.

8.5   Trigonometric equations are frequently used in practical situations.

Without the use of your calculator solve the following equations for (a) all values and (b) the fundamental values (non-negative values less than $360°$). **(8.1, 8.2, 8.3, 8.4)**

1. $2\sin\theta = \sqrt{3}$

2. $3\sec^2\theta - 4 = 0$

3. $2\sin^2\theta + \sin\theta - 1 = 0$

4. $2\sin^2\theta + \sqrt{3}\sin\theta = 0$

5. $2\sin 3\theta - 1 = 0$

6. $\cos(2\theta - 60°) = -\dfrac{1}{2}$

7. $\sec\theta - 3 + 2\cos\theta = 0$

8. $\cot\theta = \cos\theta$

Without the use of your calculator solve the following equations for (a) all values and (b) the fundamental values (non-negative values less than 2π radians). **(8.1, 8.2, 8.3, 8.4)**

9. $2\sin x + \sqrt{3} = 0$

10. $2\cos^2 x - 1 = 0$

11. $\sqrt{3}\cot^2 x - 4\cot x + \sqrt{3} = 0$

12. $3\csc^3 x - 4\csc x = 0$

13. $3\tan 2x - \sqrt{3} = 0$

14. $2\cos\left(4x + \dfrac{1}{3}\pi\right) = -1$

15. $\cot x = \tan^2 x$

16. $2\sin^2 x + \cos x = 2$

Using your calculator, if necessary, solve the following equations for (a) all values and (b) the fundamental values (non-negative values less than $360°$). Round off to the nearest hundredth. **(8.1, 8.2, 8.3, 8.4)**

17. $3\sin\theta = 1$

18. $2\tan^2\theta - 3 = 0$

19. $2\sin^2\theta + 3\sin\theta - 2 = 0$

20. $\csc^2\theta - 2\csc\theta - 4 = 0$

21. $4\cos 3\theta - 3 = 0$

22. $3\sin(2\theta - 24°) - \sqrt{5} = 0$

23. $2\cot^2\theta - 3 = \csc^2\theta$

24. $\cos 2\theta + 5\cos^2\theta = 2$

Using your calculator, if necessary, solve the following equations for (a) all values and (b) the fundamental values (non-negative values less than 2π radians). Round off to four decimal places. **(8.1, 8.2, 8.3, 8.4)**

25. $\sqrt{3}\sin x + 1 = 0$

26. $5\cos^2 x - 4 = 0$

27. $3\sin^2 x + 5\sin x + 2 = 0$

28. $3\tan^2 x + 4\tan x - 3 = 0$

29. $6\tan 3x + 5 = 0$

30. $5\sin^2\left(3x + \dfrac{1}{3}\pi\right) = 4$

31. $4\sin 2x = 3\cos x$

32. $\cos^2 x + \sin x = 0$

Using your graphing calculator, estimate all the solutions in the interval $0 \le x < 2\pi$ for the following equations correct to four decimal places. Find the solution to test the reasonability of the solution found by non-graphical means. **(8.1, 8.2, 8.3, 8.4)**

33. $3\sin^2\left(3x - \dfrac{1}{4}\pi\right) = 4$

34. $\sin^2 x - \cos x = 0$

35. $\sin^2 x + 2\sin x - 3 = 0$

36. $4\cos 2x = 3\sin x$

Using the appropriate formula, solve the following problem. **(8.5)**

37. A ray of light passes through a certain medium which has an index of refraction of 1.35. If the angle of incidence is 36°, find the angle of refraction.

38. A rifle bullet with an initial velocity of 225 feet per second struck the ground 1220 feet from the place the gun was fired. Find the angle with the horizontal at which the gun was fired.

39. Find the angle formed by the intersection of the lines $y = \dfrac{2}{3}x - 6$ and $y = -\dfrac{4}{7}x + 2$.

Chapter 9

Law of Sines and Law of Cosines

Contents

Preview

Chapter 2 demonstrated problems that could be solved using right triangle trigonometry. There are many other applications of trigonometry where it is necessary to solve an oblique triangle.

   This chapter will develop the law of sines which lets us solve any triangle when one angle and two sides are known, or when two angles and a side are known. If one of the known angles is a right angle, the law of sines reduces to the definition of the sine of an angle in a right triangle.

   This chapter will develop the law of cosines to solve triangles where two sides and the angle between them are known or where three sides are known. It will then exhibit applications of this useful law to navigation, surveying, and engineering.

9.1   Derivation of the Law of Sines

In this chapter we will learn how to find the missing sides and angles of a triangle. We will start by dividing any triangle into two right triangles.

Consider any triangle:

Case I—Any Triangle Where γ Is an Acute Angle

□ Figure 9.1

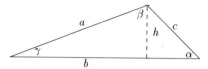

Let h be the altitude of the triangle. Notice that h divides this triangle into two right triangles.

$$\sin \gamma = \frac{h}{a} \qquad \text{and} \qquad \sin \alpha = \frac{h}{c}$$

Solving each equation for h,

$$h = a \sin \gamma \qquad h = c \sin \alpha$$

Since we have two expressions for h,

$$a \sin \gamma = c \sin \alpha$$

or rewriting,

$$\frac{a}{\sin \alpha} = \frac{c}{\sin \gamma}$$

Case II—Any Triangle Where γ Is an Obtuse Angle

□ Figure 9.2

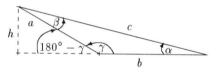

Again, let h be the altitude of the triangle.

$$\sin \alpha = \frac{h}{c} \qquad \text{and} \qquad \sin(180° - \gamma) = \frac{h}{a}$$

but because

$$\sin(180° - \gamma) = \sin \gamma \qquad \text{identity (29)}$$

$$\sin \gamma = \frac{h}{a}$$

Solving the equations $\sin \alpha$ and $\sin \gamma$ for h,

$$h = c \sin \alpha \qquad h = a \sin \gamma$$

Substituting,

$$c \sin \alpha = a \sin \gamma$$
$$\frac{c}{\sin \gamma} = \frac{a}{\sin \alpha}$$

This relationship is true for any triangle. If we consider side a to be the base of the triangles above we can also derive

$$\frac{c}{\sin \gamma} = \frac{b}{\sin \beta}$$

These two equations are called the law of sines.

9.1 Law of Sines

For any triangle $\quad \dfrac{a}{\sin \alpha} = \dfrac{b}{\sin \beta} = \dfrac{c}{\sin \gamma}$

Within any triangle (acute, obtuse, or right) the ratio of any side to the sine of the angle opposite it is the same for all three angles.

The law of sines simplifies finding missing parts of triangles that are not right triangles.

Example 1 □ If $\alpha = 40°, \beta = 60°, b = 30$, find $a, c,$ and γ.

□ Figure 9.3

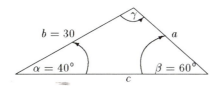

Draw your sketch as close to scale as practical.

Using the law of sines,

$$\frac{a}{\sin \alpha} = \frac{b}{\sin \beta}$$

$$\frac{a}{\sin 40°} = \frac{30}{\sin 60°}$$

$$a = \frac{30(\sin 40°)}{\sin 60°}$$

$$a \approx 22$$

Because

$$\alpha + \beta + \gamma = 180°$$

$$40° + 60° + \gamma = 180°$$

$$\gamma = 80°$$

Applying the law of sines again:

$$\frac{b}{\sin \beta} = \frac{c}{\sin \gamma}$$

$$\frac{30}{\sin 60°} = \frac{c}{\sin 80°}$$

$$\frac{30(\sin 80°)}{\sin 60°} = c$$

$$34 \approx c$$

Why not use $\dfrac{a}{\sin \alpha}$ to find side c?

We could, but, if we use the original data instead, we will avoid compounding round-off errors.

□

The equations in the law of sines relate four parts of a triangle. Therefore, if we know any angle and the side opposite it, along with one other part of the triangle, we can find the remaining three parts.

9.1 Derivation of the Law of Sines

257

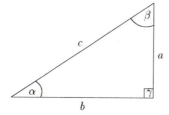

Does the law of sines work for right triangles?

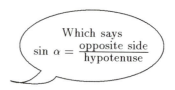

Remember, the law of sines works for any triangle. But in this case, it reduces to the definition of the sin θ.

□ Figure 9.4

If $\gamma = 90°$

$$\frac{c}{\sin \gamma} = \frac{a}{\sin \alpha}$$

becomes
$$\frac{c}{\sin 90°} = \frac{a}{\sin \alpha}$$
$$\frac{c}{1} = \frac{a}{\sin \alpha}$$
$$\sin \alpha = \frac{a}{c}$$

Which says
$$\sin \alpha = \frac{\text{opposite side}}{\text{hypotenuse}}$$

Using Your Calculator

Order of Operations
In elementary algebra an order of operations was defined. Briefly, the order is:
Operations within Parentheses
Exponents
Multiplication and Division
Addition and Subtraction

By this order of operations $2 + 3 \cdot 4 = 14$ and not 20. You should try your calculator to see that this rule is followed.

Press 2 $\boxed{+}$ 3 $\boxed{\times}$ 4 → you should get 14

Your calculator probably treats the trigonometric functions as operations within parentheses. The practical effect of this is that you can enter numbers into your calculator pretty much in the order in which they are written.

To evaluate $6 \cdot \sin 30°$ you could press $\boxed{\text{SIN}}$ 30 to get 0.5 followed by $\boxed{\times}$ 6 $\boxed{\text{ENTER}}$ to get 3.0.

But, your calculator will suspend other operations while it evaluates trigonometric functions.

This means you can evaluate $6 \boxed{\times} \sin 30°$ by pressing 6 $\boxed{\times}$ $\boxed{\text{SIN}}$ 30 $\boxed{\text{enter}}$ to get 3.0.

Try evaluating $6 \cdot \sin 30° + 4$ by pressing 6 $\boxed{\times}$ $\boxed{\text{SIN}}$ 30 $\boxed{+}$ 4 $\boxed{\text{enter}}$.

You should get 7.0 because the calculator evaluates the sine of the number in the display.

Hence, in the sequence above, $\sin 30°$ is used instead of $\sin 34°$.

Chapter 9 Law of Sines and Law of Cosines

Example 2 □ If $\alpha = 138°$, $a = 210$, and $b = 150$, solve the triangle.

Draw a sketch.

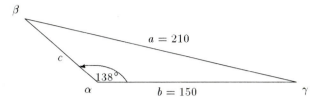

By the law of sines,

$$\frac{b}{\sin \beta} = \frac{a}{\sin \alpha}$$

$$\frac{150}{\sin \beta} = \frac{210}{\sin 138°}$$

$$\frac{150 \sin 138°}{210} = \sin \beta$$

$$29° \approx \beta$$

Now γ can be computed.

$$\gamma = 180° - (138° + 29°)$$
$$\gamma = 13°$$

Use the law of sines again to find side c.

$$\frac{c}{\sin \gamma} = \frac{a}{\sin \alpha}$$

$$\frac{c}{\sin 13°} = \frac{210}{\sin 138°}$$

$$c = \frac{(210)(\sin 13°)}{\sin 138°}$$

$$c \approx 71$$

□

The following information refers to a triangle, ABC. Find the missing parts. Remember, your answer can be no more accurate than the least accurate data given in the problem.

1. $\beta = 109.2°, \gamma = 35.6°, b = 8.56$

2. $\beta = 59.3°, \gamma = 77.5°, c = 0.0619$

3. $\alpha = 98.2°, \beta = 56.3°, b = 18.18$

4. $\beta = 132.4°, \gamma = 22.8°, c = 26.7$

5. $\alpha = 65.24°, \beta = 24.67°, a = 3.034$

6. $\beta = 52.64°, \gamma = 37.36°, b = 24.76$

7. $\alpha = 117.4°, \gamma = 10.8°, c = 106.5$

8. $\alpha = 37.54°, \beta = 75.36°, c = 0.0268$

9. $\beta = 100.2°, \gamma = 42.3°, a = 0.305$

10. $\alpha = 31.56°, \beta = 38.45°, c = 72.2$

11. $\beta = 26.7°, \gamma = 85.2°, a = 369$

12. $\alpha = 36.9°, \gamma = 72.4°, c = 3.04$

13. $\alpha = 102.6°, \gamma = 26.5°, b = 46.4$

14. $\beta = 75.26°, \gamma = 42.58°, a = 0.0842$

15. $\alpha = 29.34°, \beta = 35.61°, c = 384.2$

16. $\alpha = 79.46°, \gamma = 22.35°, b = 1.864$

17. $\alpha = 8.9°, \gamma = 15.4°, a = 22.8$

18. $\beta = 113.2°, \gamma = 53.9°, c = 18.9$

19. $\alpha = 43.9°, \beta = 101.2°, a = 8.09$

20. $\alpha = 59.67°, \gamma = 110.92°, b = 0.0219$

 Group Writing Activity

Explain, without using the law of sines, why the following information is not valid for the triangle ABC.

$$\beta = 96.8° \qquad a = 41.5 \text{ meters} \qquad b = 22.3 \text{ meters}$$

9.2 The Ambiguous Case

Did you ever try to put up a tent with a center pole that was too high for the tent? If so, you probably noticed that the pole would not stand straight up. It would, however, take a position on either side of the center of the tent.

This is an example of how it is possible to form two different triangles with the same three sides. This situation is called the ambiguous case.

Example 3 □ The side of a tent has a length c. It makes an angle α with the floor. A pole supports the tent. How should the pole be so that it will stand perpendicular to the floor?

□ Figure 9.6

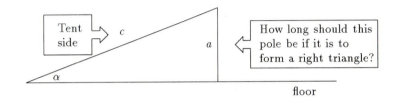

The idea is to form a right triangle.

$$\sin \alpha = \frac{a}{c}$$
$$a = c \sin \alpha$$

Correct length to form a right triangle

If the pole is shorter than $c \sin \alpha$, it will not reach the floor and will not form a triangle.

□ Figure 9.7

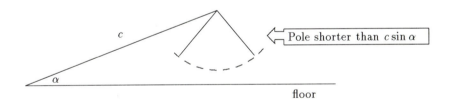

If the pole is longer than $c \sin \alpha$, it is possible to form two triangles, but neither will be a right triangle.

□ Figure 9.8

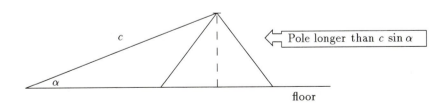

This is called the ambiguous case, because if we apply the law of sines to solve such a triangle there are two possible solutions. □

Example 4 □ If angle $\alpha = 27°$, side $a = 5.0$, and side $c = 9.0$, find the remaining parts of the triangle.

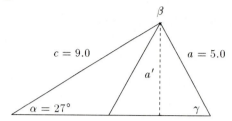

Compute the critical length for the side opposite angle α and call it a'.

$$
\begin{aligned}
a' &= c \sin \alpha \\
&= 9 \sin 27° \\
a' &\approx 4.0
\end{aligned}
$$

> This is not the length of side a. It is the altitude of the triangle. Since a is longer than this critical length, two triangles will be formed.

By the law of sines,

$$
\begin{aligned}
\frac{5}{\sin 27°} &= \frac{9}{\sin \gamma} \\
\sin \gamma &= \frac{9(\sin 27°)}{5} \\
\sin \gamma &\approx 0.8172 \\
\gamma &\approx 55° \qquad \text{(to the nearest degree)}
\end{aligned}
$$

One possible triangle looks like this:

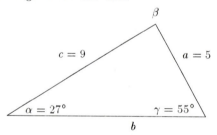

To find angle β,

$$
\begin{aligned}
\beta &\approx 180° - (27° + 55°) \\
\beta &\approx 98°
\end{aligned}
$$

To find side b in this example use the law of sines again:

$$
\begin{aligned}
\frac{b}{\sin \beta} &= \frac{a}{\sin \alpha} \\
\frac{b}{\sin 98°} &= \frac{5}{\sin 27°} \\
b &= \frac{5 \sin 98°}{\sin 27°} \\
&\approx 11
\end{aligned}
$$

Chapter 9 Law of Sines and Law of Cosines

To find the other possible triangle we need another angle with the same sine as 55°. That angle is the supplement of 55°.

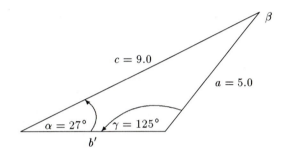

□ Figure 9.11

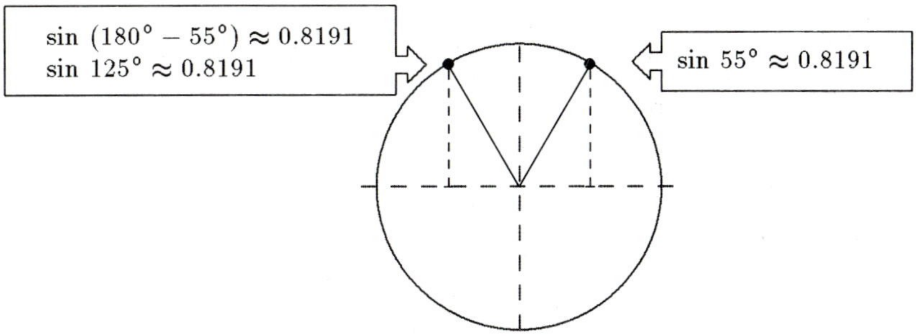

$\sin (180° - 55°) \approx 0.8191$
$\sin 125° \approx 0.8191$

$\sin 55° \approx 0.8191$

The second possible triangle looks like this:

□ Figure 9.12

β

$c = 9.0$

$a = 5.0$

$\alpha = 27°$ $\gamma = 125°$

b'

Using β' to identify the second possible value of angle β,

$$\beta' \approx 180° - (27° + 125°)$$
$$\approx 28°$$

To find side b' use the law of sines again,

$$\frac{b'}{\sin \beta'} = \frac{a}{\sin \alpha}$$
$$b' = \frac{a(\sin \beta')}{\sin \alpha}$$
$$\approx \frac{5(\sin 28°)}{\sin 27°}$$
$$b' \approx 5.2$$

□

Other Possibilities

The ambiguous case may occur in a triangle where one angle, the side opposite it, and another side are known. However, this condition does not always lead to an ambiguous case.

If $a < c \sin \alpha$:

□ Figure 9.13

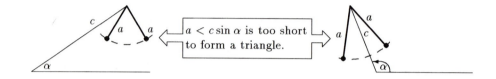

If $\alpha < 90°$ and $a = c \sin \alpha$:

□ Figure 9.14

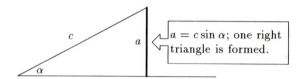

There are some possibilities where it is not necessary to compute the altitude in order to know the number of triangles possible.

□ Figure 9.15

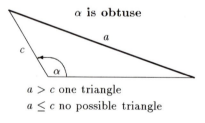

α is obtuse

$a > c$ one triangle
$a \leq c$ no possible triangle

α is acute

□ Figure 9.16

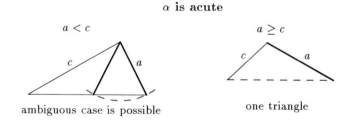

Chapter 9 Law of Sines and Law of Cosines

Example 5 □ If angle $\alpha = 35°$, side $a = 4.0$, and side $c = 10$, find the remaining parts of the triangle.

□ Figure 9.17

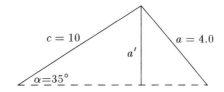

Compute the critical length for the side opposite angle α.

$$a' = c \sin \alpha$$
$$= 10 \sin 35°$$
$$a' \approx 5.7$$

This is the critical length. It is the altitude of a triangle with a 35° angle and a side c that equals 10.

□

How many triangles are possible?

None. Since side a is less than the altitude of the triangle, it will never reach the base. Hence, no triangle will be formed.

Problem Set 9.2

Find all solutions, if there are any, to each of the following triangles. Remember, your answers can be no more accurate than the least accurate data given in the problem.

1. $b = 45$, $c = 106$, $\gamma = 76°$
2. $b = 42$, $a = 10$, $\alpha = 32°$
3. $b = 30$, $c = 40$, $\beta = 20°$
4. $b = 29.13$, $c = 56.2$, $\gamma = 32.1°$
5. $b = 8.27$, $a = 7.68$, $\alpha = 45.2°$
6. $b = 2.4$, $c = 2.6$, $\beta = 78.2°$
7. $c = 35.6$, $b = 41.6$, $\gamma = 28.2°$
8. $b = 21.9$, $c = 42.3$, $\gamma = 115.4°$
9. $a = 12.8$, $b = 18.6$, $\alpha = 59.2°$
10. $a = 19.7$, $c = 42.7$, $\alpha = 21.3°$
11. $a = 17.8$, $b = 11.4$, $\alpha = 46.4°$
12. $b = 46.2$, $c = 52.9$, $\beta = 116.2°$
13. $a = 0.784$, $c = 0.692$, $\alpha = 58.3°$
14. $b = 42.3$, $a = 51.2$, $\beta = 110.1°$
15. $a = 9.2$, $c = 8.7$, $\gamma = 63.4°$
16. $a = 0.0786$, $b = 0.0942$, $\beta = 78.92°$
17. $a = 62.93$, $c = 75.68$, $\alpha = 115.62°$
18. $a = 79.34$, $c = 52.83$, $\gamma = 26.92°$

Group Writing Activity

In any triangle, the sum of the lengths of any two sides must be greater than the length of the third side. Explain why this is true.

9.3 Applications of the Law of Sines

The law of sines allows us to extend applications to situations other than a right triangle. Chapter 1 calculated the diameter of the earth. Now we can use this information and the law of sines to find the distance to the moon.

First we must find the distance between two points on the earth's surface that are 20° apart.

Example 6 □ If you could tunnel straight through the earth, what would the distance be between two places on the same line of longitude that are separated by 20° of latitude on the earth's surface?

From Chapter 1 we know the polar diameter of the earth is about 7900 miles. A radius is half a diameter, therefore the radius of the earth is 3950 miles.

Because the triangle shown is isosceles,

$$\alpha = \beta = 80°$$

Applying the law of sines,

□ Figure 9.18

$$\frac{x}{\sin 20°} = \frac{3950 \text{ miles}}{\sin 80°}$$

$$x = \frac{3950 \text{ miles}}{\sin 80°} \cdot \sin 20°$$

$$x \approx 1370 \text{ miles}$$

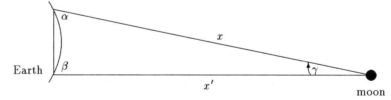

□

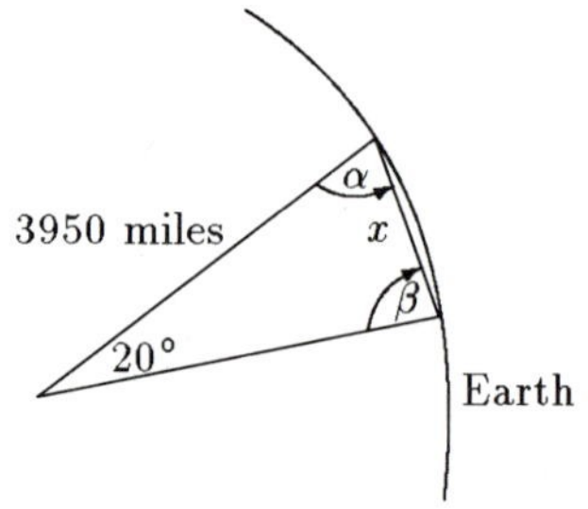

Example 7 □ Now use the information from Example 6 to find the distance to the moon.

□ Figure 9.19

Assuming we pick two spots 20° apart on the earth's surface at the proper time of the year and latitude so that $x = x'$, α and β will be measured to be 89.836° each.

Since $\alpha = \beta = 89.836°$ and the sum of the angles in a triangle is 180°, γ must be $\alpha = 180° - 2(89.836°) = 0.328°$.

$$\frac{x}{\sin \beta} \approx \frac{1370}{\sin \gamma}$$

$$x \approx 1370 \frac{\sin \beta}{\sin \gamma}$$

$$\approx 1370 \frac{\sin 89.836°}{\sin 0.328°}$$

$$\approx 240,000 \text{ miles}$$

The commonly accepted mean distance from the earth to the moon is 238,857 miles. □

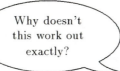

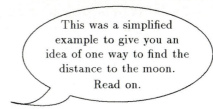

1. Several assumptions were made in setting up this problem. One of the assumptions is that the earth is a perfect sphere.

2. With astronomical distances, even the 8-digit accuracy of the calculator can lead to discrepancies of many miles.

3. Actual angular measurements would be made to greater accuracy than the figures given here.

Example 8 □ An airplane measured the angle to a radio beacon as 25° left of course. Ten miles later the pilot reads the angle to the beacon as 40° left of course. How far away is the beacon at the time of the second reading?

□ Figure 9.20

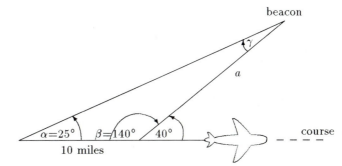

$$\gamma = 180° - (25° + 140°)$$
$$= 15°$$

Applying the law of sines,

$$\frac{a}{\sin 25°} = \frac{10 \text{ miles}}{\sin 15°}$$
$$a = \frac{(\sin 25°)10}{\sin 15°}$$
$$a \approx 16 \text{ miles to the nearest mile}$$

□

Example 9 □ While building a ski lift, a surveyor measured the angle of elevation to the base of a 20-foot tower as 30°. The angle of elevation to the top of the tower was 37°. How far away was the base of the tower?

□ Figure 9.21

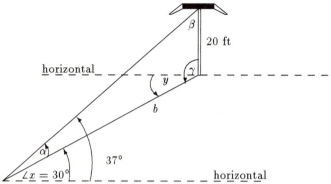

Examine the angles between the horizontal and a line from A to the base of the tower.

$$\angle x \; = \; 30° \qquad \text{Given.}$$
$$\angle x \; = \; \angle y \qquad \text{If two parallel lines are cut by a transversal, alternate interior}$$
$$\qquad \qquad \qquad \text{angles are equal.}$$
$$\angle y \; = \; 30°$$

Because the tower makes a 90° angle with the horizontal,

$$\gamma \; = \; 30° + 90°$$
$$\gamma \; = \; 120°$$

The difference in angle of elevation between the top and bottom of the tower is 7°. Call this angle α.

$$\text{Angle } \beta \; = \; 180° - (7° + 120°)$$
$$\qquad \qquad = \; 53°$$

Applying the law of sines,

$$\frac{20 \text{ ft}}{\sin 7°} \; = \; \frac{b}{\sin 53°}$$
$$\frac{20(\sin 53°)}{\sin 7°} \; = \; b$$
$$130 \text{ ft} \; \approx \; b \text{ to the nearest 10 feet}$$

□

Example 10 □ The Wild Winds Tent Company has found that if a tent rope is placed at a 45.0° angle to the ground, it will provide maximum support for the tent. How far away from the base of a 6.0 foot tent wall should an 8.0 foot rope be placed for maximum support in a strong wind?

□ Figure 9.22

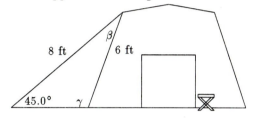

First we need to find another angle.

$$\frac{8.0}{\sin \gamma} = \frac{6.0}{\sin 45.0°}$$

$$\frac{8.0(\sin 45.0°)}{6.0} = \sin \gamma$$

$$0.9428 \approx \sin \gamma$$

$$\gamma \approx 70.5°$$

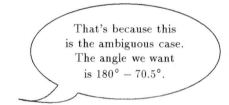

In the picture γ is greater than 90°.

That's because this is the ambiguous case. The angle we want is $180° - 70.5°$.

The proper value for γ is

$$180 - 70.5° \approx 109.5°$$

Then

$$\beta \approx 180° - (109.5° + 45°)$$

$$\approx 25.5°$$

and the proper length is x.

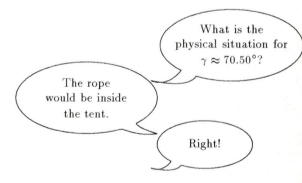

What is the physical situation for γ ≈ 70.50°?

The rope would be inside the tent.

Right!

$$\frac{x}{\sin 25.5°} \approx \frac{6}{\sin 45.0°}$$

$$x \approx \frac{6(\sin 25.5°)}{\sin 45.0°}$$

$$x \approx 3.7 \text{ feet}$$

□

Example 11 □ A boat left a dock and sailed for 5 miles on a course of N30°W. The boat then turned to a course of S20°W and sailed for 10 miles. At this time the navigator measured the angle back to the dock to be N50°E. How far away is the dock?

□ Figure 9.23

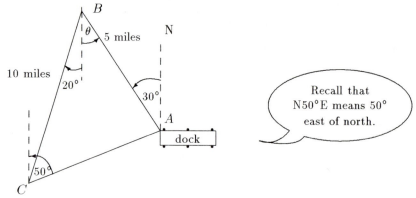

First we need to find some angles in the interior of the triangle, because parallel lines cut by a transversal form equal alternate interior angles, we know ∠θ.

$$\angle \theta = 30°$$
$$\angle CBA = 20° + 30°$$
$$= 50°$$

Similarly the angle between $\overline{BC}$ and north is 20°.

Therefore,

$$\angle BCA = 50° - 20°$$
$$= 30°$$

The remaining angle of the triangle is

$$\angle CAB = 180° - (50° + 30°)$$
$$= 100°$$

Now the law of sines can be used to find side $\overline{CA}$ which we refer to as side b.

$$\frac{b}{\sin B} = \frac{a}{\sin A}$$
$$b = \frac{a \sin B}{\sin A}$$
$$= \frac{10 \sin 50°}{\sin 100°}$$
$$b \approx 7.8 \text{ miles}$$

□

Chapter 9 Law of Sines and Law of Cosines

A Pointer About the Ambiguous Case

You need consider the possibility of the ambiguous case only when you are given an angle, the side opposite that angle, and one other side.

For identification let

α = given angle

a = side opposite α

b = the other side, which must be one side of angle α

There are only three cases to consider:

Case I — α is less than 90° and $a < b$

This is the case where the ambiguous case is possible.

□ Figure 9.24

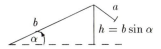 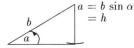 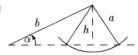

1) $a < h$
   no triangle possible

2) $a = h$
   one triangle
   a right triangle

3) $h < a < b$
   two triangles
   ambiguous case

Case II — α is less than 90° and $a > b$

□ Figure 9.25

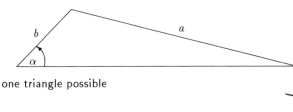

one triangle possible

No need to check $h = b \sin \alpha$ here because if $a > b$ there is only one point where side a can touch the base.

Case III — α is greater than 90°

□ Figure 9.26

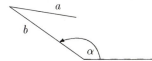

$a \leq b$
no triangle possible

$a > b$
one triangle possible

Problem Set 9.3

Solve the following problems. Express your answers to the degree of accuracy you can reasonably expect.

1. To measure the height of a vertical tree growing on a hillside, a forest ranger found the slope of the hillside to be 15°. From the foot of the tree he walked 200 feet down the slope and with a transit determined the angle of elevation of the top of the tree to be 43°. How tall was the tree to the nearest foot?

2. Two forest rangers stationed 5.6 miles apart at points A and B in a mountain range observe the same illegal campfire at point C some distance away. They measure angles CAB and CBA to be 46.0° and 55.0° respectively. How far is each ranger from the campfire to the nearest tenth of a mile?

3. A vertical telegraph pole standing alongside an inclined highway casts a 52.0 foot shadow down the incline directly along the highway when the angle of elevation of the sun is 55°. If the height of the pole is 65.0 feet, find the angle of inclination of the highway.

4. A balloonist flying at a constant rate of 5 miles per hour observed a peak directly ahead to have an angle of depression of 43°. Thirty-six minutes later, after passing directly over the peak she looked back and found the angle of depression to the peak to be 56°. To the nearest 100 feet, how high above the peak did she fly?

5. Two angles of a triangular plot of ground are 36.2° and 58.6°, and the side between them is found to be 215 meters. How many meters of fencing are needed to enclose the plot of land? (Give your answer to three significant digits.)

6. In a hilly area, a distance of 120 feet was measured down the straight slope of a hill from the base of a tree. From this point the angles of elevation of the top and foot of the tree are 37° and 22° respectively. To the nearest foot how tall was the tree?

7. Two boys desiring to estimate the height of a nearby radio tower measured the angle of elevation of the tower at their house and found it to be 52° . They took a second measurement from the second-story window and found the angle of elevation to be 44°. They next measured the window to be 24 feet above the ground. To the nearest foot what was the height of the radio tower?

8. A couple of scouts, desiring to use some of the mathematics they learned the previous school year, decided to estimate the distance between two trees. In their campsite was one tree, which they labeled "Tree A," and across a river was a second tree, which they labeled "Tree B." To complete the job they measured the distance from "Tree B" to a third tree, "Tree C," on the same side of the river to be 120 meters. With care they measured angle ACB to be 25° and angle CAB to be 110°. What was the distance from Tree A to Tree B? (Give answer to the nearest meter.)

9. A father and daughter at a picnic decided to fly a kite. They wondered how high the kite was flying over a lake. To find the height they measured the angle of elevation of the kite and found it to be 23°. While the daughter held the kite the father walked 100 paces (each pace about 3 feet in length) directly under the kite string to the edge of the lake where he took a second measurement of the kite's elevation. Here he found it to be 53°. What was the height of the kite above the lake?

10. From a lighthouse a boat was sighted at sea with a bearing of N12°W. The boat was traveling due east at 12 knots. Fifteen minutes later it had a bearing of N27°E. To the nearest nautical mile, how far was the boat from the lighthouse when last sighted? (Note: one knot = one nautical mile per hour.)

Chapter 9 Law of Sines and Law of Cosines

9.4   Derivation of the Law of Cosines

The law of sines requires that we know at least one angle and the side opposite that angle. There are times when we know only three sides, or two sides and the angle between them. The law of cosines lets us solve the triangle in these cases.

Consider the case of an air traffic controller using radar. The radar can give the distance and angle from the antenna to an aircraft. In order to determine the flight path of the aircraft, a second measurement of distance and direction from the antenna is taken seconds later. The distance and direction traveled by the aircraft in the interval are needed to determine the speed and direction of the airplane. The radar provides a computer with two distances and the angle between these distances. The computer computes the sides of a triangle opposite the known angle.

To derive the law of cosines consider any triangle ABC:

□ Figure 9.27

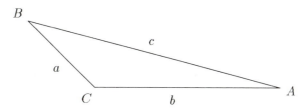

If we place a coordinate system at point C with the x-axis along side b, we have the figure below.

□ Figure 9.28

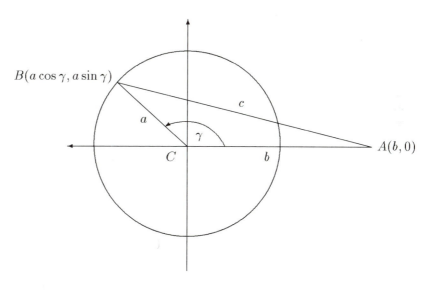

The distance formula tells us the distance c from B to A is

$$
\begin{aligned}
c^2 &= (a\cos\gamma - b)^2 + (a\sin\gamma - 0)^2 \\
&= a^2\cos^2\gamma - 2ab\cos\gamma + b^2 + a^2\sin^2\gamma \qquad \text{squaring}
\end{aligned}
$$

Rearranging the right side,

$$
\begin{aligned}
c^2 &= a^2\cos^2\gamma + a^2\sin^2\gamma + b^2 - 2ab\cos\gamma \\
&= a^2(\cos^2\gamma + \sin^2\gamma) + b^2 - 2ab\cos\gamma \qquad \text{factor } a^2 \\
&= a^2 + b^2 - 2ab\cos\gamma \qquad \text{identity (12)}
\end{aligned}
$$

9.4 Law of Cosines

For any triangle,
$$c^2 = a^2 + b^2 - 2ab\cos\gamma$$

In words, "the square of any side of a triangle is equal to the sum of the squares of the other two sides minus twice the product of the other two sides times the cosine of the angle between them."

Other forms of the law of cosines are:

$$a^2 = b^2 + c^2 - 2bc\cos\alpha$$
$$b^2 = a^2 + c^2 - 2ac\cos\beta$$

Notice that if $\gamma = 90°$ the law of cosines becomes the Pythagorean Theorem.

$$c^2 = a^2 + b^2 - 2ab\cos 90°$$
$$= a^2 + b^2 - 2ab(0)$$
$$c^2 = a^2 + b^2$$

Example 12 □ If $a = 10.0, b = 20.0$, and $\gamma = 30.0°$, solve for the remaining parts of the triangle.

□ Figure 9.29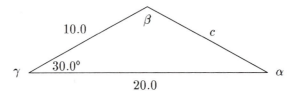

The law of cosines is
$$c^2 = a^2 + b^2 - 2ab\cos\gamma$$

Substituting

$$c^2 = 10.0^2 + 20.0^2 - 2(10.0)(20.0)\cos 30.0°$$
$$c^2 = 100 + 400 - 400\cos 30.0°$$
$$c^2 \approx 500 - 346.4$$
$$c^2 \approx 153.6$$
$$c \approx 12.4$$

Use the value of c stored in the calculator and the law of sines to find one of the other angles.

$$\frac{\sin\beta}{b} = \frac{\sin\gamma}{c}$$
$$\sin\beta = \frac{b\sin\gamma}{c}$$
$$\approx \frac{20\sin 30°}{12.4}$$
$$\approx 0.8065$$

Using a calculator,

$$\beta \approx 53.8°$$

Chapter 9 Law of Sines and Law of Cosines

Hold it!
β doesn't look
anything like
53.79°.

That's right. You can't use your
calculator without thinking. There are two
angles with a sine of 0.8069. $180° - 53.79°$ is the
other. Both angles are less than $180°$ and
therefore could be part of a triangle.

By looking at the sketch, we see the correct value for β is

$$\beta \approx 180° - 53.8°$$
$$\beta \approx 126.2°$$

Now it's possible to find the third angle.

$$\alpha \approx 180° - (126.2° + 30.0°)$$
$$\alpha \approx 23.8°$$

$\square$

Example 13 $\square$ Find the angles of a triangle with sides of 10, 15, 20.

$\square$ Figure 9.30

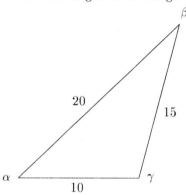

First find γ using the law of cosines.

$$c^2 = a^2 + b^2 - 2ab\cos\gamma$$
$$20^2 = 10^2 + 15^2 - 2(10)(15)\cos\gamma$$
$$400 = 100 + 225 - 300\cos\gamma$$
$$\frac{+75}{-300} = \cos\gamma$$
$$-0.2500 = \cos\gamma$$
$$105° \approx \gamma$$

To find another angle use either the law of sines or the law of cosines. We will use the law of cosines to illustrate its use.

$$15^2 = 10^2 + 20^2 - 2(10)(20)\cos\alpha$$
$$225 = 100 + 400 - 400\cos\alpha$$
$$\frac{-275}{-400} = \cos\alpha$$
$$0.6875 = \cos\alpha$$
$$47° \approx \alpha$$

The third angle is

$$\begin{aligned} \beta &\approx 180° - (105° + 47°) \\ \beta &\approx 28° \end{aligned}$$

□

Example 14 □ Find side b in a triangle with $a = 5, c = 9, \alpha = 27°$.

First make a sketch.

The ambiguous case is possible here.

□ Figure 9.31

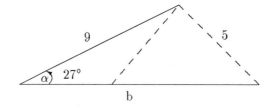

Use the law of cosines to find side b.

$$\begin{aligned} a^2 &= b^2 + c^2 - 2bc \cos \alpha \\ 5^2 &= b^2 + 9^2 - 2b(9) \cos 27° \\ 25 &\approx b^2 + 81 - 18b(0.89) \\ 25 &\approx b^2 + 81 - 16b \\ 0 &\approx b^2 - 16b + 56 \\ b &\approx \frac{16 \pm \sqrt{(-16)^2 - 4(1)(56)}}{2} \\ b &\approx \frac{16 \pm \sqrt{256 - 224}}{2} \\ b &\approx \frac{16 \pm \sqrt{32}}{2} \\ b &\approx \frac{16 \pm 5.66}{2} \\ b &\approx \frac{21.66}{2} \quad \text{or} \quad b \approx \frac{10.34}{2} \\ b &\approx 11 \quad \text{or} \quad b \approx 5 \end{aligned}$$

□

Is there any way to avoid the ambiguous case?

No. There are two possible triangles whether we like it or not. Both the law of sines and the law of cosines reveal the two possibilities.

If you know two sides and the angle between them the law of cosines is not ambiguous.

Chapter 9 Law of Sines and Law of Cosines

Pointers About Finding a Second Angle

We can use the law of cosines with two sides and the angle between them. In this case, after we have applied the law of cosines, we know the three sides and one angle. To solve the triangle we must still find a second angle. The law of sines requires the least arithmetic to find the value of an angle. But when you use the law of sines you must be aware that even though there is only one triangle possible there are two angles with the same value for the sine. You must select the proper angle.

There are three alternatives:

1. Avoid the difficulty by using the law of cosines to find the second angle. It will find the correct angle directly.

2. Select the proper angle by drawing a fairly accurate sketch. This is how we detected the problem in Example 12.

3. A third way to resolve the difficulty is to notice that in a triangle the largest side is opposite the largest angle.

Use the triangle from Example 12.

After we first used the law of cosines, we knew three sides and one angle.

□ Figure 9.32

From the law of sines we got $\beta \approx 53.8°$. Had we gone from this point we could have calculated:

$$\alpha = 180° - (\beta + \gamma)$$
$$\approx 180° - (53.8° + 30°)$$
$$\approx 96.2°$$

Is $\alpha = 96.2°$ possible?

We have a contradiction.

Notice that the side opposite α is 10, which is the shortest side of the triangle. But α is supposedly the largest angle of the triangle. Therefore side a should be the largest side.

The contradiction arises because we accepted the incorrect value of β. We can detect the contradiction because the largest angle is not opposite the largest side. If you choose to find a second angle by using the law of sines you will have to verify your answer.

If you use the law of cosines to find the second angle, the correct value is automatic. Next we illustrate how the law of cosines finds the angle directly by using the information available after the first application of the law of cosines in Example 12. At this point we know that $\gamma = 30.0°, a = 10.0, b = 20.0, c \approx 12.4$.

Write the law of cosines for angle β.

$$b^2 = a^2 + c^2 - 2ac \cos \beta$$

Solving this equation for $\cos \beta$ yields

$$\cos \beta \;\; = \;\; \frac{b^2 - (a^2 + c^2)}{-2ac}$$

$$= \;\; \frac{a^2 + c^2 - b^2}{2ac}$$

Substituting the values of a, b, c yields

$$\cos \beta \;\; \approx \;\; \frac{10.0^2 + 12.4^2 - 20.0^2}{2(10.0)(12.4)}$$

$$\cos \beta \;\; \approx \;\; \frac{253.76 - 400}{248}$$

$$\cos \beta \;\; \approx \;\; -0.5897$$

$$\beta \;\; \approx \;\; 126.1°$$

$a^2 + c^2 - b^2$ is a negative number, therefore $\cos \beta$ is negative.

When $\cos \beta$ is negative, β must be $90° < \beta < 180°$.

And the calculator automatically returns the correct value.

Chapter 9 Law of Sines and Law of Cosines

Using Your Scientific Calculator

One useful key on your calculator is the MEMORY key. Most scientific calculators have at least one memory. If your calculator has more than one memory, we strongly suggest that you spend time with your manual to learn how to use them.

Your calculator probably has two keys. One of these keys stores the number on the display in memory; it is probably labeled $\boxed{\text{STO}}$ or $\boxed{\text{X} \rightarrow \text{M}}$. To retrieve a value from memory there is a key labeled $\boxed{\text{RCL}}$ or $\boxed{\text{RM}}$ for recall memory.

  Try: 2 + 3 $\boxed{\text{STO}}$ or $\boxed{\text{X} \rightarrow \text{M}}$. A small m should appear on the display.
  Now press: $\boxed{+}$. The value 5 should appear.
  Now press: $\boxed{\text{RCL}}$ or $\boxed{\text{RM}}$. 3 should appear from memory.

Notice the 3 was stored, not the result because the $\boxed{\text{STO}}$ or $\boxed{\text{X} \rightarrow \text{M}}$ key moves the contents of the display to the memory immediately. It does not perform any suspended operations.

  If you want to store an intermediate value in memory, press the $\boxed{=}$ key to get the intermediate value on the display, then press $\boxed{\text{STO}}$.

  The memory is like a tape recording; using it does not destroy the stored value. Each time you press $\boxed{\text{RM}}$ or $\boxed{\text{RCL}}$, the current value in memory will return to the display. It usually isn't necessary to clear a value from memory because the next value you enter into memory replaces the previous value. Your calculator probably also has a $\boxed{\text{M}+}$ or $\boxed{\text{SUM}}$ key.

  This key adds the current display value to the value in memory. If it looks as if nothing happened it's because the value in memory, not the value on the display, was changed. Depending on the brand of your calculator try:
5 $\boxed{\text{STO}}$ 6 $\boxed{\text{SUM}}$ or 5 $\boxed{\text{X} \rightarrow \text{M}}$ 6 $\boxed{\text{M}+}$
Either sequence should produce an 11 in memory. To see it on the display, press $\boxed{\text{RCL}}$ or $\boxed{\text{RM}}$.

Using Your Graphing Calculator

Your graphing calculator has the capability of storing the result of an operation in a specified location in memory. To accomplish this

1. Perform the operation.

2. Press $\boxed{\text{STO}}$.

3. Select the letter you wish to store the number in.

4. Press $\boxed{\text{ENTER}}$.

This value may now be used at any time by pressing $\boxed{\text{ALPHA}}$ and then the appropriate letter.

  Ex: Store 10 + 7 in A.
    1. Press 10 + 7.

    2. Press $\boxed{\text{STO}}$.

    3. Press A.

    4. Press $\boxed{\text{ENTER}}$.

  To use that stored value try this example:
Press 5 + $\boxed{\text{ALPHA}}$ A
Press $\boxed{\text{ENTER}}$
The answer that appears on your screen is 22.

Assume the information given below is exact. Find the length of the unknown side to the same number of places as the given side.

1. $a = 56, b = 40, \gamma = 24°$ 2. $b = 44, c = 62, \alpha = 28°$

3. $a = 20.4, c = 10.8, \beta = 42.1°$ 4. $a = 22.4, b = 36.5, \gamma = 106°$

5. $a = 0.416, c = 0.684, \beta = 135.4°$ 6. $b = 6.6, c = 4.5, \alpha = 72.1°$

Solve the following triangles completely to the accuracy of the given data.

7. $a = 75.9, b = 91.3, \gamma = 48.9°$ 8. $b = 1.68, c = 3.04, \alpha = 63.9°$

9. $a = 0.0138, b = 0.0364, \gamma = 98.47°$ 10. $a = 27.9, b = 36.8, \gamma = 118.9°$

11. $a = 0.864, c = 0.582, \beta = 110.4°$ 12. $b = 4.36, c = 15.2, \alpha = 15.9°$

Find the specified angle.

13. Find angle α if $a = 5.9, b = 8.6, c = 10.4$

14. Find angle β if $a = 26, b = 34, c = 18$

15. Find angle γ if $a = 0.0179, b = 0.0423, c = 0.0362$

Find all three angles in the following triangles.

16. $a = 36, b = 27, c = 32$ 17. $a = 12.1, b = 15.2, c = 18.3$

18. $a = 5.6, b = 9.4, c = 12.8$ 19. $a = 0.932, b = 0.645, c = 0.846$

20. $a = 1286, b = 987, c = 1092$ 21. $a = 5.36, b = 4.44, c = 7.83$

22. $a = 36, b = 46, c = 56$ 23. $a = 0.0934, b = 0.0531, c = 0.0796$

9.5 Applications of the Law of Cosines

There are many applications of the law of cosines. Sometimes the law of cosines is used alone to determine a value. Frequently the law of cosines is used to find one side or angle; then the law of sines is used to find the remaining parts of the triangle.

Example 15 □ When measured from the earth, the moon subtends an angle of 0.518°. The mean distance from earth to the moon is 238,857 miles. Find the diameter of the moon.

□ Figure 9.33

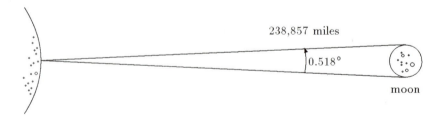
238,857 miles
0.518°
moon

We know two sides and the included angle. Therefore we can apply the law of cosines.

$$c^2 = a^2 + b^2 - 2ab \cos \gamma$$

$$
\begin{aligned}
c^2 &= 238857^2 + 238857^2 - 2(238857)(238857)\cos(0.518°) \\
c^2 &\approx 5.7052666 \times 10^{10} + 5.7052666 \times 10^{10} - 1.1410533 \times 10^{11}(0.9999591) \\
c^2 &\approx 4663230 \\
c &\approx 2159 \text{ miles is the diameter of the moon.}
\end{aligned}
$$

$\square$

Example 16 $\square$ An engineer is building a railway tunnel through a mountain. The engineer knows the distances from the peak of the mountain to the entrance and to the exit of the tunnel. The angles of depression from the peak are also measured as shown in the diagram. Determine the length of the tunnel and its angle of inclination.

$\square$ Figure 9.34

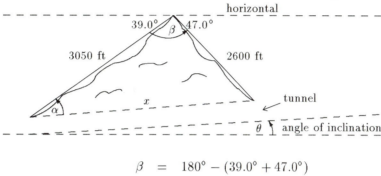

$$
\begin{aligned}
\beta &= 180° - (39.0° + 47.0°) \\
&= 94.0°
\end{aligned}
$$

By the law of cosines,

$$
\begin{aligned}
x^2 &= 3050^2 + 2600^2 - 2(3050)(2600)\cos 94.0° \\
&\approx 16062500 - 15860000(-0.0697) \\
&\approx 16062500 - (-1106338) \\
x^2 &\approx 17168838 \\
x &\approx 4140 \text{ feet}
\end{aligned}
$$

Now that a side and the angle opposite it are known, angle α can be found by using the law of sines.

$$
\begin{aligned}
\frac{2600}{\sin \alpha} &\approx \frac{4140}{\sin 94.0°} \\
\frac{2600(\sin 94.0°)}{4140} &= \sin \alpha \\
0.6265 &\approx \sin \alpha \\
38.8° &\approx \alpha
\end{aligned}
$$

That hill is too steep for a train to climb.

True; 38.8° is the angle between the base of the tunnel and the line of sight for the peak, but the angle of inclination of the tunnel is $39° - 38.8°$.

To find the angle of inclination:

Angle between the horizontal and line of sight to the peak $\approx 39.0°$

Angle between the railroad tracks and the line of sight of the peak $\approx 38.8°$

Difference = Angle of inclination of track $\approx 0.2°$

$\square$

9.5 Applications of the Law of Cosines

Navigation

A ship or an airplane navigating on the surface of the earth gets its basic information about direction of travel from a magnetic compass. The course of an airplane or ship is given as an angle measured from the north-south line.

Historically, ships have measured their courses as an acute angle from the north-south line toward east or west.

N30°E means an angle of 30° from the north-south line toward east.

Here are some other examples of courses measured in this system.

□ Figure 9.35

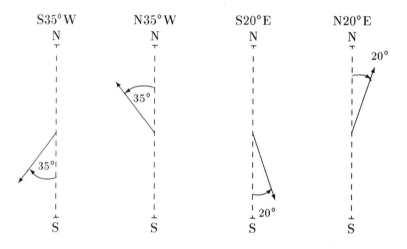

In aviation, directions are measured clockwise from north.

□ Figure 9.36

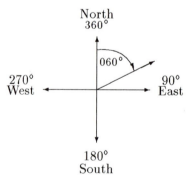

To avoid confusion on the radio, headings are given in three digits; zero-six-zero means at an angle of 60° clockwise from north.

In this system due east is 90°, south is 180°, and west is 270°. North is usually referred to as 360° rather than 0°.

Chapter 9 Law of Sines and Law of Cosines

Here are some other examples of headings as used in aviation.

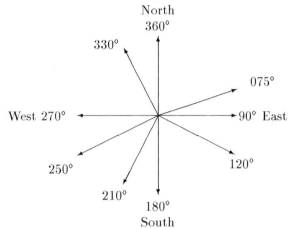

□ Figure 9.37

The principal system of air navigation in use today is based on a series of very high frequency omnidirectional ranges. These radio stations, called VORs, broadcast a series of rays radiating outward in all directions. A ray emanating from one of these navigation aids outward is called a radial. The 270° radial is a ray from the navigation aid toward the west.

□ Figure 9.38

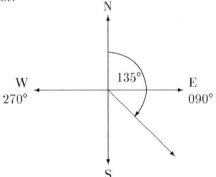

The 135° radial is a ray from the navigation aid toward the southeast.

To use trigonometry to find distances and directions, it is usually necessary to convert angles in navigational measurement to angles in standard trigonometric position. Remember, an angle in standard position in trigonometry is measured counterclockwise from the x-axis.

A sketch will help.

□ Figure 9.39

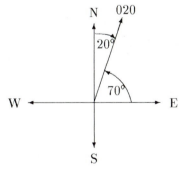

A course of 020° or N20°E must be expressed as 70° to use the angle as the argument of a trigonometric function.

The computer associated with air traffic control radar tracks aircraft in the sky by solving a problem similar to the next example, except that the time between the two readings is much smaller.

Example 17 □ A radar beam detects an aircraft at a range of 20 miles on a bearing of 040°. Thirty seconds later the radar detects the same aircraft crossing the 042° bearing from the radar site at a distance of 21 miles. What is the speed and direction of the aircraft?

To draw a sketch we will distort the actual size of the angle between the two readings.

□ Figure 9.40

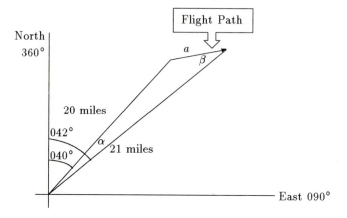

By the law of cosines:

$$
\begin{aligned}
a^2 &= b^2 + c^2 - 2bc \cos \alpha \\
&= 20^2 + 21^2 - 2(20)(21) \cos 2° \\
a &\approx 1.23 \text{ miles}
\end{aligned}
$$

The aircraft traveled 1.23 miles in 30 seconds; therefore, its speed was 2.46 miles per minute or about 148 miles per hour. Using 1.23 miles as an approximation for side a, we can compute the course of the aircraft using either the law of sines or a second application of the law of cosines.

Using the law of cosines a second time we will compute angle β.

$$
\begin{aligned}
b^2 &\approx a^2 + c^2 - 2ac \cos \beta \\
20^2 &\approx (1.23)^2 + 21^2 - 2(1.23)(21) \cos \beta \\
400 &\approx 1.51 + 441 - 51.66 \cos \beta \\
\frac{-42.51}{-51.66} &\approx \cos \beta \\
0.8229 &\approx \cos \beta \\
35° &\approx \beta
\end{aligned}
$$

The course of the airplane is the direction of side a in the diagram above. One way to get this direction is to imagine you are a pilot flying along the 42° bearing from the station. To go in the same direction as the airplane you would have to turn to the right through an angle β.

□ Figure 9.41

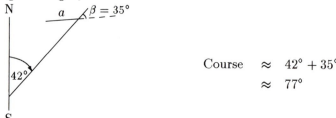

Course $\approx 42° + 35°$

$\approx 77°$

The aircraft is traveling on a course of 77° at a speed of 148 miles per hour. □

Chapter 9 Law of Sines and Law of Cosines

Example 18 □ Using distance measuring equipment (DME), a pilot determines that when it passes the 90° radial from a navigation aid, the aircraft is 80 nautical miles away. Ten minutes later, as it crosses the 125° radial the plane is 70 nautical miles away. What is the ground speed and course of the airplane?

□ Figure 9.42

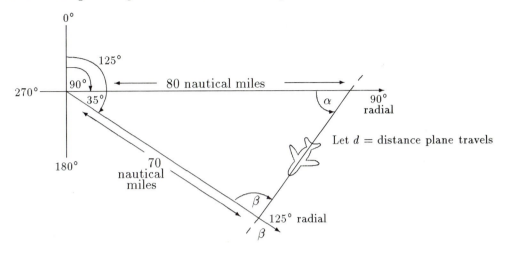

Since we know two sides and the included angle, we can use the law of cosines.

$$
\begin{aligned}
d^2 &= 70^2 + 80^2 - 2(70)(80)\cos(35°) \\
d^2 &\approx 2125 \\
d &\approx 46 \text{ nautical miles}
\end{aligned}
$$

If the airplane went 46 nautical miles in 10 minutes, its ground speed is

$$
\begin{aligned}
\frac{s \text{ nautical miles}}{60 \text{ minutes}} &\approx \frac{46 \text{ nautical miles}}{10 \text{ minutes}} \\
\frac{s}{60} &\approx \frac{46}{10} \\
s &\approx \frac{46(60)}{10} \\
&\approx 276 \text{ knots}
\end{aligned}
$$

To find the course of the airplane we must find one more interior angle of the triangle. We will use the law of sines to find angle β.

$$
\begin{aligned}
\frac{\sin \beta}{80} &\approx \frac{\sin 35°}{46} \\
\sin \beta &\approx \frac{80\,(\sin 35°)}{46} \\
\sin \beta &\approx 0.9975 \\
\beta &\approx 86°
\end{aligned}
$$

β is the angle between the course of the airplane and the 125° radial. To express the course as an angle measured from north we must add β to 125°.

9.5 Applications of the Law of Cosines   285

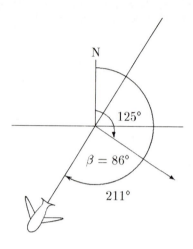

□ Figure 9.43

The course of the aircraft is
$125° + 86° = 211°$

□

Problem Set 9.5A

Solve the following. Express your answers to the degree of accuracy you can reasonably expect.

1. A new water pipeline was to replace one that had been installed years earlier. Since the terrain was rocky, the old line ran due west for 650 feet then N60°W for another 750 feet. It was decided to cut through the rocks with modern equipment and replace the old line with a straight one. How much pipe (to the nearest 10 feet) is needed to lay a straight pipeline?

2. A light airplane is flying at 180 miles per hour with a heading of 065°. One hour later a second airplane flying at 240 miles per hour leaves the same airport with a heading of 350°. To the nearest mile, how far apart are the airplanes one and one-half hours after the second airplane leaves the airport?

3. A pilot flew his airplane at a constant speed of 220 miles per hour with a heading of 120°. After one hour of flying he changed the direction of his course to 072°. He continued in this direction for one and one-half hours to reach his destination. How much farther was this flight than a straight-line flight to his destination?

4. A fishing boat adrift at sea indicated its position as 18 miles S36°E from a coast guard station. A coast guard patrol boat indicated its position as 15 miles S8°W of the coast guard station. To the nearest mile, how far was the patrol boat from the fishing boat?

5. Town A is 56 miles directly west of town B. To go from town A to town B, a person must travel 36 miles in a northeasterly direction then 48 miles in a southeasterly direction. To the nearest degree, what are the directions of the two highways leading from town A to town B?

Chapter 9 Law of Sines and Law of Cosines

6. Three circles of radius 3 cm, 5 cm, and 9 cm are all tangent to one another. Find the angles, to the nearest tenth of a degree, formed by the lines joining their centers.

7. A triangular plot of land is bounded on each side by busy streets. If the lengths of sides of the plot of land are 70.00 meters, 80.00 meters, and 100.00 meters, find the angles, to the nearest hundredth of a degree, that the streets make with each other.

8. Two airplanes leave the same airport at the same time. The planes are traveling at speeds of 140 miles per hour and 180 miles per hour respectively. After two and one-half hours the airplanes are 650 miles apart. To the nearest degree, what is the angle between their courses of flight?

Problem Set 9.5B

In actual practice, problems do not come in neat sets labeled "law of sines" or "law of cosines applications." Analyze the following problems and after choosing the appropriate law find their solutions. Solve the following triangles. Your answer can be no more accurate than the least accurate data given in the problem.

1. $\beta = 62.38°, \gamma = 49.24°, b = 29.46$

2. $\alpha = 126.9°, \beta = 39.8°, b = 54.6$

3. $a = 389.2, c = 475.3, \gamma = 18.42°$

4. $b = 8.13, a = 15.6, \beta = 39.2°$

5. $\alpha = 99.99°, \gamma = 28.76°, c = 386.5$

6. $\beta = 149.6°, \gamma = 18.4°, a = 2.36$

7. $c = 1.400, a = 2.150, \gamma = 43.92°$

8. $a = 793, c = 638, \gamma = 32.8°$

9. $a = 12.03, c = 13.46, \alpha = 52.89°$

10. $\alpha = 67.23°, \beta = 86.56°, c = 3.981$

11. $\alpha = 16.93°, \gamma = 102.56°, b = 183.9$

12. $b = 10.8, c = 8.94, \beta = 106.3°$

13. A light airplane on a navigational training flight flew at a constant speed of 150 miles per hour over a VOR on a course of 130° for 20 minutes. The pilot then flew a course of 010° until the airplane crossed the 090° radial from the VOR. At that time, how far was the aircraft from the VOR? (Round your answer to the nearest mile.)

14. A balloonist flying over an island observed that the angle of depressions of the north and south ends of the island were 32.00° and 47.00° respectively. Using special radar equipment he determined that his distance from the south end of the island was 2825 feet and from the north end was 3880 feet. What is the distance from the north end to the south end of the island?

15. Using distance measuring equipment (DME), a pilot determines that, when passing the 060° radial from a navigation aid, the aircraft is 90 nautical miles away. Continuing on a straight course another 50 nautical miles she determines that the aircraft is 110 miles away from the navigation aid. What is the radial reading at the moment? (Note: there are two possible positions for the airplane.)

16. To avoid some thunder clouds, the pilot of a small aircraft changed his course by 15°. He flew in this direction a total of 135 miles; at this time he was able to turn toward his destination but he was still 100 miles from it. If the average speed of the airplane was 125 miles per hour, how much more time did it take to reach his destination than if he could have flown directly to it instead of having to fly around the thunder storm clouds? (Give your answer to the nearest tenth of a minute.)

17. Two ships leave a harbor at the same time. One ship is traveling N36°E at 12 miles per hour; the other ship travels at 15 miles per hour. After three and one-half hours, they are 46 miles apart. In which direction is the second ship traveling? (Note: There are two possible answers, give both answers.)

18. A 75-foot powerline pole stands vertically on a hillside. The hill has an inclination of 18° with the horizontal. How long is the guy wire that is attached to the top of pole and anchored directly up the hill 44 feet from the base of the powerline pole?

19. Winds aloft are usually much stronger than winds close to the ground. Therefore, in tracking a weather balloon you cannot assume a constant speed from the time it left the ground. It is reasonable, however, to assume a constant speed and direction for 1 minute of flight. At 10 minutes after launch the weather station's radar indicates that the balloon is on a bearing of 240° from the station at an angle of elevation of 30° and a distance of 5.0 miles from the station. At 11 minutes after launch it was 5.6 miles away on the same bearing at a 25° angle of elevation. What is the speed of the balloon? (Assume that the balloon has reached maximum height and any change in the balloon's position is due to the wind.)

9.6   Area of a Triangle

The area of any triangle is given by: $A = \dfrac{1}{2}bh$. In the case of a right triangle, the value of h is one of the legs of the triangle. There are three cases where we can develop a formula for the area of an oblique triangle.

Case I — Two Sides and an Included Angle Are Known

□ Figure 9.44

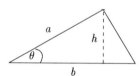 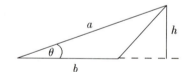

In both figures above, $h = a \sin \theta$.

$$\text{Because} \qquad A = \frac{1}{2}bh$$

$$A = \frac{1}{2}ba \, \sin \theta$$

In general, the area of any triangle is one-half the product of any two sides times the sine of the included angle.

Example 19 □ Find the area of a triangle with $b = 20$ ft, $c = 15$ ft, and $\alpha = 30°$.

We know two sides and the angle between them; the area of the triangle is one-half the product of the sides and the sine of the included angle.

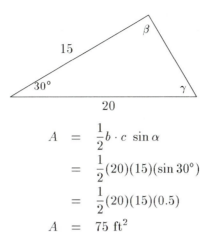

$$A = \frac{1}{2}b \cdot c \, \sin\alpha$$

$$= \frac{1}{2}(20)(15)(\sin 30°)$$

$$= \frac{1}{2}(20)(15)(0.5)$$

$$A = 75 \text{ ft}^2$$

□

Case II — Three Angles and One Side Are Known

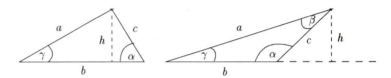

In this case we use the law of sines to find the value of a second side and substitute that value in the formula derived in Case I. Assuming that we know side b of the triangle pictured above we can find side a by using the law of sines.

$$\frac{a}{\sin\alpha} = \frac{b}{\sin\beta}$$

$$a = \frac{b\sin\alpha}{\sin\beta}$$

Substituting this value of a in the formula for the area of a triangle developed in Case I,

$$A = \frac{1}{2}ba\sin\gamma$$

becomes

$$A = \frac{1}{2}b\left(\frac{b\sin\alpha}{\sin\beta}\right)\sin\gamma$$

which simplifies to

$$A = \frac{1}{2}b^2\frac{\sin\alpha\sin\gamma}{\sin\beta}$$

Example 20 □ Find the area of a triangle with angles of 50°, 60°, and 70° if the longest side of the triangle is 25 feet.

In any triangle, the angle opposite the longest side is always the largest angle. In this example, if $b = 25$ ft and angle $\beta = 70°$, the area of the triangle is

$$A = \frac{1}{2}b^2 \frac{\sin\alpha \sin\gamma}{\sin\beta}$$

$$= \frac{1}{2}(25^2)\frac{\sin 50° \sin 60°}{\sin 70°}$$

$$\approx 221 \text{ ft}^2 \text{ to the nearest ft}^2$$

□

Case III — Three Sides Are Known

It is also possible to find the area of a triangle if we know only the lengths of all three sides.

□ Figure 9.47

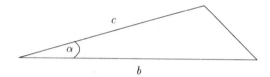

Start with the formula for the area of a triangle when two sides and the included angle are known.

$$A = \frac{1}{2}bc\sin\alpha$$

$$A^2 = \frac{1}{4}b^2c^2\sin^2\alpha \qquad \text{squaring both sides}$$

$$= \frac{1}{4}b^2c^2(1 - \cos^2\alpha) \qquad \text{substituting}$$

$$= \frac{1}{2}bc(1 + \cos\alpha)\frac{1}{2}bc(1 - \cos\alpha) \qquad \text{factoring}$$

Next use the law of cosines to express $\cos\alpha$.

Because $\quad a^2 = b^2 + c^2 - 2bc\cos\alpha$

$$\cos\alpha = \frac{a^2 - b^2 - c^2}{-2bc} \qquad \text{solving for } \cos\alpha$$

$$\cos\alpha = \frac{b^2 + c^2 - a^2}{2bc} \qquad \text{multiplying by } \left(\frac{-1}{-1}\right)$$

Continuing by substituting for $\cos\alpha$,

$$A^2 = \frac{1}{2}\left[bc\left(1 + \frac{b^2 + c^2 - a^2}{2bc}\right)\right] \cdot \frac{1}{2} \cdot \left[bc\left(1 - \frac{b^2 + c^2 - a^2}{2bc}\right)\right]$$

$$= \frac{1}{2}\left[bc\left(\frac{2}{2}\right) + \frac{bc(b^2 + c^2 - a^2)}{2bc}\right] \cdot \frac{1}{2} \cdot \left[bc\left(\frac{2}{2}\right) - \frac{bc(b^2 + c^2 - a^2)}{2bc}\right]$$

$$= \frac{1}{2}\left[\frac{2bc + b^2 + c^2 - a^2}{2}\right] \cdot \frac{1}{2} \cdot \left[\frac{2bc - b^2 - c^2 + a^2}{2}\right]$$

$$= \frac{1}{2}\left[\frac{b^2 + 2bc + c^2 - a^2}{2}\right] \cdot \frac{1}{2} \cdot \left[\frac{a^2 - (b^2 - 2bc + c^2)}{2}\right]$$

$$= \left[\frac{(b + c)^2 - a^2}{4}\right] \cdot \left[\frac{a^2 - (b - c)^2}{4}\right] \qquad \begin{array}{l}\text{factoring and writing as}\\\text{a difference of 2 squares}\end{array}$$

$$= \frac{[(b + c) + a][(b + c) - a][a - (b - c)][a + (b - c)]}{4 \cdot 4} \qquad \begin{array}{l}\text{factoring the differences of}\\\text{squares}\end{array}$$

$$= \frac{[b+c+a]}{2} \cdot \frac{[b+c-a]}{2} \cdot \frac{[a+c-b]}{2} \cdot \frac{[a+b-c]}{2}$$

To make this expression easier to remember, we define the semi-perimeter as

$$s = \frac{1}{2}(a+b+c)$$

Then notice that

$$\begin{aligned}
s - a &= \frac{a+b+c}{2} - a \\
&= \frac{a+b+c}{2} - \frac{2a}{2} \\
&= \frac{b+c-a}{2}
\end{aligned}$$

which is the second factor above.

Similarly $\quad s - b = \dfrac{a+c-b}{2}$

and

$$s - c = \frac{a+b-c}{2}$$

When you make these substitutions, the expression for A^2 becomes

$$A^2 = s(s-a)(s-b)(s-c)$$

Taking the square root of both sides, and writing in an order easier to remember, yields:

$$A = \sqrt{s(s-a)(s-b)(s-c)}$$

The formula above is referred to as Heron's formula.

In his book, *An Introduction to the History of Mathematics*, Howard Eves attributes a geometrical development of this formula to Heron of Alexandria, who he believes lived in the second half of the first century A.D.

Example 21 □ Find the area of a triangle with sides 3 feet, 4 feet, and 5 feet.

This is a right triangle with legs of 3 feet and 4 feet; therefore its area is

$$\begin{aligned}
A &= \frac{1}{2}bh \\
&= \frac{1}{2}(3)(4) \\
&= 6 \text{ square feet}
\end{aligned}$$

Applying Heron's formula should yield the same result.

$$\begin{aligned}
s &= \frac{1}{2}(a+b+c) \\
&= \frac{1}{2}(3+4+5) \\
&= 6
\end{aligned}$$

Then:

$$\begin{aligned}
A &= \sqrt{s(s-a)(s-b)(s-c)} \\
&= \sqrt{6(6-3)(6-4)(6-5)} \\
&= \sqrt{6(3)(2)(1)} \\
A &= 6
\end{aligned}$$

□

Example 22 □ Find the area of a triangular yacht sail that has sides 10 feet, 13 feet, and 15 feet.

Using Heron's formula:

$$s = \frac{1}{2}(a + b + c)$$
$$= \frac{1}{2}(10 + 13 + 15)$$
$$= 19 \text{ feet}$$

$$A = \sqrt{s(s-a)(s-b)(s-c)}$$
$$= \sqrt{19(19 - 10)(19 - 13)(19 - 15)}$$
$$= \sqrt{19 \cdot 9 \cdot 6 \cdot 4}$$
$$= \sqrt{4104}$$
$$= 64 \text{ square feet to the nearest square foot}$$

□

Problem Set 9.6

Find the area of the following triangles. Your answer can be no more accurate than the least accurate data given in the problem.

1. $\alpha = 40°$, $b = 18$, $c = 21$
2. $\beta = 75.2°$, $a = 12.3$, $c = 16.9$
3. $\gamma = 105.3°$, $a = 126$, $b = 91.5$
4. $\alpha = 25.6°$, $\beta = 42.9°$, $c = 13.4$
5. $\beta = 18.9°$, $\gamma = 58.7°$, $b = 9.34$
6. $\alpha = 69.41°$, $\gamma = 75.72°$, $a = 23.49$
7. $a = 15$, $b = 12$, $c = 19$
8. $a = 89$, $b = 93$, $c = 65$
9. $a = 1.463$, $b = 2.831$, $c = 3.102$
10. $a = 0.786$, $b = 0.956$, $c = 0.549$

Chapter 9   Key Ideas

9.1 Law of sines

1. For any triangle $\dfrac{a}{\sin \alpha} = \dfrac{b}{\sin \beta} = \dfrac{c}{\sin \gamma}$

2. These equations are called the law of sines.

3. The law of sines is used to solve any triangle when two angles and a side are given.

9.2
1. If two sides and an angle opposite one of them is given, then law of sines is used.

2. The condition where two sides and an angle opposite one of them is known may lead to the ambiguous case. For example, given angle α, sides a and b:

If $\alpha < 90°$ and $a > b$, then there is one triangle.

If $\alpha < 90°$ and $a = b$, then there is one triangle.

If $\alpha < 90°$ and $a < b$, then there are three possibilities:

(1) no triangle, (2) one triangle, (3) two triangles.

If $\alpha > 90°$ and $a \leq b$, then there is no triangle.

If $\alpha > 90°$ and $a > b$, then there is one triangle.

9.3 The law of sines allows us to extend applications to situations without a right triangle.

9.4 **1.** The law of cosines is used in any triangle when two sides and the included angle are given or when three sides are given.

2. The three forms of the law of cosines are:

$$
\begin{aligned}
a^2 &= b^2 + c^2 - 2bc \cos \alpha \\
b^2 &= a^2 + c^2 - 2ac \cos \beta \\
c^2 &= a^2 + b^2 - 2ab \cos \gamma
\end{aligned}
$$

3. Given three sides of the triangle, the angles can be found by using the law of cosines in the following forms:

$$
\begin{aligned}
\cos \alpha &= \frac{b^2 + c^2 - a^2}{2bc} \\
\cos \beta &= \frac{a^2 + c^2 - b^2}{2ac} \\
\cos \gamma &= \frac{a^2 + b^2 - c^2}{2ab}
\end{aligned}
$$

9.5 The law of cosines allows us to extend applications to situations without a right triangle.

9.6 **1.** The area of any triangle is found by using one of the following formulas.

a. Given two sides and the included angle:

$$
\begin{aligned}
A &= \frac{1}{2}ab \sin \gamma \\
A &= \frac{1}{2}ac \sin \beta \\
A &= \frac{1}{2}bc \sin \alpha
\end{aligned}
$$

b. Given one side and any two angles:

$$
\begin{aligned}
A &= a^2 \frac{\sin \beta \sin \gamma}{2 \sin \alpha} \\
A &= b^2 \frac{\sin \alpha \sin \gamma}{2 \sin \beta} \\
A &= c^2 \frac{\sin \alpha \sin \beta}{2 \sin \gamma}
\end{aligned}
$$

c. Given three sides:

$$
A = \sqrt{s(s-a)(s-b)(s-c)}
$$

$$
\text{Where: } s = \frac{1}{2}(a + b + c)
$$

Chapter 9 Review Test

The following data refers to a triangle ABC. Find the missing parts. **(9.1)**

1. $\beta = 25.7°$, $\gamma = 80.3°$, $b = 22.3$ **2.** $\gamma = 68.4°$, $\beta = 54.7°$, $b = 293$

3. $\beta = 138.4°$, $\gamma = 16.1°$, $a = 29.3$

Find all solutions to each of the following triangles. **(9.2)**

4. $a = 26.4$, $c = 31.4$, $\alpha = 72.2°$ **5.** $a = 65.3$, $b = 203$, $\alpha = 8.4°$

Find the length of the side not given. **(9.4)**

6. $a = 8.60$, $b = 7.20$, $\gamma = 24.8°$ **7.** $b = 29.8$, $c = 41.9$, $\alpha = 112.9°$

8. $a = 0.859$, $c = 0.436$, $\beta = 78.3°$

Solve the following triangles completely. **(9.1)**, **(9.4)**

9. $a = 7.27$, $b = 9.06$, , $\gamma = 47.2°$ **10.** $b = 309$, $c = 458$. $\alpha = 132.7°$

Find all three angles in the following triangle. **(9.1)**, **(9.4)**

11. $a = 1.86$, $b = 7.09$, $c = 5.93$

Solve the following problem. **(9.5)**

12. Two airplanes leave an airport at the same time. The first flies 180 miles per hour in the direction 200°. The second plane flies 220 miles per hour along a course to the left of the first airplane. After two and one-half hours the airplanes are 175 miles apart. What is the direction of the second airplane?

Solve the following problems. **(9.3)**

13. An electrical cable was stretched from a pole on level ground to a pole on a hillside. The angle of elevation measured from the base of the pole on level ground to the base of the pole on the hillside was found to be 31.4°. The distance from the pole on level ground to the base of the hill was 78.9 feet and the distance from the base of the hill to the pole on the hillside was 115 feet. How long was the span covered by the cable?

14. Two observers stationed 10 km apart at stations A and B observe a fire at point C. They measure angles CAB and CBA to be 43° and 52° respectively. How far is each observer from the fire?

15. To determine the width of a small body of water, a surveyor stood at a point 1600 feet from one end of a pond and 4800 feet from the other end. He read the bearing of the first end to be N4°E and the bearing of the second end to be N46°E. Find the length of the pond.

Find the area of the following triangles. **(9.6)**

16. $a = 9.23$, $\beta = 75.6°$, $\gamma = 45.0°$ **17.** $a = 5.384$, $b = 7.209$, $c = 4.018$

Chapter 10

Vectors

Contents

Preview

This chapter will introduce the concept of a vector. Vectors are essential to scientists and engineers when describing forces of motion. They are a basic tool in engineering, architecture, navigation, optics, electronics, and acoustics.

After this chapter has defined vectors, it will show you how to add vectors using the trigonometry you have already learned. In the second part of the chapter you will learn how to use vectors in applied problems.

10.1   Addition of Vectors

Vectors represent quantities that have both magnitude and direction; for example, force and distance moved. Scalars are quantities with magnitude only; for example, volume and temperature. Since motion is a vector quantity, it makes sense to say twenty miles north. Volume, on the other hand, is a scalar quantity. A scalar, such as volume, has amount only; it does not have direction. You can refer to 6 quarts, but 6 quarts north is meaningless.

The difference between vectors and scalars is apparent in addition. Two quarts and three quarts is five quarts no matter how you combine them. But consider force, which is a vector. You and a friend each have a rope attached to a boat, as pictured below.

□ Figure 10.1

150 pounds

100 pounds

If you pull with a force of 100 pounds and your friend pulls with a force of 150 pounds, the combination of your efforts will be 250 pounds *only* if you both pull in the same direction. If you pull forward and your friend pulls in the opposite direction, the result of your combined efforts will be 50 pounds in the direction of your friend.

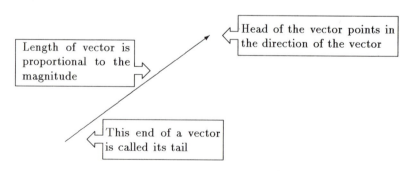

What if I pull forward and my friend pulls sideways?

As you can guess, the boat will move somewhere between the two of you. Vectors describe the motion.

Graphical Vectors

Because a vector is a quantity with amount and direction, we can use a line segment with an arrowhead on a graph to represent a vector. The direction of the segment is the direction of the vector and the length of the segment is a scalar representation of the amount of the vector.

□ Figure 10.2

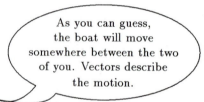

Length of vector is proportional to the magnitude

Head of the vector points in the direction of the vector

This end of a vector is called its tail

Here is how vectors could represent some of the possible ways you and your friend could pull on a boat.

□ Figure 10.3

□ Figure 10.4

□ Figure 10.5

□ Figure 10.6

□ Figure 10.7

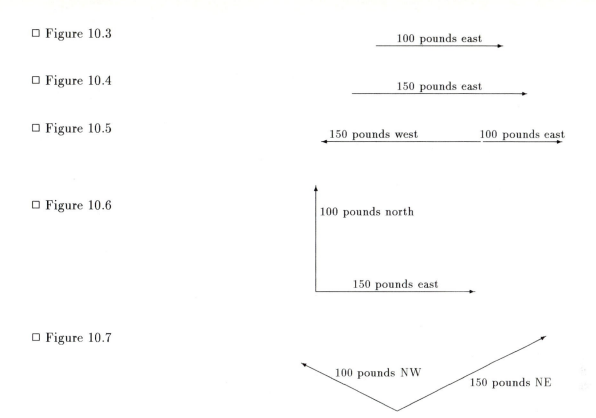

Obviously we need a better method of indicating direction than north, east, south, and west. Usually we use degrees from the positive x-axis according to the conventions of trigonometry.

Resultant of Two Vectors

The sum of two vectors acting on an object is called the resultant. One way to find the resultant is to think of each vector acting separately in sequence. The resultant will be the final effect.

Example 1 □ If you walk 3 blocks east and 4 blocks north, what will be the magnitude of the vector sum?

□ Figure 10.8

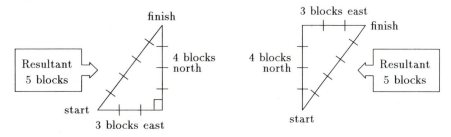

In both cases the magnitude of the resultant is five blocks.

The order in which you add two vectors doesn't matter; the result is the same. You wind up in the same spot if you first walk east, then walk north or if you walk north and then east. □

10.1 Addition of Vectors

The sum or resultant of two geometrical vectors is found by placing the vectors head to tail. The sum is the vector from the tail of the starting vector to the head of the ending vector.

Example 2 □ To tie a boat to a dock, two sailors pull on ropes. One pulls with 100 pounds of force perpendicular to the dock, the other pulls with 150 pounds force at an angle of 45° to the dock. What is the magnitude and direction of the resultant force?

□ Figure 10.9

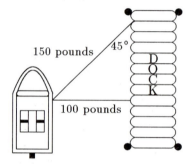

If we "slide" the 150-pound vector parallel to itself so that its tail is at the head of the 100-pound vector, we have a picture like Figure 10.10, where $\overrightarrow{R}$ indicates the vector resultant of the two forces.

□ Figure 10.10

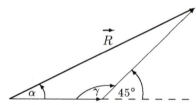

To find the magnitude and direction of the resultant, we must solve the triangle above for $\overrightarrow{R}$ and α.

$$\begin{aligned} \gamma &= 180° - 45° \\ &= 135° \end{aligned}$$

The law of cosines will give $\left|\overrightarrow{R}\right|$.

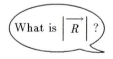

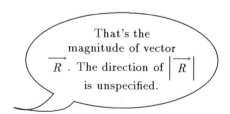

Chapter 10 Vectors

In this text a vector quantity is represented either as a variable with an arrow over it or a variable in boldface type. Because you cannot use boldface in handwriting, be sure to write an arrow over any vector.

$$\left| \overrightarrow{R} \right|^2 = 150^2 + 100^2 - 2(150)(100)\cos(135°)$$

$$\approx 53713$$

$$\left| \overrightarrow{R} \right| \approx 232 \text{ pounds}$$

$$\frac{\sin 135°}{232} \approx \frac{\sin \alpha}{150}$$

$$\frac{\sin(135°)(150)}{232} \approx \sin \alpha$$

$$0.4572 \approx \sin \alpha$$

$$27° \approx \alpha$$

The resultant is $\overrightarrow{R} = 232$ pounds at 27° from the 100-pound force. □

Parallelogram Method of Adding Vectors

Notice in Example 2 that the same value of $\overrightarrow{R}$ can be found by using the diagonal of a parallelogram.

□ Figure 10.11

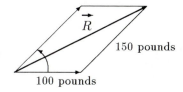

150 pounds

$\overrightarrow{R}$

100 pounds

Geometry Reminder

Parallelograms
The sum of the interior angles of a parallelogram is 360°. The opposite sides of a parallelogram are equal and parallel. Each pair of opposite angles of a parallelogram is equal.

Example 3 □ What is the resultant of a 40-lb and a 30-lb force acting on a sled if the two forces are separated by 30°?

□ Figure 10.12

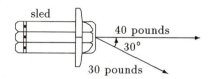

Make a force parallelogram.

□ Figure 10.13

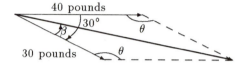

From the geometry reminder:

$$30° + 30° + \theta + \theta = 360°$$
$$30° + \theta = 180°$$
$$\theta = 150°$$

Use the law of cosines to find $\left| \overrightarrow{R} \right|$.

$$\left| \overrightarrow{R} \right|^2 = 30^2 + 40^2 - 2(30)(40)\cos 150°$$

$$\left| \overrightarrow{R} \right|^2 \approx 4578$$

$$\left| \overrightarrow{R} \right| \approx 67.7$$

Use the law of sines to find the direction of $\left| \overrightarrow{R} \right|$.

If we call the angle between the resultant and the 30-lb force β,

$$\frac{\sin \beta}{40} = \frac{\sin 150°}{\left| \overrightarrow{R} \right|}$$

$$\sin \beta \approx \frac{40(0.5)}{67.7}$$
$$\sin \beta \approx 0.2957$$
$$\beta \approx 17°$$

The resultant is a 68-lb force at an angle of 17° from the 30-lb force. The angle of the resultant from the 40-lb force is:

$$30° - 17° \approx 13°$$

□

Chapter 10 Vectors

Problem Set 10.1

Solve the following.

1. If a person drives 10 miles N45°W and then 10 miles straight north, how far is the person from the starting point? What is the bearing of the final position with respect to the starting position?

2. A sailboat sailed 4.8 miles due north then 6.9 miles N15.6°E. Find its distance and bearing from its starting point.

3. Two tugs are pulling a barge, one with a force of 900 pounds, and the other with a force of 1250 pounds. The angle between the tugs is 26°. Find the magnitude of the resultant force and the angle between the resultant force and each of the tugs.

4. The resultant of two forces acting at an angle of 62° with respect to each other is 90 kilograms. If one of the forces is 65 kilograms, find the other force and the angle it makes with the resultant.

5. Two forces, one 62 pounds and the other 39 pounds, have a resultant force of 82 pounds. Find the angle between the two forces.

6. An airplane flew 300 miles on a course of 060°; it then turned left 40° and flew for 200 miles. How far and in what direction was it from its starting point?

7. Two airplanes leave an airport at the same time. One travels in the direction 038° at 140 mph, and the other travels in the direction 350° at 180 mph. After two hours, how far apart are the airplanes?

8. An airplane flew on a course of 110° at 180 mph for $1\frac{1}{2}$ hours. It then changed its course and flew on a course of 072° at 160 mph for two hours. At this time how far was it from its starting point, and what was its bearing from the starting point?

9. A blimp flew at 45 mph in the direction 160°. After two hours it changed its course and flew at 40 mph for $2\frac{1}{2}$ hours in the direction 305°. How far was it from its starting point, and what was its bearing from its starting point?

10. Two sailboats leave a dock at the same time. One sails in the direction S25°W at 15 knots, and the other sails N40°W at 12 knots. After three hours, how far apart are the sailboats?

11. Two forces of 150 pounds and 180 pounds are acting at the same point, with an angle of 56° between them. What is the magnitude of the resultant force, and what angle does it make with the 150-pound force?

12. One force of 92 pounds makes an angle of 74° with another force. How strong must the second force be so that the second force will make an angle of 28° with the resultant? What will be the magnitude of the resultant?

10.2 Geometric Resolution of Vectors

The methods shown in section 10.1 can be cumbersome, particularly if more than two vectors are to be added. Resolving vectors into components makes the task easier.

Resolving Vectors into Components

Example 4 □ Consider an airplane flying at 100 miles per hour on a course 40° to the north of east.

□ Figure 10.14

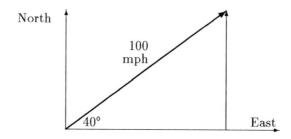

We can think of the airplane's velocity as being made up of two parts or components, an easterly part and a northerly component.

□ Figure 10.15

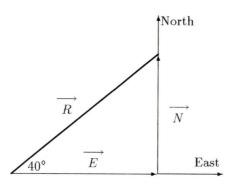

$$\text{By definition} \quad \cos 40° = \frac{\left| \overrightarrow{E} \right|}{100}$$

$$100 \, \cos 40° = \left| \overrightarrow{E} \right|$$

$$76.6 \text{ mph east} \approx \overrightarrow{E}$$

To find the component of the aircraft's speed in the direction of north, solve for the other leg of the triangle.

$$\sin 40° = \frac{\left| \overrightarrow{N} \right|}{100}$$

$$100 \, \sin 40° = \left| \overrightarrow{N} \right|$$

$$64.3 \text{ mph north} \approx \overrightarrow{N}$$

The motion of the airplane can be resolved into two components, 76.6 mph east and 64.3 mph north.

□

Chapter 10 Vectors

Navigation Examples

A Reminder

Recall that in navigation directions are measured from north in a clockwise direction. The course of the airplane in Example 4 would be $90° - 40° = 050°$. Other courses are shown below.

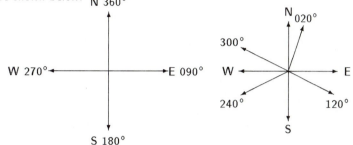

Because navigation headings are measured from north in a clockwise or negative direction, and calculators measure angles counterclockwise from the x-axis (East), we need an efficient way to convert aviation angles to calculator angles.

Think of north as $+90°$ in calculator terms. Then aviation angles tell you how much rotation there is in the negative direction from $+90°$. A heading of $020°$ is $90° - 20° = 70°$ on your calculator.

$$\text{Aviation} \longrightarrow \text{Conversion} \longrightarrow \text{Calculator}$$
$$010° \rightarrow 90° - 10° \rightarrow +80°$$
$$120° \rightarrow 90° - 120° \rightarrow -30°$$
$$270° \rightarrow 90° - 270° \rightarrow -180°$$
$$300° \rightarrow 90° - 300° \rightarrow -210°$$

The next example will show how to find the resultant by resolving vectors into horizontal and vertical components and adding the components.

Example 5 □ A blimp is flying at a speed of 30 knots on a heading of $037°$. If the wind is blowing from $250°$ to $070°$ at 15 knots, what is the groundspeed of the blimp?

It may help to visualize the blimp in the air over western Kansas. In that part of the country the roads run north and south or east and west, one mile apart. From the air the ground looks like a big piece of graph paper.

First we will consider where the blimp would travel if there were no wind. Then we will consider the effect of the wind. First look at the blimp and ignore the wind. It is on a heading of $037°$ from the north. Therefore, the blimp is pointed at $053°$ north of east.

□ Figure 10.17

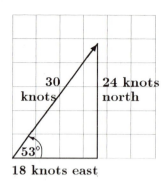

In this case the blimp is flying east at

$$\overrightarrow{E} = 30 \cos 53°$$

$$\overrightarrow{E} \approx 18 \text{ nautical miles/hr east.}$$

The blimp is flying north at

$$\overrightarrow{N} = 30 \sin 53°$$

$$\overrightarrow{N} \approx 24 \text{ nautical miles/hr north.}$$

Now consider the effect of the wind. If the blimp shut its engines off and just sat in the air, a 15-knot wind from 250° blowing toward 070° would blow the blimp along a path 20° north of east. Now find the effect of the wind on the blimp in the east direction.

The wind is blowing the blimp east at

□ Figure 10.18

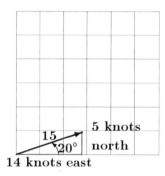

$$\overrightarrow{E} = 15 \cos 20°$$

$$\overrightarrow{E} \approx 14 \text{ nautical miles/hr east.}$$

The wind is blowing the blimp north at

$$\overrightarrow{N} = 15 \sin 20°$$

$$\overrightarrow{N} \approx 5 \text{ nautical miles/hr north.}$$

To get the combined effect of the blimp's travel through the air and the motion of the air (wind), we add all the east components, then add all the north components, and find the resultant.

□ Figure 10.19

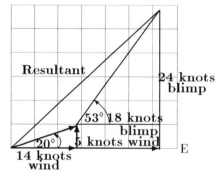

East Components		
E_{blimp}	$\approx$	18
E_{wind}	$\approx$	14
E_{total}	$\approx$	32

North Components		
N_{blimp}	$\approx$	24
N_{wind}	$\approx$	5
N_{total}	$\approx$	29

This means that the blimp will cross about 29 east-west roads per hour and will cross about 32 north-south roads per hour.

But what are the speed and direction of the blimp?

That vector is the resultant of the sum of the east travel vector plus the sum of the north travel vector.

304

Chapter 10 Vectors

To find the resultant, first find the direction of travel.

$$\tan \theta = \frac{N_{\text{total}}}{E_{\text{total}}}$$

$$\tan \theta = \frac{29}{32}$$

$$\theta \approx 42°$$

The blimp is traveling along the ground on a course 42° north of east.

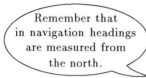

Remember that in navigation headings are measured from the north.

That means the heading is $90° - 42° \approx 048°$.

To find the speed use either $\sin \theta$ or $\cos \theta$ to find the magnitude of the resultant.

☐ Figure 10.20

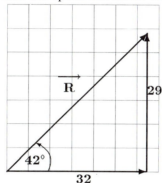

We'll use $\sin \theta$.

$$\sin(42°) \approx \frac{29}{\left| \overrightarrow{R} \right|}$$

$$\left| \overrightarrow{R} \right| \approx \frac{29}{\sin(42°)}$$

$$\left| \overrightarrow{R} \right| \approx 43 \text{ nautical miles/hr}$$

☐

To Summarize:

☐ Figure 10.21

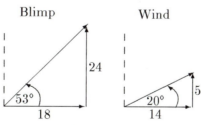

Blimp

Wind

Resultant

Example 6 ☐ An airplane is flying on a heading of 284° at 120 knots. The wind is blowing from 160° at 25 knots. What is the resultant groundspeed of the plane?

☐ Figure 10.22

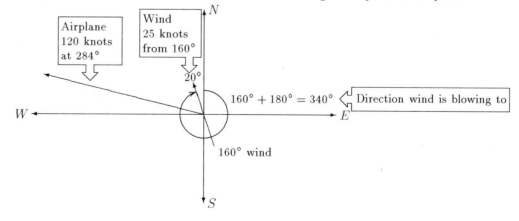

Airplane 120 knots at 284°

Wind 25 knots from 160°

$160° + 180° = 340°$

Direction wind is blowing to

160° wind

Just as a west wind blows east, a wind from 160° means the wind is blowing from 160° to 340°. The actual direction of the vector for the wind is 340°. If we consider the x-axis in the east-west direction, the angle between the wind and the x-axis is $90° - 340° = -250°$.

The component of the wind in the direction of the x-axis is

$$
\begin{aligned}
W_x &= 25\cos(-250°) \\
&\approx 25(-0.3420) \qquad \text{Note: } \cos(-250°) \text{ is negative} \\
W_x &\approx -8.6 \text{ knots}
\end{aligned}
$$

The component of the wind in the direction of the positive y-axis (north) is

$$
\begin{aligned}
W_y &= 25\sin(-250°) \\
&\approx 25(0.9397) \qquad \text{Note: } \sin(-250°) \text{ is positive} \\
W_y &\approx 23.5 \text{ knots}
\end{aligned}
$$

Next resolve the velocity of the airplane. A course of 284° means an angle of $-194°$ from the x-axis.

$$
\begin{array}{llll}
P_x &= 120\cos(-194°) & \qquad P_y &= 120\sin(-194°) \\
P_x &\approx -116.4 \text{ knots} & \qquad P_y &\approx 29.0 \text{ knots}
\end{array}
$$

$$
\begin{array}{ll}
\text{total of the } x \text{ components} & \qquad \text{total of the } y \text{ components} \\
\begin{aligned}
\text{Total}_x &= P_x + W_x \\
&\approx -116.4 + (-8.6) \\
&\approx -125.0 \text{ knots}
\end{aligned}
&
\begin{aligned}
\text{Total}_y &= P_y + W_y \\
&\approx 29.0 + 23.5 \\
&\approx 52.5 \text{ knots}
\end{aligned}
\end{array}
$$

Look at the sum of the components.

□ Figure 10.23

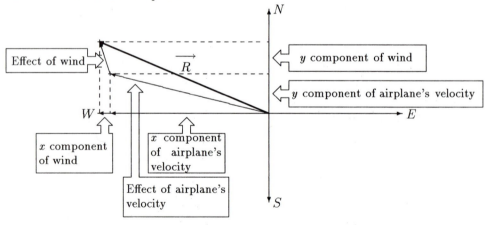

The tangent of the resultant velocity of the airplane is

$$
\begin{aligned}
\tan\theta &= \frac{\text{Total}_y}{\text{Total}_x} \\
&\approx \frac{52.5 \text{ knots}}{-125.0 \text{ knots}} \\
&\approx -0.4200 \\
\theta &\approx -22.8°
\end{aligned}
$$

Chapter 10 Vectors

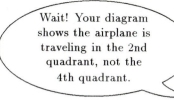

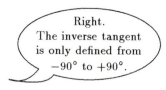

If you look at the signs of the sum of the x-coordinates and the sum of the y-coordinates, you can determine the quadrant of $\overrightarrow{R}$. Negative x and positive y means $\overrightarrow{R}$ lies in the second quadrant. The direction of $\overrightarrow{R}$ is then 22.8° toward the north from 180° which is 292.8°.

To find the magnitude of the resultant, solve for $\left|\overrightarrow{R}\right|$. Using the values for the sum of the x components and the sum of the y components we found on the last page, we can find the value of $\left|\overrightarrow{R}\right|$.

□ Figure 10.24

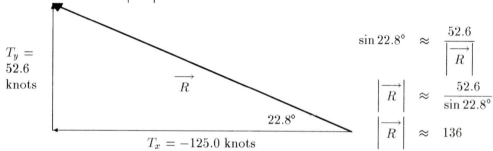

$$\sin 22.8° \approx \frac{52.6}{\left|\overrightarrow{R}\right|}$$

$$\left|\overrightarrow{R}\right| \approx \frac{52.6}{\sin 22.8°}$$

$$\left|\overrightarrow{R}\right| \approx 136$$

We did the calculations to the accuracy of the calculator. But we give the final answer to the accuracy of the data, which is optimistically to three places. The plane will travel along a ground track 23° north of west at 135 knots.

Expressed in navigational terms, this is a course of 293° at 135 knots.   □

Problem Set 10.2

Solve the following:

1. An airplane is flying at 140 miles per hour on a course 072°. What are its northerly and easterly components?

2. A rifle bullet is fired into the air with a velocity of 900 feet per second at an angle of 55° from the horizontal. Find the magnitude of the horizontal and vertical components of the velocity vector.

3. An airplane was headed in the direction 238° with an airspeed of 325 miles per hour, and the wind was blowing 50 miles per hour from the direction 016°. Find the groundspeed and the course of the airplane.

4. An airplane is flying on a heading of 130° at 180 knots. The wind is blowing from 280° toward 100° at 40 knots. What is the resultant groundspeed of the airplane, and what is its course?

10.2 Geometric Resolution of Vectors

5. An airplane is headed 023° at 175 knots. After 3.5 hours the airplane has actually gone 595 nautical miles in the direction 016°. Find the speed of the wind and the direction of the wind.

6. What heading and airspeed are required to fly 240 miles per hour due south if a wind of 50 miles per hour is blowing from 200° toward 020°?

7. In a stunt for a movie, a motorcycle travelling 100 miles per hour flies off a ramp at an angle of 37° to the horizontal. At the instant the motorcycle leaves the ramp, what are the horizontal and vertical components of its velocity?

8. A jumbo jet flying at 625 miles per hour has a heading of 310°. A 40-mph wind is blowing from 218°. Find the resultant course of the jet and its groundspeed.

9. One way we get weather data is from aircraft in flight. The wind correction angle of an aircraft in flight can provide windspeeds at upper altitudes which are particularly helpful in predicting weather. An airliner flying at an airspeed of 514 mph observes that a heading of 123° is required to actually fly a course of 116° with a speed of 550 mph over the ground. Find the speed and direction of the wind to the nearest mph and degree.

10. A ship sailing at 24 knots is headed in the direction S72.4°E. What is the easterly component and southerly component of its travel?

11. A man in a rowboat can row 4.5 miles per hour in still water. The river is flowing from north to south at 3 miles per hour. In what direction must he row to reach a point on the east bank that is S54°E from a dock on the west bank?

12. What are the initial horizontal and vertical components of the velocity vector of a rifle bullet that is fired into the air with an initial velocity of 1250 feet per second at an angle of 50° from the horizontal?

13. The direction from airport A to airport B is 128°. An airplane flying from airport A to airport B has a groundspeed of 212 miles per hour. The wind is blowing at 40 miles per hour from 340°. What is the airspeed of the plane, and what heading should it maintain to fly directly from A to B?

14. A blimp has a heading of 221° at 35 miles per hour. An 18-mph wind is blowing from 070°. Find the resultant groundspeed and the course of the blimp.

 Group Writing Activity

A sailboat in a channel is being driven by a small outboard motor, but it also has its sail up and is being pushed by the wind in a slightly different direction. If the current in the channel is in a third direction, explain how you would, using components, determine the speed of the boat and its direction.

10.3   Algebraic Resolution of Vectors

Think of the blimp in the previous section just floating in the air without its engine running. An observer at an airport on the ground might notice that the blimp traveled from 1 mile west and 3 miles south of the airport to 3 miles east and 2 miles north of the airport in one hour. We can visualize the vector representing the blimp's travel with a coordinate system.

□ Figure 10.25

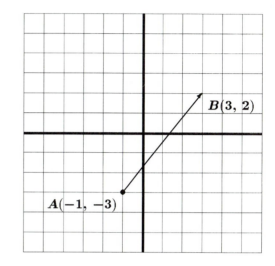

We can use the coordinate system on the right to represent the blimp's motion as a vector **AB** from $A(-1, -3)$ to $B(3, 2)$.

The change in the blimp's x position is

$$\begin{aligned}
\Delta x &= 3 - (-1) \\
\Delta x &= 4
\end{aligned}$$

The change in the blimp's y position is

$$\begin{aligned}
\Delta y &= 2 - (-3) \\
\Delta y &= 5
\end{aligned}$$

Vectors are defined as directed distances. Equivalent vectors have the same magnitude and direction. They do not have to have the same initial and terminal points. Suppose the observer released a balloon at the airport at the same time as the blimp was being tracked. The path of the balloon would have the same magnitude and direction as the blimp.

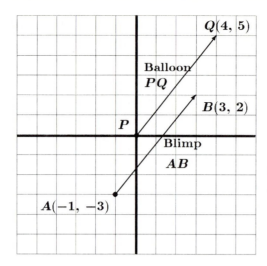

□ Figure 10.26

The vectors $\boldsymbol{PQ}$ and $\boldsymbol{AB}$ are equal. Vectors with their initial point at the origin are said to be in standard position. We can use bold letters or an arrow above the letters to indicate a quantity is a vector. Frequently we will use a single bold letter such as $\boldsymbol{u}$ or $\boldsymbol{v}$ to denote a vector. When you write vectors by hand you will need to use arrows because you cannot indicate bold with a pencil.

Vectors in standard position can be specified by the coordinates of their terminal points. Hence

$$\boldsymbol{PQ} = \langle 4, 5 \rangle$$

We use angle brackets to indicate that $\langle 4, 5 \rangle$ is a vector and not an ordered pair.

To specify the vector for the blimp, we merely translate it to the origin. This is equivalent to finding Δx and Δy from its initial to its terminal point.

The magnitude of a vector is defined using the Pythagorean Theorem:

$$\text{Magnitude of } \langle a, b \rangle = \sqrt{a^2 + b^2}$$

The magnitude of a vector is a scalar.

Algebraic Addition of Vectors

Algebraic addition of vectors is analogous to adding geometric vectors by components.

Definition 10.3A   Addition of Vectors

If $\boldsymbol{u} = \langle a, b \rangle$ and $\boldsymbol{v} = \langle c, d \rangle$ then

$$\boldsymbol{u} + \boldsymbol{v} = \langle a + c, b + d \rangle$$

Example 7 □ Add $\boldsymbol{u} = \langle 6, 8 \rangle$ and $\boldsymbol{v} = \langle 3, -4 \rangle$ and find the magnitude of the resultant.

$$\begin{aligned}
\langle 6, 8 \rangle + \langle 3, -4 \rangle &= \langle 6 + 3, 8 + (-4) \rangle \\
\boldsymbol{u} + \boldsymbol{v} &= \langle 9, 4 \rangle
\end{aligned}$$

The magnitude of $\boldsymbol{u} + \boldsymbol{v}$ is

$$\begin{aligned}
|\boldsymbol{u} + \boldsymbol{v}| &= \sqrt{9^2 + 4^2} \\
&= \sqrt{97}
\end{aligned}$$

□

Chapter 10 Vectors

Definition 10.3B Scalar Product

If k is a scalar and $\boldsymbol{u} = \langle a, b \rangle$ then the scalar product of k and $\boldsymbol{u}$ is the vector

$$
\begin{aligned}
k\boldsymbol{u} &= k \langle a, b \rangle \\
&= \langle ka, kb \rangle
\end{aligned}
$$

Definition 10.3C Unit Vector

Any vector with a magnitude of 1 is called a unit vector.

Any vector can be specified as the product of a scalar and a unit vector.

Example 8 □ Find a unit vector with the same direction as $\boldsymbol{P} = \langle -4, 5 \rangle$.

The magnitude of $\boldsymbol{P} = \langle -4, 5 \rangle$ is

$$
\begin{aligned}
\boldsymbol{P} &= \sqrt{(-4)^2 + 5^2} \\
&= \sqrt{41}
\end{aligned}
$$

To find a unit vector with the same direction as $\boldsymbol{P}$ but magnitude of 1, we divide both components of $\boldsymbol{P}$ by its magnitude:

$$
\boldsymbol{u} = \left\langle -\frac{4}{\sqrt{41}}, \frac{5}{\sqrt{41}} \right\rangle
$$

□

Special Unit Vectors

Because of their usefulness in defining the direction of any vector, we define two special unit coordinate vectors.

Definition 10.3D Unit Coordinate Vectors

$\boldsymbol{i} = \langle 1, 0 \rangle$ a unit vector in the direction of the x-axis.
$\boldsymbol{j} = \langle 0, 1 \rangle$ a unit vector in the direction of the y-axis.

As in the previous section, any vector can be resolved into components in the direction of the $\boldsymbol{i}$ and $\boldsymbol{j}$ vectors.

Example 9 □ Express $\boldsymbol{v} = \langle 3, 4 \rangle$ in terms of $\boldsymbol{i}$, $\boldsymbol{j}$ vectors.

$$
\boldsymbol{v} = 3\boldsymbol{i} + 4\boldsymbol{j}
$$

□

If $u = \langle -2, 5 \rangle$ and $v = \langle 3, -1 \rangle$, calculate each quantity below.

1. $u + v$	2. $u - v$
3. $3u - v$	4. $2u + 4v$
5. $|u|$	6. $|v|$
7. $|u + v|$	8. $|u| + |v|$

Add the following vectors and find the magnitude of the resultant.

9. $u = \langle 3, -4 \rangle$, $v = \langle 5, 7 \rangle$ 10. $u = \langle -1, 6 \rangle$, $v = \langle 2, -3 \rangle$

11. $u = \langle -2, -5 \rangle$, $v = \langle 3, 7 \rangle$ 12. $u = \langle 4, -1 \rangle$, $v = \langle -5, -2 \rangle$

Find a unit vector with the same direction as the given vector.

13. $v = \langle 2, -3 \rangle$ 14. $v = \langle 6, -4 \rangle$

15. $v = \langle -1, -5 \rangle$ 16. $v = \langle 3, 4 \rangle$

Express the following vectors in terms of i and j.

17. $v = \langle 4, -2 \rangle$ 18. $v = \langle -7, 5 \rangle$

19. $v = \langle 3, 1 \rangle$ 20. $v = \langle -12, -3 \rangle$

10.4   Work, Inclined Planes, and the Dot Product

Most of physics involves vectors in one way or the other. So far we have dealt with problems that required the addition of two vectors. Many parts of physics require the product of two vectors.

There are two vector products: the dot product and the cross product. In this text, we will only deal with dot products.

Before we start with work, we need to look at how forces can be resolved into perpendicular components.

Example 10 □ A child is pulling a wagon with a force of 20 pounds on a rope that makes an angle of 30° with the horizontal. Resolve the force on the tongue of the wagon into horizontal and vertical components.

□ Figure 10.27

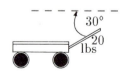

$$\sin 30° = \frac{\left|\overrightarrow{V}\right|}{20} \qquad \cos 30° = \frac{\left|\overrightarrow{H}\right|}{20}$$

$$20\ \sin 30° = \left|\overrightarrow{V}\right| \qquad 20\ \cos 30° = \left|\overrightarrow{H}\right|$$

$$20(0.5) = \left|\overrightarrow{V}\right| \qquad 17.32 \approx \left|\overrightarrow{H}\right|$$

$$\text{a force of 10 lbs vertical} = \overrightarrow{V} \qquad \text{a force of 17.32 lbs horizontal} \approx \overrightarrow{H}$$

If we choose an x-axis parallel to the horizon, we can say the force vector on the wagon is $\overrightarrow{F} = \langle 17.32, 10 \rangle$.

$\square$

Work

People usually think of work as anything that takes effort; for example, learning trigonometry or standing still holding a child. Neither of these examples fits the definition of work in physics. In physics work has two elements; first, there must be a force and second, there must be a motion.

If a force is in the same direction as the motion, then the work done by a force $\overrightarrow{F}$ applied for a distance d is the scalar

$$W = \left|\overrightarrow{F}\right| \cdot d$$

Example 11 $\square$ Find the work done lifting a 50-pound box from the floor to a table 3 feet high.

$$
\begin{aligned}
W &= \left|\overrightarrow{F}\right| \cdot d \\
W &= 50 \text{ lbs} \cdot 3 \text{ ft} \\
W &= 150 \text{ ft} \cdot \text{lbs}
\end{aligned}
$$

If you carry the 50-pound box for 5 miles at a height of 3 feet, you will not do any work by the definition. The force you are applying to hold the box is perpendicular to the direction of motion. Only force in the direction of motion counts in the definition of work. $\square$

What if the force is not applied in the direction of motion?

An extended definition of work which can be used when the force is not in the direction of motion is given below.

Definition 10.4A Work

The work done by applying a force $\overrightarrow{F}$ over a distance d is a scalar W.

$$W = \left|\overrightarrow{F}\right| \cdot \cos\theta \cdot \left|\overrightarrow{d}\right|$$

where θ is the angle between $\overrightarrow{F}$ and the direction of motion,
$\left|\overrightarrow{F}\right| \cos\theta$ is the component of F in the direction of motion

The Dot Product

In computing work, we multiply the magnitude of a vector $\overrightarrow{d}$ by the component of another vector $\overrightarrow{F}$ in the direction of $\overrightarrow{d}$. This product is called the scalar product, or dot product, of the two vectors.

Definition 10.4B Geometric Dot Product

The dot product of the vectors $\overrightarrow{A}$ and $\overrightarrow{B}$ is a scalar.

$$\overrightarrow{A} \bullet \overrightarrow{B} = \left|\overrightarrow{A}\right|\left|\overrightarrow{B}\right| \cos\theta$$

where θ is the angle between $\overrightarrow{A}$ and $\overrightarrow{B}$

We can use the dot product to define work.

Definition 10.4C Work: Using the Dot Product

If a force $\overrightarrow{F}$ causes motion over a distance $\overrightarrow{d}$, the work done is given by:

$$W = \overrightarrow{F} \bullet \overrightarrow{d}$$

This definition is equivalent to the previous definition of work.

Example 12 □ A force of 40 lbs on a rope attached at an angle of 60° with the horizontal to a block just overcomes friction and moves the block along the surface. How much work is done moving the block 100 ft along a level surface?

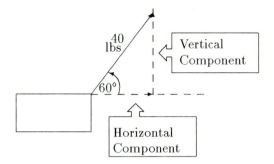

□ Figure 10.28

Because there is no vertical motion, the vertical component of the force does no work. The only motion is in the horizontal direction.

The work is then

$$W = \left| \overrightarrow{F} \right| \cdot \cos\theta \cdot \left| \overrightarrow{d} \right|$$
$$W = 40 \text{ lb} \cdot \cos 60° \cdot 100 \text{ ft}$$
$$W = 40 \text{ lb}(.5) \cdot 100 \text{ ft}$$
$$W = 2000 \text{ ft} \cdot \text{lbs}$$

□

Algebraic Definition of the Dot Product

It is possible to define the dot product using algebraic notation of vectors.

Definition 10.4D Algebraic Dot Product

The dot product of the vectors $\boldsymbol{v} = \langle a, b \rangle$ and $\boldsymbol{u} = \langle c, d \rangle$ is the scalar

$$\boldsymbol{v} \bullet \boldsymbol{u} = ac + bd$$

We will show how this definition is equivalent to the geometric definition of the dot product shortly, but first let's illustrate how to find the dot product algebraically.

Example 13 □ Find the dot product of $\boldsymbol{P} = \langle 2, 7 \rangle$ and $\boldsymbol{Q} = \langle 3, -4 \rangle$.

$$\begin{aligned} \boldsymbol{P} \bullet \boldsymbol{Q} &= 3 \cdot 2 + (-4)(7) \\ &= 6 - 28 \\ &= -22 \end{aligned}$$

A negative dot product indicates that the angle between the two vectors is greater than 90° and less than or equal to 180°. If the dot product is zero, the two vectors are perpendicular (orthogonal) to each other. □

The next example is done two ways to illustrate that the algebraic dot product produces the same result as the geometric dot product.

Example 14 □ Find the work done moving a cart 40 ft up a 15° incline if it requires a 100-lb force on a rope attached to the cart at a 45° angle to the horizontal to move the cart.

□ Figure 10.29

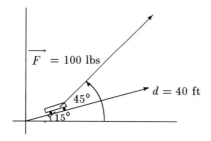

Work is defined as $W = \overrightarrow{F} \bullet \overrightarrow{d}$. We will first express $\overrightarrow{F}$ and $\overrightarrow{d}$ as algebraic vectors:

$\overrightarrow{F}$ = 100 lbs at 45° from the horizontal

$\overrightarrow{F_x}$ = 100 lbs cos 45° $\overrightarrow{F_y}$ = 100 lbs sin 45°

$\left| \overrightarrow{F_x} \right|$ ≈ 70.7 lbs $\left| \overrightarrow{F_y} \right|$ ≈ 70.7 lbs

$\boldsymbol{F}$ ≈ $\langle 70.7, 70.7 \rangle$ The dimension of the force vector is lbs.

$\overrightarrow{d}$ = 40 ft in a direction 15° above the horizontal

$\overrightarrow{d_x}$ = 40 ft cos 15° $\overrightarrow{d_y}$ = 40 ft sin 15°

$\left| \overrightarrow{d_x} \right|$ ≈ 38.6 ft $\left| \overrightarrow{d_y} \right|$ ≈ 10.4 ft

$\boldsymbol{d}$ ≈ $\langle 38.6, 10.4 \rangle$ The dimension of the distance vector is ft.

Now applying the definition of work:

$$
\begin{aligned}
W &= \boldsymbol{F} \bullet \boldsymbol{d} \\
&\approx \langle 70.7, 70.7 \rangle \bullet \langle 38.6, 10.4 \rangle \\
&\approx [(70.7)(38.6) + (70.7)(10.4)] \, \text{ft} \cdot \text{lbs} \\
&\approx 3464 \, \text{ft} \cdot \text{lbs}
\end{aligned}
$$

If we repeat this calculation leaving the expressions for the vector components in the $\boldsymbol{i}$ and $\boldsymbol{j}$ directions as trigonometric expressions, we can see that both definitions for the dot product are equivalent.

$$
\begin{aligned}
W &= \boldsymbol{F} \bullet \boldsymbol{d} \\
&= \langle 100 \cos 45°, 100 \sin 45° \rangle \, \text{lbs} \bullet \langle 40 \cos 15°, 40 \sin 15° \rangle \, \text{ft} \\
&= [100 \cos 45° \cdot 40 \cos 15° + 100 \sin 45° \cdot 40 \sin 15°] \, \text{ft} \cdot \text{lbs} \\
&= 100 \cdot 40 [\cos 45° \, \cos 15° + \sin 45° \, \sin 15°] \, \text{ft} \cdot \text{lbs} \\
&= 100 \cdot 40 \cos 30° \, \text{ft} \cdot \text{lbs} \qquad\qquad \text{identity 18 } \cos(\alpha - \beta) \\
W &\approx 3464 \, \text{ft} \cdot \text{lbs}
\end{aligned}
$$

The work done is the product of the magnitudes of the two vectors times the cosine of the angle between them.   □

Definition 10.4E Angle Between Two Vectors

We can use the definition of the dot product to find the angle between two vectors. Because

$$\overrightarrow{A} \bullet \overrightarrow{B} = \left|\overrightarrow{A}\right|\left|\overrightarrow{B}\right| \cdot \cos\theta$$

$$\cos\theta = \frac{\overrightarrow{A} \bullet \overrightarrow{B}}{\left|\overrightarrow{A}\right|\left|\overrightarrow{B}\right|}$$

Example 15 □ Find the angle between $\overrightarrow{P} = \langle -2, 3 \rangle$ and $\overrightarrow{Q} = \langle 4, 5 \rangle$.

$$\cos\theta = \frac{\overrightarrow{P} \bullet \overrightarrow{Q}}{\left|\overrightarrow{P}\right|\left|\overrightarrow{Q}\right|}$$

$$= \frac{(-2)(4) + (3)(5)}{\sqrt{(-2)^2 + (3)^2} \cdot \sqrt{(4)^2 + (5)^2}}$$

$$\cos\theta \approx 0.3032$$

$$\theta \approx 72.35°$$

□

The Projection of $\overrightarrow{A}$ on $\overrightarrow{B}$

The projection of $\overrightarrow{A}$ on $\overrightarrow{B}$ is

$$\text{Proj}_B A = \left|\overrightarrow{A}\right|\cos\theta \qquad \text{where } \theta \text{ is the angle between } \overrightarrow{A} \text{ and } \overrightarrow{B}$$

Substituting the previous formula for $\cos\theta$:

$$\text{Proj}_B A = \frac{\left|\overrightarrow{A}\right|\left(\overrightarrow{A} \bullet \overrightarrow{B}\right)}{\left|\overrightarrow{A}\right| \cdot \left|\overrightarrow{B}\right|}$$

$$= \frac{\overrightarrow{A} \bullet \overrightarrow{B}}{\left|\overrightarrow{B}\right|}$$

Example 16 □ Find the projection of $\overrightarrow{P} = \langle -2, 3 \rangle$ on $\overrightarrow{Q} = \langle 4, 5 \rangle$.

$$\text{Proj}_Q P = \frac{\langle -2, 3 \rangle \bullet \langle 4, 5 \rangle}{\sqrt{4^2 + 5^2}}$$

$$= \frac{-8 + 15}{\sqrt{41}}$$

$$\text{Proj}_Q P \approx 1.09$$

You may verify that this is correct by multiplying the magnitude of $\overrightarrow{P}$ by the cosine of the angle between $\boldsymbol{P}$ and $\boldsymbol{Q}$ as found in the previous example. □

Problem Set 10.4

Write the following vectors in algebraic notation.

1. 100 lbs at 37°

2. 70 lbs at 30°

3. 200 lbs at 127°

4. 80 lbs at −30°

If $u = \langle 3, 4 \rangle$ and $v = \langle -5, 6 \rangle$ calculate each quantity below.

5. $u + v$

6. $u - v$

7. $3u - v$

8. $2u + 4v$

9. $|u|$

10. $|v|$

11. $|u + v|$

12. $|u| + |v|$

13. $u \bullet v$

14. $v \bullet u$

15. Find the angle between u and v.

16. Find the projection of v on u.

17. If vector A has a magnitude of 120 lbs at an angle of 53° from vector B which has a magnitude of 15 ft, find $A \bullet B$.

18. X is 20° above the horizontal with a magnitude of 150 lbs, and Y which has a magnitude of 200 lbs is 70° below the horizontal. Find $X \bullet Y$.

19. A father is pulling his son in a wagon with a force of 40 pounds on a rope that makes an angle of 36° with the horizontal. Resolve the forces on the rope into horizontal and vertical components.

20. An elephant is pulling a 300-lb tent pole at a circus site. One end of a rope is tied to the pole and the other to the elephant. The rope makes an angle of 40° with the horizontal. Find the horizontal and vertical components of the force. How much work is done by the elephant if it pulls the tent pole 100 ft?

21. Two delivery men carry a desk into an office. They each support half the weight, but they pull sideways against each other to hold the desk away from their legs. If each man pulls with a force of 125 pounds at an angle of 15° from the vertical, how heavy is the desk and how hard do the men pull against each other? How much work does each man do in lifting the desk 6 inches off the ground?

22. A 3000-pound car is standing on a ramp that has a 20° incline with the horizontal. To the nearest pound, how much force is needed to keep the car from rolling down the ramp and what is the force perpendicular to the ramp? How much work is required to move the car 3 feet up the ramp?

23. A 200-lb man in a wheelchair is on a ramp that makes an angle of 15° with the horizontal. Find the force required to keep the wheelchair from rolling down the ramp. How much work is needed to move the man 25 ft up the ramp?

□ Figure 10.

□ Figure 10.

□ Figure 10.

24. To close her drapes at home, a woman pulls on a cord with a constant force of 20 pounds, at a constant angle of 75° with the horizontal. Find the work done to close the drapes on a 6-foot-wide window.

25. A gardener pushes a lawnmower with a force of 45 pounds on the handle that makes an angle of 40° with the horizontal. How much work is done by the gardener if she has to make 20 passes at a 60-foot-wide yard.

Group Writing Activity

To find the dot product of two vectors $u = \langle 3, -4 \rangle$ and $v = \langle 2, 5 \rangle$ we multiply the components.

$$\begin{aligned} u \bullet v &= (3)(2) + (-4)(5) \\ &= -14 \end{aligned}$$

If we express u and v in terms of i and j we get

$$u = 3i - 4j \qquad \text{and} \qquad v = 2i + 5j$$

Multiply $(3i - 4j)(2i + 5j)$ like binomials to get

$$6i^2 + 15i \bullet j - 8j \bullet i - 20j^2$$

How do you justify this being equal to -14? In other words, explain why $i^2 = j^2 = 1$ and $i \bullet j = j \bullet i = 0$.

Group Writing Activity

How would you write the dot product to account for vectors that are three-dimensional? That is,

$$u = \langle a, b, c \rangle \qquad\qquad v = \langle x, y, z \rangle$$

Explain the geometric interpretation of the dot product for these two vectors.

10.1 **1.** A vector is a quantity with amount and direction.

2. The resultant of the sum of two vectors is formed by joining the head of one vector to the tail of another. The vector sum is the vector from the tail of the starting vector to the head of the ending vector.

3. If two vectors are used as two adjacent sides of a parallelogram, the diagonal of the parallelogram gives the value of the resultant of the sum of the vectors.

10.2 **1.** Some vectors can be resolved into a north-south component and an east-west component.

2. Some vectors can be resolved into horizontal and vertical vectors.

3. The resultant of two or more vectors can be calculated by adding the horizontal components of each and by adding the vertical components of each.

4. In air navigation, direction is measured from the north line in a clockwise position. To determine the actual direction and groundspeed of an airplane, you can resolve the vector representing the airspeed and the vector representing the wind speed into perpendicular components, usually along the x-axis and y-axis. The resultant vector can be determined using the fact that the sum of two or more vectors is the sum of their components.

10.3 **1.** Vectors with their initial point at the origin are vectors in standard position. Such vectors can be specified by the coordinates of their terminal points.

2. The magnitude of a vector is calculated using the Pythagorean Theorem.

3. A unit vector is a vector with magnitude 1.

4. i is a unit vector in the direction of the x-axis.

5. j is a unit vector in the direction of the y-axis.

10.4 **1.** To determine the effect of an object on an inclined plane, resolve the vertical force into a component perpendicular to the plane and a component parallel to the inclined plane. The component perpendicular to the plane determines the force exerted against the plane. The component parallel to the plane determines the force needed to hold an object on the inclined plane.

2. The dot product $\vec{A} \bullet \vec{B}$ of two vectors $\vec{A}$ and $\vec{B}$ is $\left|\vec{A}\right| \cdot \left|\vec{B}\right|$ times the cosine of the angle between $\vec{A}$ and $\vec{B}$.

3. Work is a scalar equal to $\vec{F} \bullet \vec{d}$.

4. The angle between two vectors can be found using the dot product.

5. The projection of $\vec{A}$ on $\vec{B}$ is the magnitude of $\vec{A}$ times the cosine of the angle between $\vec{A}$ and $\vec{B}$.

Chapter 10 Review Test

Add the following vectors and find the magnitude of the resultant. **(10.3)**

1. $u = \langle 7, -3 \rangle$, $v = \langle -2, 5 \rangle$ 2. $u = \langle 3, 6 \rangle$, $v = \langle 1, -4 \rangle$

Find a unit vector with the same direction as the given vector. **(10.3)**

3. $v = \langle 3, -4 \rangle$ 4. $v = \langle -5, -3 \rangle$

Express the following vectors in terms of i and j. **(10.3)**

5. $v = \langle -3, 1 \rangle$ 6. $v = \langle 2, 7 \rangle$

Write the following vectors in algebraic notation. **(10.4)**

7. 65 lbs at $35°$ 8. 100 lbs at $-45°$

If $u = \langle 7, -2 \rangle$ and $v = \langle 3, -4 \rangle$, calculate each quantity below. **(10.3)**, **(10.4)**

9. $4u + 2v$ 10. $|u - v|$ 11. $u \bullet v$

12. Find the angle between u and v.

13. Find the projection of v on u.

14. If vector A has a magnitude of 95 lbs at an angle of $70°$ from vector B which has a magnitude of 25 ft, find $A \bullet B$.

15. Two forces of 40 kilograms and 50 kilograms act on an object at an angle of $65°$ between them. Find the magnitude of the resultant force and, to the nearest degree, the angle the resultant makes with the 40-kilogram force. **(10.1)**

16. The resultant of two forces acting at an angle of $116°$ with respect to each other is 85 pounds. If one of the forces is 90 pounds, find the other force and the angle it makes with the resultant. **(10.1)**

17. Two forces, one 25 kilograms and the other 37 kilograms, have a resultant of 44 kilograms. To the nearest degree find the angle between the two forces. **(10.1)**

18. Two college students decided to move their desk. One pushed the desk with a force of 95 pounds toward the north, while the other pushed the desk toward the east with a force of 74 pounds. What was the resultant force and in what direction did the desk move? How much work was done to move the desk 5 feet? **(10.4)**

19. An airplane is flying on a $280°$ course at 180 miles per hour. What are the northerly and westerly components? **(10.2)**

20. A cabin cruiser is towing a rowboat. The tow line makes an angle of $34°$ with the horizontal, and the tension on the line is 60 pounds. What are the horizontal and vertical forces on the rowboat? How much work is done by the cabin cruiser to pull the rowboat 3000 feet? **(10.2)**

21. An inclined ramp is used to get heavy appliances into a store. The ramp is inclined to the horizontal at an angle of 25°. How much force is needed to keep a 450-pound refrigerator from sliding down the ramp? What is the perpendicular force against the ramp? How much work is required to move the refrigerator 6 feet up the ramp? **(10.4)**

22. Two boys are pushing an object along the floor. One is pushing with a force of 75 pounds in the direction S36°W and the other is pushing with a force of 90 pounds in the direction S24°E. In what direction is the object moving? **(10.2)**

23. A gondolier pushes his pole with a force of 30 pounds at an angle of 36°. How much work is required to move the gondola 5000 feet through the canals? **(10.4)**

24. Find the work required to pull a barge from the shore along a river 300 feet if a force of 100 pounds is applied at an angle of 17° with the edge of the river. **(10.4)**

25. A motorboat crossed a river flowing from north to south. The current is flowing at 8 miles per hour and the boat is moving at 15 miles per hour. To the nearest degree, in what direction must the boat go in order to reach a point across the river in the direction S78°E from the starting point? **(10.2)**, **(10.4)**

26. An airplane is headed in the direction 328° at 260 miles per hour and the wind is blowing from 044° at 46 miles per hour. Find the groundspeed of the airplane and its resultant course. **(10.2)**

Chapter 11

Complex Numbers

Contents

Preview

This chapter will review complex numbers that you may recall from algebra. The second section will show you how to represent complex numbers in trigonometric form. This representation of complex numbers is particularly useful to electrical engineers working with alternating-current circuits. In the third section of this chapter, you will learn how DeMoivre's Theorem is used to reduce the work required to find powers and roots of complex numbers.

11.1 Algebraic Operations with Complex Numbers

This first section is a review of addition, subtraction, multiplication, and division of complex numbers. If you have recently completed a strong intermediate algebra course you may go directly to Section 11.2.

To find the square root of negative numbers, we must extend our set of numbers. To do this, we define a new system of numbers where $\sqrt{-1} = i$.

> How can
> you say that out
> of the blue?

> The same
> way we defined $1 \div 4$
> to be $\frac{1}{4}$ or $-x$ as the number
> you add to x to get zero
> for an answer.

Definition 11.1A Imaginary Unit

The imaginary unit is a number called i where

$$i^2 = -1 \quad \text{and} \quad i = \sqrt{-1}$$

Definition 11.1B Pure Imaginary Numbers

Pure imaginary numbers are numbers of the form bi

$$\text{where} \quad b \ \in \ \text{reals} \quad (\in \text{ is read "is an element of")}$$
$$i \ = \ \sqrt{-1}$$

> Does
> imaginary
> mean they don't
> exist?

> Imaginary is
> a poor word choice,
> but we're stuck with it.
> Read on.

If you think of imaginary as meaning a product of the mind, then these numbers exist as any other numbers exist. When was the last time you touched -2 or $+2$ for that matter?

Because the rules for exponents apply to imaginary numbers also, some properties follow immediately from the definition of i.

$$
\begin{array}{llll}
\text{Since} & i^2 & = & -1 \\
\text{then} & i^3 & = & i^2 \cdot i \quad \text{and} \quad i^4 = i^2 \cdot i^2 \\
& & = & -i \qquad\qquad\qquad\;\; = (-1)(-1) \\
& & & \qquad\qquad\qquad\qquad\;\; = +1
\end{array}
$$

Example 1 □ Evaluate i^{35}.

$$
\begin{aligned}
i^{35} &= i^{32+3} \\
&= i^{32} \cdot i^3 \\
&= (i^4)^8 \cdot i^3 \\
&= 1^8 \cdot i^3 \\
&= -i
\end{aligned}
$$

Why call i^{35} i^{32+3}?

Because 32 is a multiple of 4, and $i^4 = 1$.

It also follows if $a > 0$,

then

$$
\begin{aligned}
\sqrt{-a} &= \sqrt{a(-1)} \\
&= \sqrt{a}\sqrt{-1} \\
&= \sqrt{a}\, i \qquad \text{by definition, } \sqrt{-1} = i
\end{aligned}
$$

□

Example 2 □ Find the following square roots.

$$
\begin{aligned}
\text{a. } \sqrt{-25} &= \sqrt{25}\sqrt{-1} \\
&= 5i \\
\text{b. } \sqrt{-16} &= \sqrt{16}\sqrt{-1} \\
&= 4i \\
\text{c. } \sqrt{-20} &= \sqrt{4}\sqrt{5}\sqrt{-1} \\
&= 2\sqrt{5}\, i \\
\text{d. } \sqrt{-18} &= \sqrt{9}\sqrt{2}\sqrt{-1} \\
&= 3\sqrt{2}\, i
\end{aligned}
$$

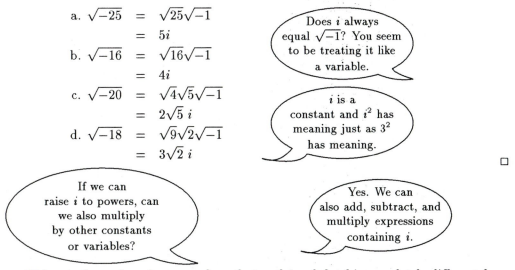

Does i always equal $\sqrt{-1}$? You seem to be treating it like a variable.

i is a constant and i^2 has meaning just as 3^2 has meaning.

□

If we can raise i to powers, can we also multiply by other constants or variables?

Yes. We can also add, subtract, and multiply expressions containing i.

This set of pure imaginary numbers that we have defined is completely different from the set of real numbers.

There is, however, a set of numbers that includes both. It is the complex number system.

Definition 11.1C   The Complex Number System

The complex number system is the set of numbers of the form

$$a + bi$$

$$
\begin{aligned}
\text{where} \quad a, b &\in \text{ real numbers} \\
i &= \text{ imaginary unit}
\end{aligned}
$$

a is called the real part of $a + bi$.
bi is called the pure imaginary part of $a + bi$.

In complex numbers of the form $a + bi$,

If $a \neq 0$, $b = 0$ $a + bi$ is a real number

If $a = 0$, $b \neq 0$ $a + bi$ is a pure imaginary number

If $a \neq 0$, $b \neq 0$ $a + bi$ is an imaginary number

Definition 11.1D   Equality of Complex Numbers

$$a + bi = c + di$$

if, and only if, $a = c$ and $b = d$.

For addition and subtraction of complex numbers we add or subtract the real and the imaginary parts separately.

Definition 11.1E   Addition of Complex Numbers

$$(a + bi) + (c + di) = (a + c) + (b + d)i$$

Example 3 □ Subtract $(4 - 3i) - (-10 + 5i)$.

$$
\begin{aligned}
(4 - 3i) - (-10 + 5i) &= (4 + 10) + (-3 - 5)i \\
&= 14 - 8i
\end{aligned}
$$

□

To multiply two complex numbers, use the same procedure as for multiplication of two binomials.

Definition 11.1F   Product of Two Complex Numbers

$$
\begin{aligned}
(a + bi)(c + di) &= ac + bci + adi + bdi^2 \\
&= ac + (bc + ad)i - bd \\
&= (ac - bd) + (bc + ad)i
\end{aligned}
$$

Example 4 □ Multiply $(3 + 4i)(2 - 5i)$.

$$
\begin{aligned}
(3 + 4i)(2 - 5i) &= 6 + 8i - 15i - 20i^2 \\
&= 6 - 7i - 20(-1) \\
&= 26 - 7i
\end{aligned}
$$

□

Example 5 □ Multiply $(2 + \sqrt{-3})(4 - \sqrt{-3})$.

$$
\begin{aligned}
(2 + \sqrt{-3})(4 - \sqrt{-3}) &= (2 + \sqrt{3}\,i)(4 - \sqrt{3}\,i) \\
&= 8 + 4\sqrt{3}\,i - 2\sqrt{3}\,i - 3i^2 \\
&= 8 + 2\sqrt{3}\,i + 3 \\
&= 11 + 2\sqrt{3}\,i
\end{aligned}
$$

□

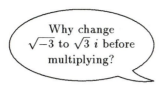

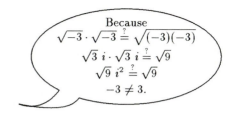

Always change from a negative number under the radical sign to i before multiplying or dividing complex numbers.

To divide a complex number by a real number, we use the fact that division by reals is defined as multiplication by the multiplicative inverse.

Example 6 □ Divide $\dfrac{(6+4i)}{2}$.

$$
\begin{aligned}
\frac{6+4i}{2} &= (6+4i)\frac{1}{2} \\
&= \frac{6}{2} + \frac{4}{2}i \qquad \text{distributive property} \\
&= \underline{\qquad}
\end{aligned}
$$

$3 + 2i$

□

To divide a complex number by a complex number, use a process similar to rationalizing denominators:

Example 7 □ Divide $5 + 10i$ by $5i$.

$$
\begin{aligned}
\frac{5+10i}{5i} &= \frac{5+10i}{5i} \cdot \frac{i}{i} \\
&= \frac{5i+10i^2}{5i^2} \\
&= \frac{5i-10}{-5} \\
&= -i+2 \\
&= 2-i
\end{aligned}
$$

□

Definition 11.1G Conjugate of a Complex Number

Each complex number has a conjugate.

$$
\begin{aligned}
\text{If} \quad z &= a + bi \quad \text{then} \\
\text{the conjugate of } z \text{ is} \quad \overline{z} &= a - bi
\end{aligned}
$$

Example 8 □ Divide $5 + 4i$ by $1 + 2i$.

To divide, multiply both the numerator and the denominator by the conjugate of the denominator.

$$
\begin{aligned}
\frac{5 + 4i}{1 + 2i} &= \frac{5 + 4i}{1 + 2i} \cdot \frac{(1 - 2i)}{(1 - 2i)} \\
&= \frac{5 - 6i - 8i^2}{1 - 4i^2} \\
&= \frac{5 - 6i + 8}{1 + 4} \\
&= \frac{13 - 6i}{5} \\
&= \frac{13}{5} - \frac{6}{5}i.
\end{aligned}
$$

□

Problem Set 11.1

Find the following square roots.

1. $\sqrt{-9}$
2. $\sqrt{-25}$
3. $-\sqrt{-36}$
4. $\sqrt{-125}$
5. $-\sqrt{-18}$
6. $-\sqrt{-8}$
7. $\sqrt{-75}$
8. $-\sqrt{-98}$

Evaluate each of the following.

9. i^5
10. $-i^{19}$
11. i^{77}
12. $-i^{103}$
13. i^{87}
14. i^{73}
15. $(-i)^{58}$
16. $(-i)^{101}$

Simplify the following.

17. $(3 + 2i) + (-5 + 6i)$
18. $(-4 + 8i) - (7 - 12i)$
19. $(18 + \sqrt{-8}) + (-8 - 3\sqrt{-2})$
20. $(-6 - \sqrt{-36}) - (7 + 6i)$
21. $(16 - 3\sqrt{-8}) - (-4 - 2\sqrt{-32})$
22. $(-7 + 4\sqrt{-12}) + (8 - 5\sqrt{-48})$
23. $3i(7 - 8i)$
24. $-8i(12 + 3i)$
25. $\sqrt{-25}(7 - \sqrt{-49})$
26. $(3 - 4i)^2$
27. $(\sqrt{5} + 2i)(\sqrt{5} - 3i)$
28. $(\sqrt{27} - \sqrt{-2})(\sqrt{3} + \sqrt{-8})$
29. $4i(8 - \sqrt{-25})$
30. $-9i(7 - 4\sqrt{-12})$
31. $\dfrac{9 - 24i}{3}$
32. $\dfrac{18 - \sqrt{-12}}{2}$
33. $\dfrac{-\sqrt{3} - 2i}{6i}$
34. $\dfrac{39}{2 - i}$
35. $\dfrac{12i}{3 + 2i}$
36. $\dfrac{3i}{5 - \sqrt{-4}}$
37. $\dfrac{5 + 3i}{2 - 2i}$
38. $\dfrac{3 - 4i}{5 + 2i}$
39. $\dfrac{8i}{2 - \sqrt{2}\,i}$
40. $\dfrac{-3}{3 - \sqrt{-4}}$
41. $\dfrac{6i}{8 - \sqrt{-12}}$
42. $\dfrac{5i}{10 + \sqrt{-75}}$
43. $\dfrac{7 - \sqrt{-18}}{8 - \sqrt{-32}}$
44. $\dfrac{3 + \sqrt{-48}}{5 - \sqrt{-27}}$
45. $\dfrac{2\sqrt{2} - \sqrt{-3}}{\sqrt{2} + \sqrt{-3}}$
46. $\dfrac{\sqrt{3} - 3\sqrt{-2}}{2\sqrt{3} - \sqrt{-2}}$

11.2 Trigonometric and Polar Representation of Complex Numbers

Graphical Representation of Complex Numbers

To graph complex numbers, use a graph similar to the Cartesian coordinate system for real numbers. The graph used is called the complex plane. This plane has a horizontal real axis and a vertical imaginary axis.

□ Figure 11.1

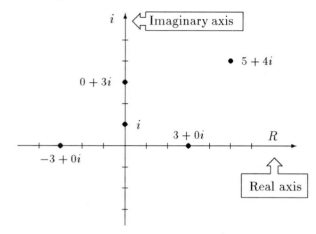

Definition 11.2A   Modulus of a Complex Number

The modulus or absolute value of a complex number, z, is the distance between z and the origin of the complex plane.

$$\text{If } z = a + bi, \text{ then } |z| = \sqrt{a^2 + b^2}$$

□ Figure 11.2

The modulus of a complex number is similar to the magnitude of a vector from the origin to the point on the plane representing the complex number.

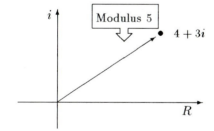

Example 9 □ Find the modulus of $z = 2 - 7i$.

$$
\begin{aligned}
|z| &= \sqrt{2^2 + (-7)^2} \\
&= \sqrt{4 + 49} \\
&= \sqrt{53}
\end{aligned}
$$

□

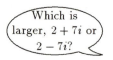

Which is
larger, $2 + 7i$ or
$2 - 7i$?

We can't
order complex
numbers.

The form $a + bi$ is called the rectangular form of a complex number. Another useful way to express a complex number is trigonometric form. In trigonometric form, the real and imaginary parts of the number are represented as functions of the modulus, which is usually called r, and of θ, which is the angle between r and the positive real axis in the complex plane.

Definition 11.2B   Trigonometric Form of Complex Numbers

A complex number $z = a + bi$ in trigonometric form is

□ Figure 11.3

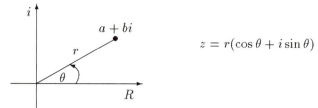

$$z = r(\cos \theta + i \sin \theta)$$

where $r = \sqrt{a^2 + b^2}$. Where $\tan \theta$ is defined, it is convenient to define θ as the angle with $\tan \theta = \dfrac{b}{a}$ in the same quadrant as $a + bi$.

Example 10 □ Write $1 + \sqrt{3}i$ in trigonometric form.

A sketch may help:

□ Figure 11.4

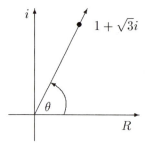

$$
\begin{aligned}
r &= \sqrt{1^2 + (\sqrt{3})^2} \\
&= \sqrt{1 + 3} \\
&= 2 \\
\tan \theta &= \frac{\sqrt{3}}{1} \\
\tan \theta &= \sqrt{3} \\
\theta &= 60°
\end{aligned}
$$

In trigonometric form,

$$1 + \sqrt{3}i = 2(\cos 60° + i \sin 60°)$$

□

Chapter 11 Complex Numbers

In general, to change a complex number from rectangular form to trigonometric form, do the following:

□ Figure 11.5

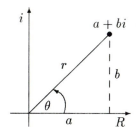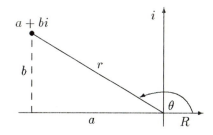

1. Find the magnitude of r by the Pythagorean Theorem.

$$r = \sqrt{a^2 + b^2}$$

2. Find θ.

$$\tan \theta = \frac{b}{a}$$

θ will be in the same quadrant as $a + bi$

3. Find a.

$$\text{Because } \frac{a}{r} = \cos \theta, \qquad a = r \cos \theta$$

4. Find b.

$$\text{Because } \frac{b}{r} = \sin \theta, \qquad b = r \sin \theta$$

Then by substitution,
$$
\begin{aligned}
a + bi &= r \cos \theta + r \sin \theta i \\
&= r(\cos \theta + i \sin \theta)
\end{aligned}
$$

Example 11 □ Write $-1 - \sqrt{3}i$ in trigonometric form.

$$
\begin{aligned}
r &= \sqrt{(-1)^2 + (-\sqrt{3})^2} \\
&= \sqrt{1 + 3} \\
r &= 2 \\
\tan \theta &= \frac{-\sqrt{3}}{-1} \\
&= \sqrt{3} \\
&= 60°
\end{aligned}
$$

Wait a minute. This number is in the third quadrant.

Right. The tangent is positive in two quadrants. We must know into which quadrant the angle falls to properly evaluate it.

In this case,

□ Figure 11.6

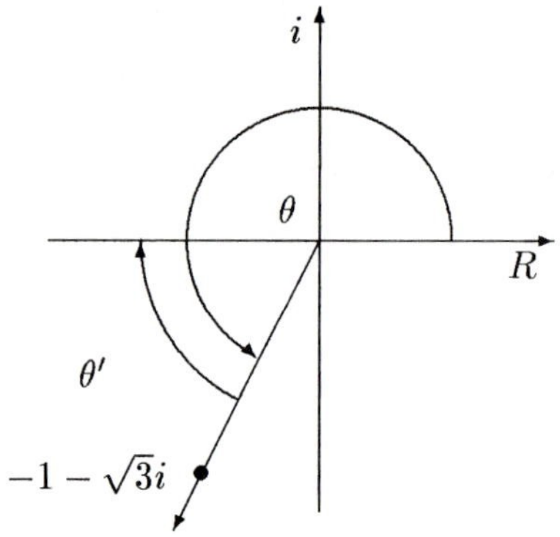

think of $\arctan \left| \dfrac{b}{a} \right| = \theta'$ as a reference angle.

What we found above is that $\theta' = 60°$.

The figure above shows θ is in the third quadrant.

Hence:

$$\begin{aligned} \theta &= 180° + 60° \\ &= 240° \end{aligned}$$

Therefore,

$$-1 - \sqrt{3}i = 2(\cos 240° + i \sin 240°)$$

□

Example 12 □ Write $-2i$ in trigonometric form.

$-2i$ is equivalent to $0 - 2i$.

Draw a sketch.

□ Figure 11.7

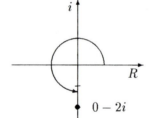

$$\begin{aligned} r &= \sqrt{0^2 + (-2)^2} \\ &= 2 \end{aligned}$$

From the sketch it is obvious that $\theta = 270°$.

So that's why definition 11.2B qualified how to find θ.

Yes; θ is the angle with $\tan \theta = \frac{b}{a}$ in the same quadrant as $a + bi$ in all cases where $\tan \theta$ is defined.

You could also say that θ is the angle where

$\sin \theta = \dfrac{b}{\sqrt{a^2 + b^2}}$ and $\cos \theta = \dfrac{a}{\sqrt{a^2 + b^2}}$

Normally we just use the tangent function to find θ' and determine the value of θ from the quadrant of $a + bi$.

For this case $\theta = 270°$ and $r = 2$.

Therefore,

$$-2i = 2(\cos 270° + i \sin 270°)$$

□

Complex numbers are of particular importance to electrical engineers who frequently need to add, subtract, multiply, and divide complex numbers. As you have probably observed, addition of complex numbers in rectangular form is straightforward; multiplication can be tedious. Fortunately, multiplication of complex numbers written in trigonometric form is relatively easy.

Example 13 □ Multiply $2(\cos 30° + i \sin 30°)$ by $5(\cos 60° + i \sin 60°)$.

Treat this as a product of two binomials: $2(\cos 30° + i \sin 30°) \cdot 5(\cos 60° + i \sin 60°)$

$$= 10[\cos 30° \cos 60° + \cos 30° \, i \sin 60° + i \sin 30° \cos 60° + i^2 \sin 30° \sin 60°]$$

Rearranging and replacing i^2 with -1,

$$= 10[\cos 30° \cos 60° - \sin 30° \sin 60° + i \cos 30° \sin 60° + i \sin 30° \cos 60°]$$

$$= 10[\cos(30° + 60°) + i \sin(30° + 60°)]$$

| This is the product | This is the sum of the two angles |
| of the 2 magnitudes | |

$$= 10[\cos 90° + i \sin 90°] \hspace{4cm} □$$

Definition 11.2C Product of Two Complex Numbers in Trigonometric Form

The product of the complex numbers

$$a(\cos \alpha + i \sin \alpha) \cdot b(\cos \beta + i \sin \beta) = a \cdot b[\cos(\alpha + \beta) + i \sin(\alpha + \beta)]$$

Example 14 □ Find the product of $(1 + \sqrt{3}i)(2\sqrt{3} + 2i)$ using both rectangular and trigonometric forms.

In rectangular form:

$$
\begin{aligned}
(1 + \sqrt{3}i)(2\sqrt{3} + 2i) &= 2\sqrt{3} + 2i + \sqrt{3}i \cdot 2\sqrt{3} + \sqrt{3}i \cdot 2i \\
&= 2\sqrt{3} + 2i + 2 \cdot 3i + 2\sqrt{3}i^2 \\
&= 2\sqrt{3} - 2\sqrt{3} + 2i + 6i \\
&= 0 + 8i
\end{aligned}
$$

In trigonometric form:

First convert each number in rectangular form to the equivalent trigonometric form.

$$
\begin{aligned}
\text{For} \quad & 1 + \sqrt{3}i \\
r &= \sqrt{1^2 + (\sqrt{3})^2} \\
r &= 2 \\
\tan \theta &= \frac{\sqrt{3}}{1} \\
\theta &= 60°
\end{aligned}
$$

Therefore, $1 + \sqrt{3}i = 2(\cos 60° + i\sin 60°)$.

$$
\begin{aligned}
\text{For} \qquad 2\sqrt{3} + 2i \\
&= \sqrt{(2\sqrt{3})^2 + 2^2} \\
&= \sqrt{4 \cdot 3 + 4} \\
r &= 4 \\
\tan\theta &= \frac{2}{2\sqrt{3}} \\
\theta &= 30°
\end{aligned}
$$

Therefore, $2\sqrt{3} + 2i = 4(\cos 30° + i\sin 30°)$.

$$
\begin{aligned}
\text{The product} \quad 2(\cos 60° \quad + \quad i\sin 60°) \cdot 4(\cos 30° + i\sin 30°) \\
&= 8[\cos(60° + 30°) + i\sin(60° + 30°)] \\
&= 8[\cos 90° + i\sin 90°] \\
&= 8[0 + i] \\
&= 8i
\end{aligned}
$$

□

The product is the same no matter in which form the numbers are written.

Electrical engineers use an abbreviated notation for the trigonometric form of a complex number. It is called polar form.

Definition 11.2D Polar Form of a Complex Number

In polar form the complex number $r(\cos\theta + i\sin\theta)$ is written $r\,\underline{/\theta}$.

That sure reduces writing.

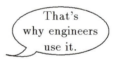
That's why engineers use it.

To multiply two numbers in polar form just multiply the magnitudes and add the angles.

Example 15 □ Multiply $3\,\underline{/20°}\cdot 4\underline{/30°}$.

Write your answer in rectangular form.

$$
\begin{aligned}
3\underline{/20°} \cdot 4\underline{/30°} &= 12\underline{/50°} \\
&= 12(\cos 50° + i\sin 50°) \\
&= 12\cos 50° + 12i\sin 50° \\
&\approx 7.7136 + (9.192)i
\end{aligned}
$$

□

Example 16 □ Multiply $5\,\underline{/230°}\cdot 2\underline{/320°}$.

Leave the answer in polar form with the argument expressed as an angle between $0°$ and $360°$.

$$
\begin{aligned}
5\underline{/230°} \cdot 2\underline{/320°} &= 10\underline{/550°} \\
&= 10\underline{/190°} + 360° \\
&= 10\underline{/190°}
\end{aligned}
$$

□

Example 17 □ Add $3 \underline{/20°} + 4 \underline{/30°}$.

Polar form is convenient for multiplication but not for addition. To add, convert each number to rectangular form.

$$
\begin{aligned}
3\underline{/20°} &= 3(\cos 20° + i \sin 20°) \\
&= 3\cos 20° + 3i\sin 20° \\
&\approx 2.819 + 1.026i \\
4\underline{/30°} &= 4(\cos 30° + i \sin 30°) \\
&\approx 4\cos 30° + 4i\sin 30° \\
&\approx 3.464 + 2.000i \\
3\underline{/20°} + 4\underline{/30°} &\approx (2.819 + 1.026i) + (3.464 + 2.000i) \\
&\approx 6.283 + 3.026i
\end{aligned}
$$

□

In actual applications you can usually plan ahead so that complex numbers are written in the form that is easiest for the operation you wish to perform.

Quotient of Two Complex Numbers

As you might expect, the quotient of two complex numbers written in polar form involves subtraction of the angles. We will derive the formula.

Remember, polar form is merely shorthand for trigonometric form.

$$
\frac{a\underline{/\alpha}}{b\underline{/\beta}} = \frac{a(\cos \alpha + i \sin \alpha)}{b(\cos \beta + i \sin \beta)}
$$

Multiply the numerator and denominator by the conjugate of the denominator.

$$
\begin{aligned}
&= \frac{a(\cos \alpha + i \sin \alpha)}{b(\cos \beta + i \sin \beta)} \cdot \frac{(\cos \beta - i \sin \beta)}{(\cos \beta - i \sin \beta)} \\
&= \frac{a(\cos \alpha \cos \beta - \cos \alpha \cdot i \sin \beta + i \sin \alpha \cos \beta - i^2 \sin \alpha \sin \beta)}{b(\cos^2 \beta - i^2 \sin^2 \beta)} \\
&= \frac{a(\cos \alpha \cos \beta + \sin \alpha \sin \beta + i \sin \alpha \cos \beta - i \cos \alpha \sin \beta)}{b(\cos^2 \beta + \sin^2 \beta)} \\
&= \frac{a}{b}[\cos(\alpha - \beta) + i \sin(\alpha - \beta)]
\end{aligned}
$$

Definition 11.2E Quotient of Two Complex Numbers in Polar Form

$$
\frac{a\underline{/\alpha}}{b\underline{/\beta}} = \frac{a}{b}\underline{/\alpha - \beta}
$$

Example 18 □ Find the quotient $\dfrac{10\underline{/40^\circ}}{5\underline{/70^\circ}}$.

$$\frac{10\underline{/40^\circ}}{5\underline{/70^\circ}} = \frac{10}{5}\underline{/40^\circ - 70^\circ}$$

$$= 2\underline{/-30^\circ}$$

$$2\underline{/-30^\circ} = 2\underline{/330^\circ}$$

What about -30°?

Either leave it as it is or convert it to an angle between 0° and 360°.

□

Problem Set 11.2

Find the modulus of the following complex numbers.

1. $-1 + 2i$ **2.** $\sqrt{3} - 2i$ **3.** $2\sqrt{3} + \sqrt{3}i$ **4.** $-8\sqrt{2} + 6\sqrt{2}i$

Write the following complex numbers in trigonometric form.

5. $1 - i$ **6.** $-3 + \sqrt{3}i$ **7.** $-2i$ **8.** 4

9. $5 + 12i$ **10.** $-15 + 8i$ **11.** $\sqrt{2} + \sqrt{2}i$ **12.** $\sqrt{3} - 3i$

Perform the indicated operations. Leave answers in trigonometric form with the argument expressed as a number between 0° and 360°, including 0° but not 360°.

13. $[3(\cos 30^\circ + i \sin 30^\circ)] \cdot [4(\cos 70^\circ + i \sin 70^\circ)]$

14. $[\sqrt{3}(\cos 80^\circ + i \sin 80^\circ)] \cdot [2\sqrt{2}(\cos 310^\circ + i \sin 310^\circ)]$

15. $[5\sqrt{6}(\cos 200^\circ + i \sin 200^\circ)] \cdot [3\sqrt{3}(\cos 170^\circ + i \sin 170^\circ)]$

16. $[3\sqrt{2}(\cos 220^\circ + i \sin 220^\circ)] \cdot [5\sqrt{3}(\cos 318^\circ + i \sin 318^\circ)]$

17. $[8(\cos 120^\circ + i \sin 120^\circ)] \div [4(\cos 80^\circ + i \sin 80^\circ)]$

18. $[6(\cos 320^\circ + i \sin 320^\circ)] \div [12(\cos 240^\circ + i \sin 240^\circ)]$

19. $[\sqrt{3}(\cos 280^\circ + i \sin 280^\circ)] \div [\sqrt{3}(\cos 20^\circ + i \sin 20^\circ)]$

20. $[4(\cos 80^\circ + i \sin 80^\circ)] \div [3(\cos 20^\circ + i \sin 20^\circ)]$

Perform the indicated operations. Leave your answers in polar form with the argument expressed as a number between 0° and 360°, including 0° but not 360°.

21. $3\underline{/40°} \cdot 8\underline{/120°}$ **22.** $2\sqrt{3}\underline{/240°} \cdot 3\underline{/320°}$

23. $\sqrt{10}\underline{/70°} \cdot \sqrt{5}\underline{/340°}$ **24.** $6\underline{/132°} \cdot 4\underline{/286°}$

25. $\dfrac{10\underline{/130°}}{\sqrt{5}\underline{/85°}}$ **26.** $\dfrac{6\underline{/128°}}{\sqrt{3}\underline{/242°}}$ **27.** $\dfrac{12\sqrt{5}\underline{/12°}}{9\sqrt{10}\underline{/348°}}$

28. $\dfrac{5\underline{/80°}}{3\sqrt{10}\underline{/50°}}$ **29.** $\dfrac{\sqrt{2}\underline{/120°}}{\sqrt{3}\underline{/150°}}$ **30.** $\dfrac{6\sqrt{6}\underline{/60°}}{8\sqrt{2}\underline{/370°}}$

Perform the following operations. Write your answers in rectangular form.

31. $3\underline{/120°} \cdot 4\underline{/60°}$ **32.** $\sqrt{3}\underline{/180°} \cdot \sqrt{2}\underline{/120°}$ **33.** $2\sqrt{6}\underline{/130°} \cdot \sqrt{3}\underline{/60°}$

34. $\dfrac{3\underline{/340°}}{4\underline{/220°}}$ **35.** $\dfrac{3\underline{/130°}}{2\underline{/310°}}$ **36.** $\dfrac{5\underline{/60°}}{12\underline{/150°}}$

Convert the following complex numbers to polar form. Perform the indicated operations; leave your answers in polar form. Express each argument as a number between 0° and 360°, including 0° but not 360°.

37. $(1+i)(-1-i)$ **38.** $(1-\sqrt{3}i)(-1+i)$ **39.** $(3-\sqrt{3}i)(-4+4i)$

40. $(8+8i)(-3-3\sqrt{3}i)$ **41.** $\dfrac{2\sqrt{3}-2i}{-5+5i}$ **42.** $\dfrac{3+\sqrt{3}i}{-\sqrt{3}+i}$

43. $\dfrac{\sqrt{2}-3i}{-2+\sqrt{3}i}$ **44.** $\dfrac{-3\sqrt{2}-2}{3-\sqrt{2}i}$

Group Writing Activity

Multiply the complex number $3+i$ by i, i^2, i^3. When you change each of these complex numbers to polar form, what do you observe happening to the angle? Why does this happen?

11.3 DeMoivre's Theorem

The previous section showed how working with the polar form of a complex number simplified multiplication and division. This section will show how to find powers of complex numbers by working in polar form.

Consider squaring a complex number:

$$
\begin{aligned}
z &= a+bi = r(\cos\theta + i\sin\theta) \\
z^2 &= r(\cos\theta + i\sin\theta)r(\cos\theta + i\sin\theta) \\
&= r^2(\cos^2\theta + 2i\sin\theta\cos\theta + i^2\sin^2\theta) \\
&= r^2[\cos\theta\cos\theta - \sin\theta\sin\theta + (\sin\theta\cos\theta + \sin\theta\cos\theta)i] \\
&= r^2[\cos 2\theta + i\sin 2\theta]
\end{aligned}
$$

Using a similar technique,

$$z^3 = r^3[\cos 3\theta + i\sin 3\theta]$$

Using mathematical induction, it is possible to show that this relationship is true for all integers.

11.3 DeMoivre's Theorem

For any complex number

$$z = r(\cos\theta + i\sin\theta)$$
$$z^n = r^n(\cos n\theta + i\sin n\theta)$$

In polar form,

$$\left(r\underline{/\theta}\right)^n = r^n\underline{/n\theta}$$

Example 19 □ Find $\left(\sqrt{2} + \sqrt{2}i\right)^5$.

First write $\sqrt{2} + \sqrt{2}i$ in trigonometric form:

$$r = \sqrt{\left(\sqrt{2}\right)^2 + \left(\sqrt{2}\right)^2}$$
$$r = 2$$
$$\tan\theta = \frac{\sqrt{2}}{\sqrt{2}}$$
$$\theta = 45°$$
$$z = \sqrt{2} + \sqrt{2}i = 2(\cos 45° + i\sin 45°)$$

To raise z to the fifth power,

$$z^5 = 2^5[\cos(5 \cdot 45°) + i\sin(5 \cdot 45°)]$$
$$= 32(\cos 225° + i\sin 225°)$$
$$= 32\left[\frac{-\sqrt{2}}{2} + i\left(\frac{-\sqrt{2}}{2}\right)\right]$$
$$z^5 = -16\sqrt{2} - 16\sqrt{2}i$$

□

Let's look at an example where we can predict the result.

Example 20 □ Find $(2i)^4$.

In trigonometric form,

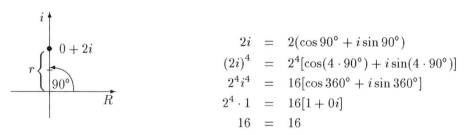

□ Figure 11.8

$$2i = 2(\cos 90° + i\sin 90°)$$
$$(2i)^4 = 2^4[\cos(4 \cdot 90°) + i\sin(4 \cdot 90°)]$$
$$2^4 i^4 = 16[\cos 360° + i\sin 360°]$$
$$2^4 \cdot 1 = 16[1 + 0i]$$
$$16 = 16$$

□

Using DeMoivre's Theorem to Find Roots

In advanced math classes it can be proven that DeMoivre's Theorem is true if n is any real number. That means we can use DeMoivre's Theorem to find roots of complex numbers. Just as a real number has two square roots, a complex number has two square roots, three cube roots, four fourth roots, and so on. To find all these roots, we must write the trigonometric form of the complex number more completely.

$$a + bi = r[\cos(\theta + 360°k) + i\sin(\theta + 360°k)]$$

Because all multiples of 360° added to
θ are equivalent angles

With this addition we can use DeMoivre's Theorem to find roots of complex numbers.

Example 21 □ Find three cube roots of -8.

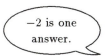

-2 is one
answer.

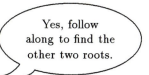

Yes, follow
along to find the
other two roots.

First write -8 in trigonometric form:

$$
\begin{aligned}
-8 &= -8 + 0i \\
&= 8[\cos(180° + 360°k) + i\sin(180° + 360°k)]
\end{aligned}
$$

□ Figure 11.9

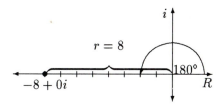

By DeMoivre's Theorem

$$(-8)^{1/3} = 8^{1/3}\left[\cos\frac{1}{3}(180° + 360°k) + i\sin\frac{1}{3}(180° + 360°k)\right]$$

$$(-8)^{1/3} = 2[\cos(60° + 120°k) + i\sin(60° + 120°k)]$$

This is the
general solution

To find all roots we now substitute $k = 0, 1, 2, 3 \ldots$.
For $k = 0$ in the general solution,

$$
\begin{aligned}
(-8)^{1/3} &= 2[\cos 60° + i\sin 60°] \quad \Leftarrow k = 0\\
&= 2\left[\frac{1}{2} + i\frac{\sqrt{3}}{2}\right] \\
&= 1 + \sqrt{3}i
\end{aligned}
$$

Is that
a cube root
of -8?

Yes,
multiply it out:
$(1 + \sqrt{3}i)^3 = -8$.

For $k = 1$ in the general solution,

$$
\begin{aligned}
(-8)^{1/3} &= 2[\cos(60° + 1 \cdot 120°) + i \sin(60° + 1 \cdot 120°)] \\
&= 2[\cos 180° + i \sin 180°] \\
&= 2[-1 + 0i] \\
&= -2, \text{ which is the root we expected.} \quad \boxed{k = 1}
\end{aligned}
$$

For $k = 2$ in the general solution,

$$
\begin{aligned}
(-8)^{1/3} &= 2[\cos(60° + 2 \cdot 120°) + i \sin(60° + 2 \cdot 120°)] \quad \boxed{k = 2} \\
&= 2[\cos 300° + i \sin 300°] \\
&= 2\left[\frac{1}{2} + i\left(-\frac{\sqrt{3}}{2}\right)\right] \\
&= 1 - \sqrt{3}i
\end{aligned}
$$

For $k = 3$ in the general solution,

$$
\begin{aligned}
(-8)^{1/3} &= 2[\cos(60° + 3 \cdot 120°) + i \sin(60° + 3 \cdot 120°)] \\
&= 2[\cos(60° + 360°) + i \sin(60° + 360°)]
\end{aligned}
$$

This is a repeat of the first solution. All other values for k will yield repeated solutions.

$\square$

The Nature and Number of Complex Roots

Plot the three cube roots of -8 on the complex plane.

$\square$ Figure 11.10

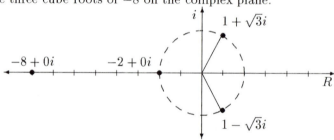

All these roots lie equally spaced on a circle of radius $= r^{1/3}$.

If we are looking for the nth root, we would find n of them equally spaced on a circle of radius $r^{1/n}$.

Group Writing Activity

Why are the roots equally spaced on a circle of radius $r^{1/n}$?

Example 22 $\square$ Solve the equation $x^4 + 2i = 2\sqrt{3}$.
This equation can be rewritten as

$$
x^4 = 2\sqrt{3} - 2i
$$

Now we can use DeMoivre's Theorem to find four fourth roots of the complex number $2\sqrt{3} - 2i$. Write $2\sqrt{3} - 2i$ in trigonometric form:

$$r = \sqrt{(2\sqrt{3})^2 + 2^2}$$
$$= \sqrt{4 \cdot 3 + 4}$$
$$r = 4$$
$$\tan\theta' = \left|-\frac{2}{2\sqrt{3}}\right|$$
$$\tan\theta' = +\frac{1}{\sqrt{3}}$$
$$\theta' = +30° \quad \text{(This is a reference angle,}$$
$$\theta \text{ is in the fourth quadrant.)}$$
$$\theta = 330°$$
$$2\sqrt{3} - 2i = 4[\cos(330° + 360°k) + i\sin(330° + 360°k)]$$
$$(2\sqrt{3} - 2i)^{1/4} = \sqrt{2}\left[\cos\frac{1}{4}(330° + 360°k) + i\sin\frac{1}{4}(330° + 360°k)\right]$$
$$= \sqrt{2}[\cos(82.5° + 90°k) + i\sin(82.5° + 90°k)]$$

For
$$k = 0$$
$$x^{1/4} \approx 1.414(\cos 82.5° + i\sin 82.5°)$$
$$\approx 0.1846 + 1.402i$$

For
$$k = 1$$
$$x^{1/4} \approx 1.414(\cos 172.5° + i\sin 172.5°)$$
$$\approx -1.402 + 0.1846i$$

For
$$k = 2$$
$$x^{1/4} \approx 1.141(\cos 262.5° + i\sin 262.5°)$$
$$\approx -0.1846 - 1.402i$$

For
$$k = 3$$
$$x^{1/4} \approx 1.414(\cos 352.5° + i\sin 352.5°)$$
$$\approx 1.402 - 0.1846i$$

□

Problem Set 11.3

Use DeMoivre's Theorem to raise the following to the indicated power. Leave the answer in polar form with the argument expressed as an angle between $0°$ and $360°$, including $0°$ but not $360°$.

1. $(2\underline{/30°})^3$ 2. $(\sqrt{2}\underline{/70°})^4$ 3. $(2\underline{/80°})^5$ 4. $(2\sqrt{3}\underline{/130°})^3$ 5. $(1 - i\sqrt{3})^2$

6. $(-\sqrt{3} - i)^4$ 7. $(-3 - 3i)^4$ 8. $(\sqrt{3} + 3i)^3$ 9. $(-3i)^4$ 10. $(-2 + 3i)^4$

Use DeMoivre's Theorem to raise the following to the indicated powers. Express each answer in rectangular form. Round off to two decimal places.

11. $(\sqrt{3}\underline{/50°})^4$ 12. $(2\underline{/140°})^3$ 13. $(2\sqrt{3} - 2i)^2$ 14. $(-1 + 2i)^4$

Use DeMoivre's Theorem to find the indicated roots. Leave your answer in polar form.

15. 3 cube roots of $-i$

16. 2 square roots of 16

17. 4 fourth roots of $-16i$

18. 5 fifth roots of -32

19. 4 fourth roots of $-3 + \sqrt{3}i$

20. 6 sixth roots of $-64 - 64\sqrt{3}i$

Use DeMoivre's Theorem to find the indicated roots. Express your answer in rectangular form. Round off to two decimal places.

21. 3 cube roots of $-27i$

22. 4 fourth roots of $-8 - 8\sqrt{3}i$

23. 2 square roots of -4

24. 4 fourth roots of -4

25. 4 fourth roots of $-3 + 4i$

26. 3 cube roots of $4 - 4\sqrt{3}i$

Solve the following equations for all real and complex roots.

27. $x^4 + 16 = 0$

28. $x^5 + 1 = 0$

29. $x^3 - 8i = 0$

30. $x^4 - 3i = \sqrt{2}$

31. $x^3 + 2i = \sqrt{3}$

32. $x^6 + 64 = 0$

Chapter 11   Key Ideas

11.1 **1.** We define $\sqrt{-1} = i$, which has the property $i^2 = -1$.

2. Numbers of the form bi are called imaginary numbers.

$$\sqrt{-a} = \sqrt{a}i$$

3. The complex number system is a set of numbers of the form $a + bi$.

$$a + bi = c + di \quad \text{if} \quad a = c \text{ and } b = d$$
$$(a + bi) + (c + di) = (a + c) + (b + d)i$$
$$(a + bi)(c + di) = (ac - bd) + (bc + ad)i$$

4. The conjugate of the complex number $z = a + bi$ is $\overline{z} = a - bi$.

11.2 **1.** The modulus of a complex number $z = a + bi$ is $|z| = \sqrt{a^2 + b^2}$.

2. The trigonometric form of the complex number $z = a + bi$ is
$z = r(\cos\theta + i\sin\theta)$ where $r = \sqrt{a^2 + b^2}$ and θ is the angle with $\tan\theta = \dfrac{b}{a}$ in the same quadrant as $a + bi$.

3. The product of two complex numbers is

$$a(\cos\alpha + i\sin\alpha) \cdot b(\cos\beta + i\sin\beta) = a \cdot b[\cos(\alpha + \beta) + i\sin(\alpha + \beta)].$$

4. In polar form, the complex number $r(\cos\theta + i\sin\theta)$ is written $r\underline{/\theta}$.

5. The product of two complex numbers in polar form is

$$a\underline{/\alpha} \cdot b\underline{/\beta} = a \cdot b\underline{/\alpha + \beta}$$

6. The quotient of two complex numbers in trigonometric form is

$$\frac{a(\cos\alpha + i\sin\alpha)}{b(\cos\beta + i\sin\beta)} = \frac{a}{b}[\cos(\alpha - \beta) + i\sin(\alpha - \beta)]$$

and in polar form is

$$\frac{a\underline{/\alpha}}{b\underline{/\beta}} = \frac{a}{b}\underline{/\alpha - \beta}$$

11.3 **1.** DeMoivre's Theorem says that for any complex number:

$$\text{if}\qquad z = r(\cos\theta + i\sin\theta),$$
$$\text{then}\qquad z^n = r^n(\cos n\theta + i\sin n\theta)$$

and in polar form:

$$\text{if}\qquad z = r\underline{/\theta},\qquad \text{then}\qquad z^n = r^n\underline{/n\theta}$$

2. To find the roots of a complex number in trigonometric form, use the complex number written as

$$a + bi = r[\cos(\theta + k \cdot 360°) + i\sin(\theta + k \cdot 360°)]$$

or in radians

$$a + bi = r[\cos(\theta + k \cdot 2\pi) + i\sin(\theta + k \cdot 2\pi)]$$

Then,

$$(a + bi)^{1/n} = r^{1/n}\left[\cos\frac{1}{n}(\theta + k \cdot 360°) + i\sin\frac{1}{n}(\theta + k \cdot 360°)\right]$$

or in radians

$$(a + bi)^{1/n} = r^{1/n}\left[\cos\frac{1}{n}(\theta + k \cdot 2\pi) + i\sin\frac{1}{n}(\theta + k \cdot 2\pi)\right]$$

3. The $n - n^{th}$ roots of a complex number in trigonometric form are equally spaced on a circle which has a radius $r^{\frac{1}{n}}$.

Chapter 11 Review Test

Perform the indicated operations. **(11.1)**

1. $(12 - \sqrt{12}i) + (3 + \sqrt{3}i)$

2. $(15 + \sqrt{-8}) - (-7 + \sqrt{-18})$

3. $(\sqrt{3} + 2i)(2\sqrt{3} - 8i)$

4. $(\sqrt{27} - \sqrt{-32})(\sqrt{3} + \sqrt{-50})$

5. $\dfrac{3 + 4i}{2 - 3i}$

6. $\dfrac{8 - \sqrt{-3}}{3 + \sqrt{-12}}$

Find the modulus of the complex numbers. **(11.2)**

7. $5\sqrt{6} - i$

8. $5 + 4i$

Perform the indicated operations. Leave your answer in trigonometric form with the argument expressed as a number between $0°$ and $360°$ ($0° \leq \theta < 360°$). **(11.2)**

9. $3(\cos 70° + i \sin 70°) \cdot 15(\cos 120° + i \sin 120°)$

10. $[3\sqrt{6}(\cos 160° + i \sin 160°)] \cdot [2\sqrt{3}(\cos 240° + i \sin 240°)]$

11. $[3\sqrt{8}(\cos 120° + i \sin 120°)] \div [2\sqrt{18}(\cos 50° + i \sin 50°)]$

12. $[5\sqrt{5}(\cos 50° + i \sin 50°)] \div [2\sqrt{10}(\cos 170° + i \sin 170°)]$

Perform the indicated operations. Leave your answer in polar form with the argument expressed as a number between $0°$ and $360°$ ($0° \leq \theta < 360°$). **(11.2)**

13. $(2\sqrt{3}\underline{/35°}) \cdot (2\sqrt{6}\underline{/165°})$

14. $(5\sqrt{5}\underline{/125°})(4\sqrt{10}\underline{/315°})$

15. $\dfrac{8\underline{/140°}}{6\underline{/70°}}$

16. $\dfrac{3\sqrt{2}\underline{/80°}}{2\sqrt{6}\underline{/140°}}$

Perform the following operations. Write answers in rectangular form. **(11.2)**

17. $5\underline{/310°} \cdot 6\underline{/350°}$

18. $\dfrac{4\sqrt{6}\underline{/130°}}{\sqrt{3}\underline{/70°}}$

Convert the following complex numbers to trigonometric form; then perform the indicated operations. Leave your answers in trigonometric form. **(11.2)**

19. $(2\sqrt{3} - 2i)(-3 - \sqrt{3}i)$

20. $(3 + 4i)(-12 + 5i)$

21. $\dfrac{3 - 3\sqrt{3}i}{-4 + 4i}$

22. $\dfrac{-5 - 3i}{4 + i}$

Use DeMoivre's Theorem to perform the indicated operations. Leave your answer in polar form. **(11.3)**

23. $(-2 - 2\sqrt{3}i)^4$

24. $(-3i)^4$

25. Find the two square roots of $4 - 4\sqrt{3}i$.

26. Find the five fifth roots of $-32 + 32i$.

Use DeMoivre's Theorem to perform the indicated operations. Express your answer in rectangular form. **(11.3)**

27. $(-3 + \sqrt{3}i)^3$

28. Find the three cube roots of $-4 - 3i$.

Solve the following equations for all real and complex roots. **(11.3)**

29. $x^6 + 1 = 0$

30. $x^4 + i = 2\sqrt{2}$

Chapter 12

Polar Coordinates

Contents

Preview

There are times when the Cartesian coordinate system is the natural way to describe a physical situation. For example, if you are laying out a grid pattern of city streets, you will probably choose a Cartesian coordinate system.

However, there are other times when a different coordinate system is more natural and hence more useful. The natural system to describe rotation is a system of polar coordinates. This system simplifies analysis of a rotating electrical generator, a rotating radar beam, or the energy from the sun as it radiates in a sphere throughout space.

This chapter will start by defining the system of polar coordinates. Then it will compare polar coordinates to rectangular coordinates and show you how to convert between the two systems. The third section of this chapter will work with some graphs generated in polar coordinates.

12.1   The Polar Coordinate System

The rectangular or Cartesian coordinate system has many useful applications. However, there are times when other systems fit the application more naturally. Consider a rotating radar antenna. The information available is an angle of rotation and a distance from the antenna site.

To establish a polar coordinate system, we start with a point 0 and extend a reference line from the point. The point is called the pole and the line which is usually chosen to lie along the x-axis is called the polar axis.   To locate an arbitrary point on the plane, draw a vector from the pole to the point. This is called the radius vector.   The location of the point is then specified by the length of the radius vector and its angle from the polar axis.

□ Figure 12.1

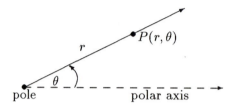

The signs of each of the polar coordinates are important. Positive θ, consistent with trigonometry, is in the counterclockwise direction. Positive r is along the terminal side of angle θ. Negative r is in the direction opposite the terminal side of θ.

Example 1 □ Plot the following in polar coordinates.

$(4, 60°), (4, -60°), (-4, 60°), (-4, -60°)$

□ Figure 12.2

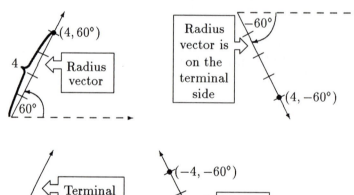

□ Figure 12.3

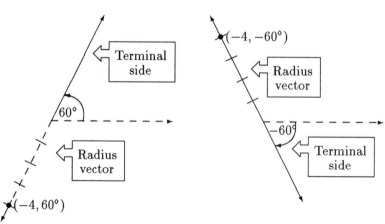

□

In a rectangular system each point corresponds to one, and only one, ordered pair. Not so in a polar coordinate system. Each point (r, θ) corresponds to $(r, \theta + k360°)$.

Even when you ignore angles with absolute values greater than 360°, there are four representations of each point in polar coordinates.

Example 2 □ Give three other ordered points to specify $(3, 45°)$.

□ Figure 12.4

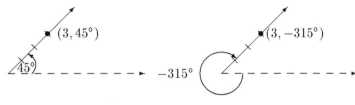

□ Figure 12.5

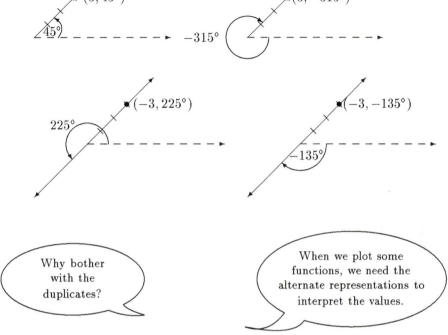

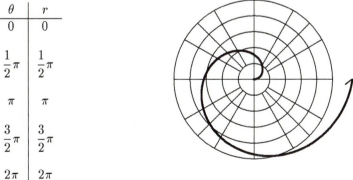

In rectangular coordinates we express y as a function of x. In polar coordinates, we usually express r as a function of θ. It is possible to plot graphs of functions expressed in polar coordinates by plotting the points (r, θ). We analyze the effect of changing θ on r to determine how the graph develops.

Example 3 □ Plot the graph of $r = \theta$.

Notice that as θ increases r increases at the same rate. The graph quickly spirals away from the pole.

□ Figure 12.6

θ	r
0	0
$\frac{1}{2}\pi$	$\frac{1}{2}\pi$
π	π
$\frac{3}{2}\pi$	$\frac{3}{2}\pi$
2π	2π

This graph is a member of a larger family of curves $r = a\theta$, $a > 0$, called Archimedes' spiral. □

Group Writing Activity

What do you think the effect of varying the value of a has on the graph of $r = a\theta$? What would happen if the restriction $a > 0$ was removed?

Example 4 □ Plot the graph of $r = \sin\theta$.

We will analyze the graph as θ passes through each quarter of a rotation. At $\theta = 0$, r is zero. The curve starts at the pole.

□ Figure 12.7

As θ increases from 0 to $\frac{1}{2}\pi$, r increases from 0 to 1.

θ	r
0	0
$\frac{1}{6}\pi$	.5
$\frac{1}{4}\pi$	.7
$\frac{1}{3}\pi$	.86
$\frac{1}{2}\pi$	1

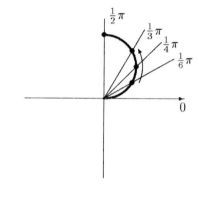

□ Figure 12.8

As θ increases from $\frac{1}{2}\pi$ to π, r decreases from 1 to 0.

θ	r
$\frac{2}{3}\pi$	.86
$\frac{3}{4}\pi$	.7
$\frac{5}{6}\pi$	.5
π	0

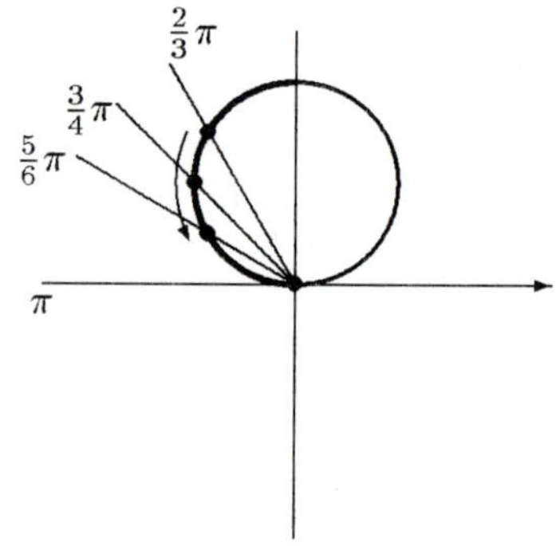

□ Figure 12.9

In this quadrant, because $\sin\theta$ is negative, r is negative.

As θ increases from π to $\frac{3}{2}\pi$, r decreases from 0 to −1.

Notice the points from $0 \le \theta \le \frac{1}{2}\pi$ are repeated.

θ	r
$\frac{7}{6}\pi$	−.5
$\frac{5}{4}\pi$	−.7
$\frac{4}{3}\pi$	−.86
$\frac{3}{2}\pi$	−1

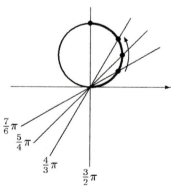

In this quadrant, r remains negative.

As θ increases from $\frac{3}{2}\pi$ to 2π, r increases from -1 to 0.

Notice the points from $\frac{1}{2}\pi \leq \theta \leq \pi$ are repeated.

θ	r
$\frac{5}{3}\pi$	$-.86$
$\frac{7}{4}\pi$	$-.7$
$\frac{11}{6}\pi$	$-.5$
2π	0

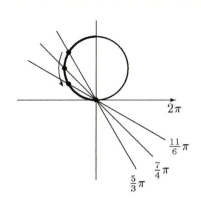

□

Example 5 □ Plot the graph of $r = 2 + 2\sin\theta$.

We will analyze the graph as θ passes through each quarter of a rotation. At $\theta = 0$, $\sin\theta = 0$, $r = 2$. The curve starts at the point $(2, 0)$.

□ Figure 12.11

As θ increases from 0 to $\frac{1}{2}\pi$, $\sin\theta$ increases from 0 to 1, therefore r increases from 2 to 4.

θ	r
0	2
$\frac{1}{6}\pi$	3
$\frac{1}{4}\pi$	3.41
$\frac{1}{3}\pi$	3.73
$\frac{1}{2}\pi$	4

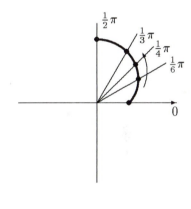

□ Figure 12.12

As θ increases from $\frac{1}{2}\pi$ to π, $\sin\theta$ decreases from 1 to 0, therefore r decreases from 4 to 2.

θ	r
$\frac{2}{3}\pi$	3.73
$\frac{3}{4}\pi$	3.41
$\frac{5}{6}\pi$	3
π	2

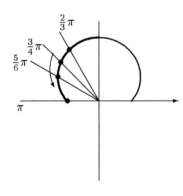

As θ increases from π to $\frac{3}{2}\pi$, $\sin\theta$ decreases from 0 to -1, therefore r decreases from 2 to 0.

θ	r
$\frac{7}{6}\pi$	1
$\frac{5}{4}\pi$	.59
$\frac{4}{3}\pi$	.27
$\frac{3}{2}\pi$	0

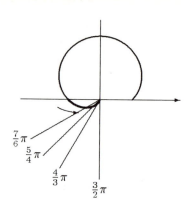

□ Figure 12.14

As θ increases from $\frac{3}{2}\pi$ to 2π, $\sin\theta$ increases from -1 to 0, therefore r increases from 0 to 2.

θ	r
$\frac{5}{3}\pi$	.27
$\frac{7}{4}\pi$	.59
$\frac{11}{6}\pi$	1
2π	2

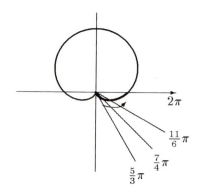

□

This curve is called a cardioid.

What happens if we substitute $\cos\theta$ for $\sin\theta$?

We get the same cardioid rotated by $\frac{1}{2}\pi$.

□ Figure 12.15

$r = 2 + 2\cos\theta$
yields the graph

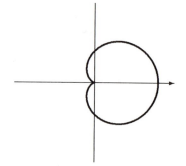

Cardioids are a subset of a larger class of curves called limacons.

Chapter 12 Polar Coordinates

Limacons

Equations of the form

$$
\begin{aligned}
r &= a + b \sin \theta \\
r &= a - b \sin \theta \\
r &= a + b \cos \theta \\
r &= a - b \cos \theta
\end{aligned}
$$

produce a class of curves called limacons. The ratio of a to b determines the shape.

In equations of the form $r = a + b \cos \theta$

if $\quad \dfrac{|a|}{|b|} = 1 \quad$ the curve will be a cardioid.

if $\quad \dfrac{a}{b} \geq 0 \quad$ the bulk of the limacon will be on the right of the pole.

if $\quad \dfrac{a}{b} < 0 \quad$ the bulk of the limacon will be on the left of the pole.

Group Writing Activity

Instead of exploring the graph of $r = a + b \cos \theta$ as above, determine what effect varying a and b has on the graph of $r = a + b \sin \theta$.

If $|a| < |b|$ then $\dfrac{|a|}{|b|}$ will be less than one, and the limacon curve will have an inner loop.

□ Figure 12.16

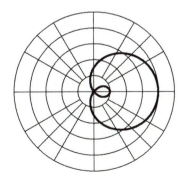

$$r = a + b \cos \theta$$

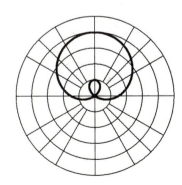

$$r = a + b \sin \theta$$

If $|a| > |b|$ then $\dfrac{|a|}{|b|}$ will be greater than one, and the limacon will have a dimple.

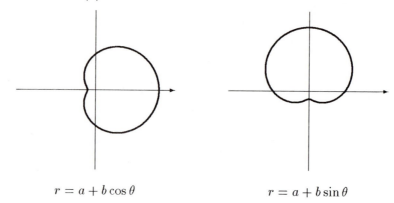

□ Figure 12.17

$$r = a + b\cos\theta \qquad\qquad r = a + b\sin\theta$$

Notice that these curves all exhibit some type of symmetry.

Symmetry in Polar Coordinates

Consider the graph of $y = x^2$ in rectangular coordinates. This curve is symmetrical about the y-axis.

□ Figure 12.18

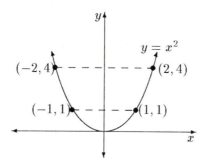

A test for symmetry about the y-axis in rectangular coordinates is: if each value of x can be replaced with $-x$ without changing the corresponding value for y, the equation is symmetric about the y-axis.

To illustrate, in $y = x^2$
if x is replaced with $-x$, the equation is
$$\begin{aligned} y &= (-x)^2 \\ &= x^2 \end{aligned}$$

Same as original equation, therefore symmetric about the y-axis

□ Figure 12.19

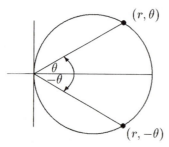

Similar tests exist for polar coordinates. Consider $r = \cos\theta$. The points (r, θ) and $(r, -\theta)$ are symmetric about the polar axis. A curve in polar coordinates is always symmetric about the polar axis if θ can be replaced with $-\theta$ without changing the value of r.

Because $\cos\theta = \cos(-\theta)$, $r = \cos\theta$ is symmetric about the polar axis.

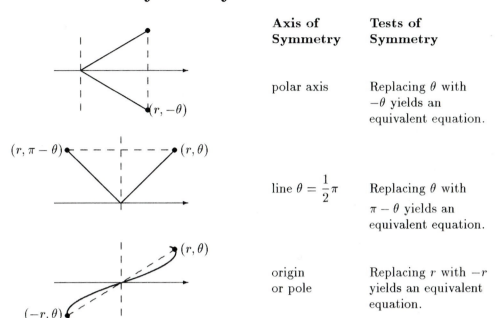

12.1 Tests for Symmetry

□ Figure 12.20

Axis of Symmetry	Tests of Symmetry
polar axis	Replacing θ with $-\theta$ yields an equivalent equation.
line $\theta = \frac{1}{2}\pi$	Replacing θ with $\pi - \theta$ yields an equivalent equation.
origin or pole	Replacing r with $-r$ yields an equivalent equation.

Do these tests always work?

If a curve passes the test, it will have the predicted form of symmetry. But some curves with the desired symmetry won't pass the test.

Example 6 □ Test the symmetry of $r = \dfrac{3}{1 - \cos\theta}$.

By replacing θ with $-\theta$ we get

$$r = \frac{3}{1 - \cos(-\theta)}$$

We recall $\cos(-\theta) = \cos\theta$

hence
$$\begin{aligned} r &= \frac{3}{1 - \cos(-\theta)} \\ &= \frac{3}{1 - \cos\theta} \end{aligned}$$

which is the original equation and therefore the graph is symmetric with respect to the polar axis. □

Using Your Graphing Calculator

Most of the newer graphing calculators allow you to watch r change as a function of θ. There are a few critical steps to consider as you use the graphing calculator in polar mode to show the graph of $r = f(\theta)$.

1. Make sure your calculator is in polar mode (consult your manual if necessary).

2. To get the proper range or viewing window you will need to set three variables: $\theta, x,$ and y.

   (a) θ determines the range of values that will be substituted into the relation $r = f(\theta)$.

   (b) $X_{\min}$ and $X_{\max}$ determine the horizontal width of the plot area.

   (c) $Y_{\min}$ and $Y_{\max}$ determine the vertical height of the plot area.

Because a graphing calculator screen is usually wider than it is high, you will need to make $Y_{\max} - Y_{\min}$ about $\frac{2}{3}$ the difference $X_{\max} - X_{\min}$ on most calculators. Some calculators have a function, ZSQUARE, on the ZOOM key to solve this problem automatically.

Problem Set 12.1

To assist you in sketching the graphs below, we have included the tracing guide for polar graphs on the right. You may find it helpful to place a sheet of plain white paper over this guide. You will be able to see enough of this guide through the sheet of paper to plot points. It is not necessary to trace the entire guide for each graph. Trace only the polar axis and $\frac{1}{2}\pi$ axis along with any other critical axes you wish to show.

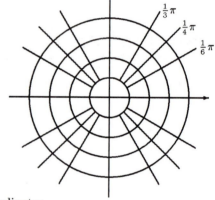

Plot the following points which are given in polar coordinates.

1. $(3, 60°)$ 2. $(-4, 45°)$ 3. $(-3, -45°)$ 4. $(2, -300°)$

5. $(-3, -150°)$ 6. $(4, 390°)$ 7. $(2, -210°)$ 8. $(3, -150°)$

Give three other ordered pairs to specify the following points. Do not use angles greater than 360°.

9. $(1, 30°)$ 10. $(5, 120°)$ 11. $(-5, 210°)$ 12. $(3, -150°)$

Make a table of values and sketch the following graphs in polar form using the tracing guide above.

13. $r = \sin \theta$ 14. $r = -\cos \theta$ 15. $r = -\sin \theta$

16. $r = 2\cos \theta$ 17. $r = 3\sin \theta$ 18. $r = -3\cos \theta$

19. $r = -2\cos \theta$ 20. $r = -3\sin \theta$ 21. $r = 2(1 - \cos \theta)$

22. $r = 3 + 2\cos \theta$ 23. $r = 3 - 2\sin \theta$ 24. $r = 3 + 3\cos \theta$

25. $r = 2 + 3\cos \theta$ 26. $r = 1 - 2\sin \theta$

Determine the symmetry of the following curves in polar form.

27. $r = -6\cos \theta$ 28. $2r = \cos \theta$ 29. $r^2 = 9\cos 2\theta$

30. $r^2 = -16\sin 2\theta$ 31. $r = 5(1 - \sin \theta)$ 32. $r = 3 + 3\cos \theta$

33. $r = \dfrac{5}{2 - 3\cos \theta}$ 34. $r = \dfrac{-6}{-5 + 4\sin \theta}$ 35. $r \sec \theta = 5$

36. $r = -6\csc \theta$

12.2 Parametric Equations

Sometimes rather than describing y as a function of x it is useful to describe both x and y as functions of a third variable such as time. This technique is called parametric equations and can be used to plot polar equations on a graphing calculator that does not have polar-coordinate graphing mode.

Think about a projectile fired from a gun at an angle θ above the horizontal with a velocity V_0.

□ Figure 12.21

There are three separate components of the motion that can be analyzed as functions of time.

1. There is a horizontal component.
   $x = V_0 \cos \theta t$
2. There is a vertical upward component.
   $y_1 = V_0 \sin \theta t$
3. There is a downward vertical motion due to gravity.
   $y_2 = -16t^2$

When we describe two variables in terms of a third variable, the third variable—time in this case—is called a parameter. The system is called a system of parametric equations.

Example 7 □ Use your graphing calculator to plot the path of a projectile fired at 300 ft/sec from a cannon pointed at 30° above the horizontal.

Put your graphing calculator in parametric mode using degrees. Then enter the following equations.

$$\begin{aligned} X_{1_T} &= 300 \cos (30°) \, t \\ Y_{1_T} &= 300 \sin (30°) \, t - 16t^2 \end{aligned}$$

Press the $\boxed{\text{RANGE}}$ key and enter

$T_{\min}$	=	0	$X_{\min}$	=	-1	
$T_{\max}$	=	10	$X_{\max}$	=	3000	
T_{step}	=	.1	X_{scl}	=	250	

$Y_{\min}$ = -100
$Y_{\max}$ = 1000
Y_{scl} = 250

Then execute the graph.

□ Figure 12.22

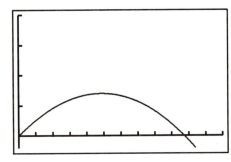

Why does your
projectile go through the
x-axis? It should stop
when it hits ground.

It will stop if it
hits ground. But the x-axis
is merely the horizontal. If
you fired off a cliff the shell
could go below the
horizontal.

If the projectile crossing the x-axis really bothers you, trace along the curve and you can find the approximate value of t when the projectile hits ground. Notice the display gives the x and y coordinates as well as t in parametric mode.

The shell will cross the horizontal between 9.3 and 9.4 seconds at a horizontal displacement of approximately 2430 ft.   □

We define parametric equations $x = f(t)$ and $y = g(t)$, where t is the parameter and f and g are continuous functions on an interval I.

To graph a set of parametric equations on a rectangular coordinate system, choose values of t and determine the resulting x and y coordinates separately for each value of t. Plot the ordered pairs (x, y) where x and y are paired for a specific t value. Join the points with a smooth curve and indicate the direction of the curve based upon the increasing values of t.

Conversion of Polar Coordinates to Parametric Equations in a Cartesian System

Superimpose a point in polar coordinates on a Cartesian system.

□ Figure 12.23

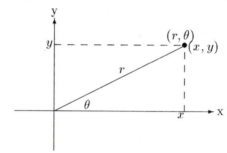

Notice the (x, y) coordinates of the point (r, θ) are given by the equations

$$x = r \cos \theta \qquad y = r \sin \theta$$

We can use this information to find parametric equations that are equivalent to a polar equation.

Example 8 □ Use parametric equations to plot a graph of $r = 2 + 3 \sin \theta$.

To find the parametric equations that represent the function, simply substitute the polar equation for r in the equations

$$x = r \cos \theta \quad \text{and} \quad y = r \sin \theta$$
$$x = (2 + 3 \sin \theta) \cos \theta \quad \text{and} \quad y = (2 + 3 \sin \theta) \sin \theta$$

To plot the graph of this function in your calculator, you may need to substitute T for θ. Set the mode of your calculator to parametric using radians.
Set the range as follows.

$$
\begin{array}{lll}
T_{\min} = 0 & X_{\min} = -6 & Y_{\min} = -2 \\
T_{\max} = 6.28 & X_{\max} = 6 & Y_{\max} = 6 \\
T_{\text{step}} = .2 & X_{\text{scl}} = .5 & Y_{\text{scl}} = .5
\end{array}
$$

Enter the following parametric equations:

$$X_{1_T} = (2 + 3\sin T)\cos T \qquad \text{and} \qquad Y_{1_T} = (2 + 3\sin T)\sin T$$

You should get this graph.

□ Figure 12.24

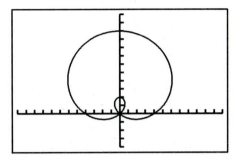

□

Problem Set 12.2

Convert the following polar equations to parametric equations. Then use your graphing calculator to plot the function. By using the trace function, examine the graph and approximate the maximum and minimum values of x and y. Also indicate if the graph appears to be symmetrical about the x-axis, y-axis, or the origin.

1. $r = 5$

2. $r = 2\theta$

3. $r = 2\sin 2\theta$

4. $r = 3\sin 3\theta$

5. $r = 5\cos 3\theta$

6. $r = 4\sin 2\theta$

7. $r^2 = 4\sin 2\theta$

8. $r^2 = 9\cos 2\theta$

9. $r = \dfrac{5}{2 - 2\sin\theta}$

10. $r = \dfrac{6}{4 - 3\cos\theta}$

Group Writing Activity

The only difference between problems 6 and 7 is that number 7 has r^2 instead of r. What effect does this have on the graph and why?

12.3　Other Graphs in Polar Coordinates

Straight Lines in Polar Coordinates

Recall from the chapter on vectors that a vector can be divided into x and y components.

$$x = r\cos\theta \qquad \text{and} \qquad y = r\sin\theta$$

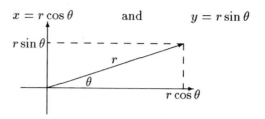

$r\sin\theta$ is equal to a constant is equivalent to saying $y = k$.

Therefore,

$$r\sin\theta = k \qquad \text{is the equation of a horizontal line.}$$

$$r\cos\theta = k \qquad \text{is the equation of a vertical line.}$$

Solving these equations for r gives the polar equations of horizontal or vertical lines.

$$r = \frac{k}{\cos\theta} \quad \longleftarrow \boxed{\text{Vertical line}}$$

$$r = \frac{k}{\sin\theta} \quad \longleftarrow \boxed{\text{Horizontal line}}$$

$ax + by = c$ is the general equation of a line in rectangular coordinates. If we substitute $x = r\cos\theta$ and $y = r\sin\theta$, we get the equation of a line in polar coordinates.

$$ar\cos\theta + br\sin\theta = c$$

The slope of this line is $-\dfrac{a}{b}$.

The y intercept is $\dfrac{c}{b}$.

Example 9 □ Write the equation of a circle centered at the origin in polar coordinates.

In rectangular coordinates the equation of a circle with its center at the origin is

$$x^2 + y^2 = a^2$$

If we substitute
$$x = r\cos\theta$$
$$y = r\sin\theta$$

in the rectangular equation of a circle, we get

$$(r\cos\theta)^2 + (r\sin\theta)^2 = a^2$$
$$r^2\cos^2\theta + r^2\sin^2\theta = a^2$$
$$r^2\left[\cos^2\theta + \sin^2\theta\right] = a^2$$
$$r^2 = a^2$$

taking the positive square root

$$r = a$$

This represents the equation of a circle at the origin in polar coordinates.　□

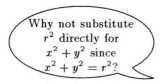

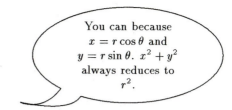

Group Writing Activity

Using the above information, write the equation of a diagonal line through the pole.

Parabolas in Polar Coordinates

A parabola is defined as the set of points that are equidistant from a fixed point (focus) and a fixed line (directrix).

□ Figure 12.26

Place the focus of the parabola at the pole of a polar coordinate system. Call the distance from the pole to the directrix p.

$$\boxed{\begin{array}{c}\text{Distance from}\\\text{focus to } P(r,\theta)\end{array}} = \boxed{\begin{array}{c}\text{Distance from } P(r,\theta)\\\text{to directrix}\end{array}}$$

$$r = p + r\cos\theta$$

Solve for r:

$$r - r\cos\theta = p$$
$$r(1 - \cos\theta) = p$$
$$r = \frac{p}{1 - \cos\theta}$$

This is the equation of a parabola in polar coordinates.

Example 10 □ Graph $r = \dfrac{2}{1 - \cos \theta}$.

The directrix is not part of the parabola; therefore, we do not draw it. Because $\cos \theta = \cos(-\theta)$, the curve is symmetric about the polar axis. We will find points from $0 < \theta < 180°$ and complete the other half of the curve using symmetry.

Notice at $\theta = 0$ the equation $r = \dfrac{2}{1 - \cos \theta}$ is not defined.

θ	0°	30°	45°	60°	90°	120°	150°	180°
$1 - \cos \theta$	0	0.13	0.29	0.5	1	1.5	1.87	2
$r = \dfrac{2}{1 - \cos \theta}$	—	14.9	6.8	4	2	1.33	1.07	1

There are no points at $\theta = 0$

Parabola crosses the polar axis halfway between the focus and the directrix.

□ Figure 12.27

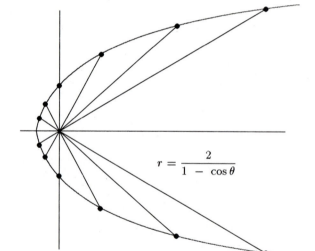

$$r = \dfrac{2}{1 - \cos \theta}$$

□

Group Writing Activity

In the polar equation

$$r = \dfrac{c}{a + b \cos \theta}$$

what effect does changing the values of a, b, and c have on the graph?
What happens if you substitute $\sin \theta$ for $\cos \theta$?

     Chapter 12 Polar Coordinates

We have graphed some figures in polar coordinates by plotting points. It might be helpful if we know approximately what a graph is going to look like before we start. Some common graphs and their equations are listed below. Use these as a guide to graph similar equations.

☐ Figure 12.28

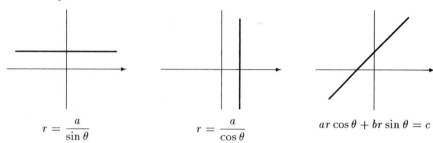

$$r = \frac{a}{\sin \theta} \qquad r = \frac{a}{\cos \theta} \qquad ar \cos \theta + br \sin \theta = c$$

☐ Figure 12.29

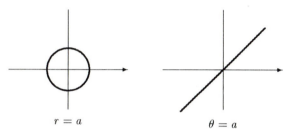
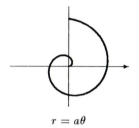

$$r = a \qquad\qquad \theta = a \qquad\qquad r = a\theta$$

☐ Figure 12.30

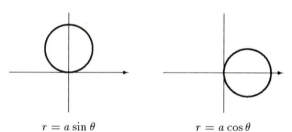
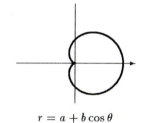

$$r = a \sin \theta \qquad r = a \cos \theta \qquad \begin{array}{c} r = a + b \cos \theta \\ \text{if } |a| = |b| \end{array}$$

☐ Figure 12.31

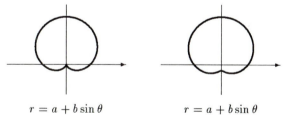
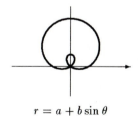

$$\begin{array}{c} r = a + b \sin \theta \\ \text{if } |a| = |b| \end{array} \qquad \begin{array}{c} r = a + b \sin \theta \\ \text{if } |a| > |b| \end{array} \qquad \begin{array}{c} r = a + b \sin \theta \\ \text{if } |a| < |b| \end{array}$$

☐ Figure 12.32

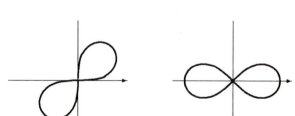

$$r^2 = a^2 \sin 2\theta \qquad r^2 = a^2 \cos 2\theta \qquad r = a \cos 2\theta$$

Group Writing Activity

Consider the polar equation

$$r^2 = a^2 \sin 2\theta$$

What happens as

1. $|a|$ is increased or decreased?

2. 2θ is replaced with -2θ?

3. 2θ is replaced with 3θ?

4. The equation is changed to $r^2 = -a^2 \sin 2\theta$?

Are these functions defined for all values of θ?

Problem Set 12.3

To assist you in sketching the graphs below we have included a tracing guide for polar graphs at the end of this problem set. Find the polar equation and sketch the curve using polar coordinates.

1. $2x + y = 6$ **2.** $x^2 + y^2 - 4x = 0$

3. $x^2 + y^2 + 6y = 0$ **4.** $x^2 + y^2 + 8x - 6y = 0$

5. $y = 4$ **6.** $x = -3$

7. $-3x + 2y = 6$ **8.** $x^2 + y^2 - 4x + 6y = 0$

Find the slope and y-intercept of the line represented by the following equations.

9. $5r \cos \theta - 3r \sin \theta = 6$ **10.** $-2r \cos \theta + 4r \sin \theta = 7$

11. $r = \dfrac{2}{3 \cos \theta - 4 \sin \theta}$ **12.** $r = \dfrac{-3}{-5 \cos \theta + 2 \sin \theta}$

Sketch the graphs of the following equations given in polar form.

13. $r = 6$ **14.** $\theta = 60°$

15. $r = -2 \sin \theta$ **16.** $r = -4 \cos \theta$

17. $r = \dfrac{3}{\sin \theta}$ **18.** $r = -\dfrac{2}{\cos \theta}$

19. $r = \dfrac{3}{1 + \cos \theta}$ **20.** $r = \dfrac{2}{1 - \sin \theta}$

21. $r = 2\theta$ **22.** $r^2 = 4 \cos 2\theta$

23. $r^2 = 9 \sin 2\theta$ **24.** $r = 3 \cos 2\theta$

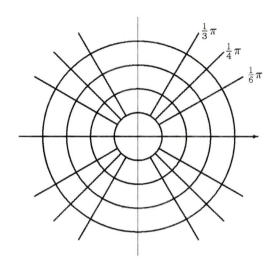

Chapter 12   Key Ideas

12.1   **1.** The polar coordinate system consists of a fixed point called the pole and a polar axis which usually is chosen to lie along the x-axis.

2. A point on the polar plane is labeled $P(r, \theta)$ where r is the length of a vector from the pole to the point, and θ is the angle the radius vector makes with the polar axis.

3. Positive θ is in the counterclockwise direction.

4. Negative θ is in the clockwise direction.

5. Negative r is in the direction opposite the terminal side of θ.

6. The graph of $r = a\theta$ is a curve called Archimedes' spiral.

7. Equations of the form:

$$r = a + b\sin\theta \qquad\qquad r = a + b\cos\theta$$
$$r = a - b\sin\theta \qquad\qquad r = a - b\cos\theta$$

are equations of limacons.

8. If $a = b$ in the above equation, the curve is a cardioid.

9. If $\dfrac{a}{b} < 1$ in the above equations, the curve is a limacon with an inner loop.

10. If $\dfrac{a}{b} > 1$ in the above equations, the curve is a limacon with a dimple.

11. Tests of symmetry for a curve in polar coordinates are:

   a.   Polar axis: Replacing θ with $-\theta$ yields an equivalent equation.

   b.   Line $\theta = \dfrac{1}{2}\pi$: Replacing θ with $\pi - \theta$ yields an equivalent equation.

   c.   Origin or pole: Replacing r with $-r$ yields an equivalent equation.

12.2 **1.** To graph a set of parametric equations, choose a value of t and determine the resulting x and y coordinates separately for each value of t.

2. To convert from the rectangular coordinate system to the polar coordinate system or vice versa, use the following formulas:

$$x = r\cos\theta$$
$$y = r\sin\theta$$
$$r = \sqrt{x^2 + y^2}$$

12.3 **1.** In polar coordinates, the equation of a vertical line is:

$$r = \frac{k}{\cos\theta}$$

2. In polar coordinates, the equation of a horizontal line is:

$$r = \frac{k}{\sin\theta}$$

3. In polar coordinates, the equation of a line is:

$$ar\cos\theta + br\sin\theta + c = 0$$

4. In polar coordinates, a circle passing through the pole with its center on the polar axis (x-axis) is:

$$r = a\cos\theta \qquad \text{or} \qquad r = -a\cos\theta$$

5. In polar coordinates, a circle passing through the pole with its center on the line $\theta = \frac{1}{2}\pi$ (y-axis) is:

$$r = a\sin\theta \qquad \text{or} \qquad r = -a\sin\theta$$

6. In polar coordinates, the equation of a circle with radius a and the center at the pole is:

$$a^2 = r^2 \qquad \text{or} \qquad r = a$$

7. A parabola in polar coordinates has an equation of the form:

$$r = \frac{p}{1 - \cos\theta}$$
$$r = \frac{p}{1 + \cos\theta}$$
$$r = \frac{p}{1 - \sin\theta}$$
$$r = \frac{p}{1 + \sin\theta}$$

Chapter 12 Review Test

Give three ordered pairs to specify the following points. Do not use angles greater than 360°. **(12.1)**

1. $(2, -120°)$

2. $(-4, 210°)$

Make a table of values and sketch the following graphs in polar form. **(12.1)**

3. $r = 4\cos\theta$

4. $r = -3\sin\theta$

5. $r = 4 - 3\cos\theta$

6. $r = 3 + 4\sin\theta$

Determine the symmetry of the following curves in polar form. **(12.1)**

7. $3r = -7\cos\theta$

8. $r^2 = 4\sin\theta$

Convert the following polar equations to parametric equations. Use your graphing calculator to plot the graph and indicate the maximum and minimum values of x and y. **(12.2)**

9. $r = 3\sin 2\theta$

10. $r = \dfrac{4}{2 - 3\cos\theta}$

Find the polar equation and sketch the curve using polar coordinates. **(12.3)**

11. $3x - 2y = 6$

12. $x^2 + y^2 = 2x$

Find the slope and y-intercept of the line represented by the following equations. **(12.3)**

13. $4r\cos\theta - 5r\sin\theta = 6$

14. $-r\cos\theta + 2r\sin\theta = -3$

Sketch the graphs of the following equations given in polar form. **(12.3)**

15. $r = 3 - 3\cos\theta$

16. $r = \dfrac{2}{1 + \sin\theta}$

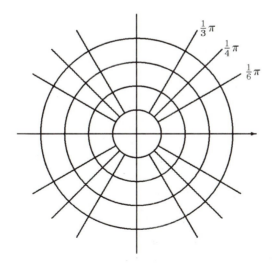

Chapters 1–12 Cumulative Review

Find the smallest non-negative angle that is coterminal with the following angles. **(1.1)**

1. $-340°$

2. $\dfrac{25}{4}\pi$

3. Express 0.800 radians in degree measure accurate to three decimal places. **(1.2)**

4. Express $221°$ in radian measure accurate to three decimal places. **(1.2)**

5. A pulley is moved by a belt. If the pulley turns through an angle of $480°$ while a point on the belt moves 19 feet, what is the radius of the pulley to the nearest tenth of a foot? **(1.2)**

6. A circular freeway on-ramp covers an arc of $135°$ with a radius of 120 ft. How long is the ramp? **(1.2)**

7. Find the linear velocity in feet per second and angular velocity in radians per second of a stone imbeded in a tire with a radius of 15 inches on a car traveling 70 miles per hour. **(1.3)**

Find the values of the six trigonometric functions of an angle θ in standard position if a point on the terminal side has the following coordinates. Consider the angle to rotate in a positive direction. **(2.1)**

8. $\left(-1, \sqrt{8}\right)$

Find the values of the six trigonometric functions for the following angles without the use of a calculator. **(2.2)**

9. $300°$

10. $\dfrac{7}{6}\pi$

Solve the following right triangle. **(2.4)**

11. $\gamma = 90°, a = 7.92, c = 12.45$

12. An approach to an overpass on a highway must be 16.8 feet high. A desirable angle of elevation for the approach is $2.2°$. How far from the overpass should the approach begin? **(2.5)**

13. Two scouts, using a compass, walked 480 paces N25°W, then turned to the left and walked 630 paces in the S65°W direction. What direction must they go and how many paces must they take to return to their starting point? **(2.5)**

14. What is the positive value of x closest to zero at the maximum of the function $y = -\dfrac{1}{2}\cos 3x$? What is the value of y at that maximum? **(3.1)**

How many periods occur in the specified domain of each function below? **(3.1, 3.2)**

15. $y = 7\cos\dfrac{1}{2}x$
    $-3\pi < x \le 3\pi$

16. $y = -3\sin\left(2x - \dfrac{1}{4}\pi\right)$
    $-\pi < x \le 6\pi$

Find the starting point, quarter point, midpoint, three-quarter point, and endpoint of a generic box. Then sketch two periods of each of the following functions.

17. $y = -3\cos\left(2x - \dfrac{1}{4}\pi\right)$ **(3.2)**

18. $y = \cot\left(\dfrac{1}{2}x + \dfrac{1}{3}\pi\right)$ **(3.4)**

Using the same trigonometric function as y_1, give the equation of the function y_2 that meets the given criteria: **(3.2)**

19. $y_1 = 3 \sin 2x$ y_2 is $\dfrac{1}{2}$ as high as y_1 with three times as many periods in any interval as y_1

20. $y_1 = \dfrac{1}{3} \cos \left(\dfrac{1}{2}x + \dfrac{1}{6}\pi \right)$ y_2 is three times as high as y_1 and is shifted $\dfrac{1}{3}\pi$ units to the right of y_1

Write the equations of the two functions that will produce each graph below. **(3.6)**

21.

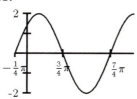

22.

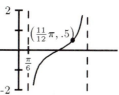

(a) Determine the domain and range of the following functions. (b) Determine if they are one-to-one functions. **(4.1)**

23. $x^2 - 2y = 3$ **24.** $y = x^3 - 3$

(a) Write the inverse of the following equations. Solve for y in terms of x. (b) Determine if the original equation is a function. (c) Determine if the inverse is a function. **(4.1)**

25. $y = 2x - 8$ **26.** $y = x^2 + 2$

27. Give the range of y values for which each of the following functions are defined. **(4.2)**

     (a) $y = \mathrm{Sin}^{-1} x$ (b) $y = \mathrm{Cos}^{-1} x$ (c) $y = \mathrm{Tan}^{-1} x$

Give the exact value of each of the following inverse functions. **(4.2)**

28. $\mathrm{Sin}^{-1} \left(-\dfrac{\sqrt{3}}{2} \right)$ **29.** $\mathrm{Cot}^{-1}(1)$ **30.** $\mathrm{Cos}^{-1} \left(-\dfrac{1}{2} \right)$

Evaluate the following. (Do not use a calculator.) **(4.2)**

31. $\tan \left[\mathrm{Cos}^{-1} \left(-\dfrac{3}{5} \right) \right]$ **32.** $\sec \left[\mathrm{Tan}^{-1} \left(-\dfrac{12}{5} \right) \right]$

Reduce the first expression to the second. **(5.1)**

33. $1 - \cot x \cos x \sin x, \sin^2 x$ **34.** $(\sin x - \csc x)^2, \cot^2 x - \cos^2 x$

Test the validity of the following identities using a graphing calculator. If the graphs coincide, prove the identity. If the equations are not identities, give a value of x that proves they are not identities. **(5.2, 5.3)**

35. $\dfrac{\cos x}{\sin x - \cos x} = \dfrac{1}{\tan x}$ **36.** $\sin^2(-x) + \dfrac{\sin^4(-x)}{\cos^2(-x)} = \tan^2 x$

37. $\dfrac{\cos(-x)\tan x - \sin(-x)}{\tan x} = 2\cos x$

Using the values of the trigonometric function for $30°, 45°,$ and/or $60°$, find the exact value. **(6.1)**

38. $\cos 75°$

Using the sum and difference identities, change the following expressions to functions of θ only. **(6.1, 6.2)**

39. $\cos(\theta - 30°)$ **40.** $\tan \left(x - \dfrac{3}{4}\pi \right)$

Chapters 1–12 Cumulative Review

Without the use of a calculator, find the following: (**6.1, 6.2**)

41. $\sin(\alpha + \beta)$, $\quad \cos \alpha = -\dfrac{\sqrt{8}}{3}$, α is in quadrant II

$\qquad \tan \beta = \dfrac{3}{4}$, β is in quadrant III

Prove the following identity. Work on one side only. (**6.1, 6.2**)

42. $\sin(\alpha - \beta) \cdot \csc \alpha = \sin \beta (\cot \beta - \cot \alpha)$

Using the double-angle formulas, find the exact value of the following. (**7.1**)

43. $\tan 240°$

Using the half-angle formulas, find the exact value of the following. (**7.2**)

44. $\sin 75°$

Prove the following identities. Work one side only. (**7.1, 7.2, 7.3**)

45. $(\csc x - \cot x)^2 = \dfrac{1 - \cos x}{1 + \cos x}$ $\qquad$ **46.** $2(1 - \sin x) = \dfrac{1 + \cos 2x}{1 + \sin x}$

Without the use of a calculator, solve the following equations for (a) all values and (b) the fundamental values $(0° \le \theta < 360°$ or $0 \le x < 2\pi)$. (**8.1, 8.2, 8.3, 8.4**)

47. $2\cos^2 \theta - \sqrt{3}\cos \theta = 0$ $\qquad\qquad \theta$ in degrees

48. $\tan x = \sin x$ $\qquad\qquad\qquad\qquad x$ in radians

49. $1 - \csc \theta + 2\sin \theta = 0$ $\qquad\qquad \theta$ in degrees

50. $\sqrt{3}\tan 3x + 1 = 0$ $\qquad\qquad\qquad x$ in radians

Using your calculator, solve the following equations for (a) all values and (b) the fundamental values $(0° \le \theta < 360°$ or $0 \le x < 2\pi)$. (**8.1, 8.2, 8.3, 8.4**)

51. $\sqrt{3}\cos \theta - 1 = 0$ $\qquad\qquad\qquad \theta$ in degrees

52. $5\sin^2 x - 3 = 0$ $\qquad\qquad\qquad\qquad x$ in radians

53. $3\tan^2 x + 5\tan x = 2$ $\qquad\qquad\qquad x$ in radians

54. $3\cos^2 3\theta + 2\cos 3\theta - 1 = 0$ $\qquad\quad \theta$ in degrees

Find the missing parts for the following oblique triangles. (**9.1**)

55. $\alpha = 29.3°$, $\beta = 67.9°$, $a = 31.9$ $\qquad\qquad$ **56.** $\beta = 42.5°$, $\gamma = 59.3°$, $a = 12.4$

Find all solutions for the following triangle. (**9.2**)

57. $a = 30.2$, $\quad b = 39.4$, $\quad \alpha = 28.4°$

58. From a balloon flying over an island, the angles of the depression of the northernmost and southernmost points were $27°$ and $53°$ respectively. If the island was 8750 feet, measured from north to south, how high was the balloon flying? (**9.3**)

For the following triangle, find the length of the side not given. (**9.4**)

59. $a = 7.94$ $\quad c = 9.23$ $\quad \beta = 43.6°$

Solve the triangle completely. (**9.4**)

60. $b = 0.869$ $\quad c = 1.34$ $\quad \alpha = 67.9°$

Find all three angles of the following triangle. (**9.4**)

61. $a = 94$ $\qquad b = 146$ $\qquad c = 65$

62. Two small boats leave a dock at the same time. One boat travels at 12 miles per hour in the direction of N67°E. The other one travels at 15 miles per hour in the direction of S75°E. After one hour and 30 minutes, how far apart are the boats? **(9.5)**

63. The resultant of two forces acting at an angle of 105° with respect to each other is 92 pounds. If one force is 80 pounds, find the other force and the angle it makes with the resultant force. **(10.1)**

64. A ramp was constructed at the entrance of an office building. The ramp is inclined at an angle of 12° to the horizontal. If a 190-pound man in a wheelchair uses the ramp, to the nearest pound, what force is necessary to keep him from rolling down the incline, and what is the perpendicular force against the plane? How much work will he do to roll the chair 20 feet up the ramp? **(10.4)**

65. An airplane is headed in the direction of 220° at 180 miles per hour. If the wind is blowing from the direction of 290° at 30 miles per hour, find the groundspeed of the airplane and its resultant course. **(10.2)**

Add the following vectors and find the magnitude of the resultant. **(10.3)**

66. $u = \langle 8, -5 \rangle$, $v = \langle -3, 2 \rangle$

67. Find a unit vector with the same direction as $v = \langle 4, -6 \rangle$. **(10.3)**

68. Find the dot product of the following vectors and the angle between them.
$u = \langle -1, 6 \rangle$, $v = \langle 2, 3 \rangle$ **(10.4)**

Simplify. **(11.1)**

69. $\left(\sqrt{8} - \sqrt{-27} \right) \cdot \left(\sqrt{2} + \sqrt{-12} \right)$

70. $\dfrac{5 + \sqrt{-50}}{2 - \sqrt{-18}}$

Find the modulus for the following complex number. **(11.2)**

71. $7 - \sqrt{6}i$

Perform the indicated operations. Leave your answer in trigonometric form with the argument expressed as a number between 0° and 360°. **(11.2)**

72. $5 \left(\cos 278° + i \sin 278° \right) \cdot 9 \left(\cos 162° + i \sin 162° \right)$

73. $5\sqrt{2} \left(\cos 224° + i \sin 224° \right) \div \sqrt{10} \left(\cos 348° + i \sin 348° \right)$

Perform the indicated operations. Write your answer in rectangular form. **(11.2)**

74. $\sqrt{6} \, \underline{/250°} \cdot \sqrt{12} \, \underline{/230°}$

75. $\dfrac{3\sqrt{2} \underline{/80°}}{\sqrt{6} \underline{/110°}}$

Convert the following complex numbers to trigonometric form, then perform the indicated operations. Leave your answer in trigonometric form $(0° \leq \theta < 360°)$. **(11.2)**

76. $\left(3\sqrt{5} - 2i \right) \cdot \left(-4 - 2\sqrt{5}i \right)$

77. $\dfrac{-6 + 8i}{3 + i}$

Use DeMoivre's Theorem to perform the indicated operations. Leave your answer in polar form $(0° \leq \theta < 360°)$. **(11.3)**

78. $\left(\sqrt{3} - 3i \right)^4$

79. Find the fourth roots of $-8\sqrt{3} - 8i$

Using polar coordinates make a table of values and sketch the following curve. **(12.1)**

80. $r = 3 - 5\cos\theta$

81. Find the Cartesian coordinates of the point $(5, 225°)$. **(12.2)**

82. Find the polar form of the coordinates of the point $(-4, 3)$. Express θ such that $(0° \leq \theta < 360°)$. **(12.2)**

Chapters 1–12 Cumulative Review

Determine the symmetry of the following curves in polar form. **(12.1)**

83. $r = -5 \sin \theta$ **84.** $r^2 = 4 \cos \theta$

Convert the following polar equation to a parametric equation. Use your graphing calculator to plot the graph and approximate the maximum and minimum values of x and y using the trace function. **(12.2)**

85. $r = \dfrac{4}{3 - 2 \sin \theta}$

Find the polar equation and sketch the rectangular curve using polar coordinates. **(12.3)**

86. $x^2 + y^2 = 2y$

87.   Find the slope and y-intercept of the line represented by $-r \cos \theta + 2r \sin \theta = -3$. **(12.3)**

Appendix

Rounding Off and Significant Figures

Some numbers are exact, others are necessarily approximations. Counting numbers, for example, are exact. Numbers derived from theory, for example π or $\sqrt{2}$, are considered to be exact even though we cannot evaluate them exactly. There are exactly four sides to a rectangle. However, measuring the four sides of a rectangle to more than four-place accuracy is either very difficult or very expensive. Quantities obtained from measurements are never exact.

The way we express the accuracy of a measurement is with significant digits.

Significant Digits

The digits known to be correct in a number obtained by a measurement are called significant digits.
The following rules apply:

1. The digits 1, 2, 3, 4, 5, 6, 7, 8, and 9 are always significant, whereas the digit 0 may or may not be significant.

2. Zeros that come between two other digits are significant, as in 103 or 20.04.

3. If the zero's only function is to place the decimal point, it is not significant, as in

$$\underset{\substack{\uparrow \\ \text{placeholders}}}{\mathbf{0.000023}} \qquad \text{or} \qquad \underset{\substack{\uparrow \\ \text{placeholders}}}{23{,}000}$$

4. If zero does more than fix the decimal point, it is significant, as in:

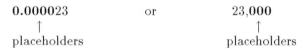

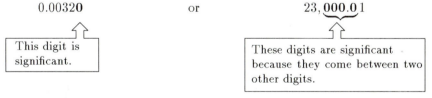

5. Rules 3 and 4 can, of course, result in certain ambiguities, such as in 23,000 (measured to the exact unit). To avoid such confusion, use scientific notation. 2.3×10^4 has two significant digits.

    2.3000×10^4 has five significant digits.

Example 1 □ The following numbers have two significant digits:
46, 0.00083, 4.0×10^1, 0.050.

□

Example 2 □ The following numbers have three significant digits:
523, 403, 4.00×10^2, 0.000800.

□

Example 3 □ The following numbers have four significant digits:
600.1, 4.000×10^1, 0.0002345.

□

When doing calculations with approximate numbers (particularly when using a calculator), it is often necessary to round off final results.

To Round Off Numbers:

Identify the digit in the place value desired. Next examine the digit to the right of this digit in the desired place accuracy.

1. If the digit on the right of the digit in the desired place accuracy is less than five, replace this digit and all the digits to its right with zero.

2. If the digit on the right of the digit in the place value desired is five or greater:

   (a) Increase the digit in the desired place accuracy by one.

   (b) Replace all digits to the right of the new digit in the desired place accuracy with zero.

Elaborate rules for computation of approximate data can be developed when necessary for some applications, such as in chemistry, but there are two simple rules that will work satisfactorily for the material presented in this textbook.

Rules for Significant Digits

Addition-subtraction: Add or subtract in the usual fashion; then round off the result so that the last digit retained is in the column farthest to the right in which ALL given numbers have significant digits.

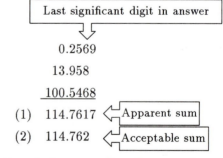

The first row (1) shows the solutions before rounding the correct number of significant digits. The second row (2) shows the solutions after rounding. The last digit in each solution in this row corresponds to the last digit that is significant in ALL of the numbers added or subtracted.

Multiplication-division: Multiply or divide in the usual fashion and then round off the results to the smaller number of significant digits found in either of the given numbers.

Examples of multiplication and division of approximate numbers:

	Multiplication		Division	
	23.55	132.0064	1300	1.039 ⟵ Acceptable
	× 90	× 4.05	1295.5892	1.03871111 ⟵ Apparent
Apparent ⟶	2119.5	534.62592	62.5⎸80974.325	405.0⎸420.678
Acceptable ⟶	2100	535		

In solving triangles in this course, we will assume a certain relationship in the accuracy of the measurement between the sides and the angles.

Accuracy in Sides	Accuracy in Angles
2 significant figures	nearest degree
3 significant figures	nearest tenth of a degree
4 significant figures	nearest hundredth of a degree

This chart means that, if the data includes one side given with two significant digits and another with three significant digits, the angle would be computed to the nearest degree. If one side is given to four significant digits and an angle to the nearest tenth of a degree, then the other sides would be given to three significant digits and the angles computed to the nearest hundredth of a degree. In general, results computed from the above table should not be more accurate than the least accurate item of the given data.

Selected Answers

Chapter 1

Problem Set 1.1A

1. 90°

3. 45°

5. 30°

7. −150°

9. 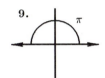 π

11. $\frac{1}{4}\pi$

13. 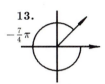 $-\frac{7}{4}\pi$

15. 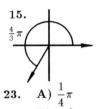 $\frac{4}{3}\pi$

17. 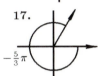 $-\frac{5}{3}\pi$

19. **A)** 45°
    B) 90°
    C) 150°
    D) 210°
    E) 225°
    F) 300°
    G) 315°

21. **A)** $\frac{1}{4}\pi$ **E)** $\frac{5}{4}\pi$
    B) $\frac{1}{2}\pi$ **F)** $\frac{5}{3}\pi$
    C) $\frac{5}{6}\pi$ **G)** $\frac{7}{4}\pi$
    D) $\frac{7}{6}\pi$

23. **A)** $\frac{1}{4}\pi$
    B) 75°

Problem Set 1.1B

1. **A)** $\frac{1}{4}\pi$ **B)** $\frac{1}{6}\pi$ **C)** $\frac{3}{4}\pi$ **D)** $-\frac{3}{4}\pi$ **E)** $-\frac{2}{3}\pi$

3. 240°   5. 140°   7. 290°   9. 155°   11. 155°

13. $\frac{4}{3}\pi$ 15. $\frac{5}{6}\pi$ 17. $\frac{3}{4}\pi$ 19. $\frac{3}{2}\pi$ 21. 0

23. 360°   25. 180°   27. 135°   29. −330°   31. −45°

33. $\frac{5}{6}\pi$ 35. $-\frac{3}{2}\pi$ 37. $\frac{5}{2}\pi$ 39. $\frac{23}{6}\pi$ 41. $-\frac{11}{4}\pi$

Problem Set 1.2

1. 1.65 cm   3. 7.1 in   5. 42 cm   7. 7.1 in   9. 3210 ft
11. 29 cm   13. 2 min   15. 1414 cm^2 17. 1518 ft^2 19. 7136 ft^2

Problem Set 1.3

1. $2.5 \frac{\text{radians}}{\text{second}}$

3. 3 seconds

5. 400 radians

7. $360 \frac{\text{inches}}{\text{minute}}$

9. $\frac{5}{6\pi}$ ft $\approx .27$ ft

11. $\frac{1}{6}\pi \frac{\text{radians}}{\text{hour}} \approx .524 \frac{\text{radians}}{\text{hour}}$

13. $\frac{25}{88}\pi$ mph $\approx .89$ mph

15. $\omega = \frac{1}{12}\pi \frac{\text{radians}}{\text{hour}} \approx 0.26 \frac{\text{radians}}{\text{hour}}$
    $v = \frac{1000}{3}\pi \frac{\text{miles}}{\text{hour}}$

17. $25.14 \frac{\text{radians}}{\text{second}} \approx 25 \frac{\text{radians}}{\text{second}}$

19. $654.5 \frac{\text{feet}}{\text{second}}$

21. 11.5 inches in diameter

23. $2 \frac{\text{revolutions}}{\text{second}}$, $10 \frac{\text{miles}}{\text{hour}}$

Review Test—Chapter 1

1. $180°$ 2. $120°$ 3. $\dfrac{11}{6}\pi$ 4. $\dfrac{1}{3}\pi$ 5. $390°$

6. $1485°$ 7. $\dfrac{11}{6}\pi$ 8. $\dfrac{7}{2}\pi$ 9. 164.93 m   10. 65.35 cm

11. 39.6 sq. cm.   12. 4595920 sq. ft.   13. 760 sq. in.   14. $2.4\ \dfrac{\text{radians}}{\text{second}}$

15. 4 seconds   16. 640 radians

17. $800\,\pi$ radians ≈ 2513 radians   18. $0.54\ \dfrac{\text{revolutions}}{\text{day}}$

19. $6785.8\dfrac{\text{meters}}{\text{minute}} \approx 6786\dfrac{\text{meters}}{\text{minute}}$ 20. $9.4\dfrac{\text{centimeters}}{\text{minute}}$

21. $300.1\dfrac{\text{revolutions}}{\text{minute}} \approx 300\dfrac{\text{revolutions}}{\text{minute}}$ 22. $171.9\dfrac{\text{revolutions}}{\text{minute}} \approx 172\dfrac{\text{revolutions}}{\text{minute}}$

Chapter 2

Problem Set 2.1

1. 5 3. 13

	$\sin\theta$	$\cos\theta$	$\tan\theta$	$\cot\theta$	$\sec\theta$	$\csc\theta$
5.	$\dfrac{2\sqrt{13}}{13}$	$-\dfrac{3\sqrt{13}}{13}$	$-\dfrac{2}{3}$	$-\dfrac{3}{2}$	$-\dfrac{\sqrt{13}}{3}$	$\dfrac{\sqrt{13}}{2}$
7.	$\dfrac{7}{25}$	$-\dfrac{24}{25}$	$-\dfrac{7}{24}$	$-\dfrac{24}{7}$	$-\dfrac{25}{24}$	$\dfrac{25}{7}$

9. 1st and 3rd   11. 2nd and 4th   13. 2nd

15. $90° < \theta < 180°$ 17. $270° < \theta < 360°$

19. $\sin\theta = -\dfrac{4}{5}, \quad \cot\theta = -\dfrac{3}{4}, \quad \sec\theta = \dfrac{5}{3}, \quad \csc\theta = -\dfrac{5}{4}$

21. $\sin\theta = -\dfrac{7}{25}, \quad \cos\theta = \dfrac{24}{25}, \quad \cot\theta = -\dfrac{24}{7}, \quad \sec\theta = \dfrac{25}{24}$

23. $\sin\theta = -\dfrac{5}{13}, \quad \cos\theta = \dfrac{12}{13}, \quad \cot\theta = -\dfrac{12}{5}, \quad \sec\theta = \dfrac{13}{12}, \quad \csc\theta = -\dfrac{13}{5}$

25. $\cos\theta = -\dfrac{1}{3}, \quad \tan\theta = 2\sqrt{2}, \quad \cot\theta = \dfrac{\sqrt{2}}{4}, \quad \sec\theta = -3, \quad \csc\theta = -\dfrac{3\sqrt{2}}{4}$

27. $\sin\theta = \dfrac{1}{5}, \quad \cos\theta = -\dfrac{2\sqrt{6}}{5}, \quad \tan\theta = -\dfrac{\sqrt{6}}{12}, \quad \cot\theta = -2\sqrt{6}, \quad \sec\theta = -\dfrac{5\sqrt{6}}{12}$

Problem Set 2.2

1. $60°$ 3. $30°$ 5. $45°$ 7. $\dfrac{1}{3}\pi$ 9. $\dfrac{1}{3}\pi$ 11. $\dfrac{1}{6}\pi$

	$\sin\theta$	$\cos\theta$	$\tan\theta$	$\cot\theta$	$\sec\theta$	$\csc\theta$
13.	$\dfrac{\sqrt{3}}{2}$	$-\dfrac{1}{2}$	$-\sqrt{3}$	$-\dfrac{\sqrt{3}}{3}$	-2	$\dfrac{2\sqrt{3}}{3}$
15.	0	-1	0	undefined	-1	undefined
17.	$-\dfrac{\sqrt{2}}{2}$	$-\dfrac{\sqrt{2}}{2}$	1	1	$-\sqrt{2}$	$-\sqrt{2}$
19.	-1	0	undefined	0	undefined	-1
21.	$-\dfrac{1}{2}$	$-\dfrac{\sqrt{3}}{2}$	$\dfrac{\sqrt{3}}{3}$	$\sqrt{3}$	$-\dfrac{2\sqrt{3}}{3}$	-2
23.	$\dfrac{1}{2}$	$-\dfrac{\sqrt{3}}{2}$	$-\dfrac{\sqrt{3}}{3}$	$-\sqrt{3}$	$-\dfrac{2\sqrt{3}}{3}$	2

Problem Set 2.3

1. $37°$ 3. $45.0°$ 5. 1.03 7. $70.5°$, 11.3 ft   9. $20.6°$, 42.7 ft

11. 108 m   13. 7.4 ft   15. 4.05 light years   17. 13.9 in   19. $6.1°$

Problem Set 2.4

1. $A = 53°$, $\quad B = 37°$, $\quad c = 4.4$
3. $A = 77.6°$, $\quad a = 70.7$, $\quad b = 15.5$
5. $B = 52°$, $\quad b = 24$, $\quad c = 31$
7. $A = 42.2°$, $\quad B = 47.8°$, $\quad a = 8.46$
9. $B = 20.6°$, $\quad a = 242$, $\quad b = 90.8$
11. $A = 35.09°$, $\quad B = 54.91°$, $\quad b = 12.90$
13. $A = 33.24°$, $\quad b = 86.61$ m, $\quad c = 103.5$ m

Problem Set 2.5

1. 65.7 ft 3. 8.0 mi 5. 77 ft 7. 3580 ft 9. 91 ft 11. 9 ft
13. 29.0 km from Running Deer; 22.7 km from Lazy Bear

Problem Set 2.6

1. $\dfrac{\sqrt{2}}{2}, -\dfrac{\sqrt{2}}{2}, -1$
3. $-\dfrac{\sqrt{3}}{2}, \dfrac{1}{2}, -\sqrt{3}$
5. $-\dfrac{\sqrt{3}}{2}, -\dfrac{1}{2}, \sqrt{3}$

7. 1
9. $\dfrac{5}{6}$
11. 1

13. $\dfrac{4}{3}\pi, -\dfrac{2}{3}\pi$
15. $\dfrac{7}{6}\pi, -\dfrac{5}{6}\pi$
17. $\dfrac{3}{2}\pi, -\dfrac{1}{2}\pi$

19. $\sin s = \dfrac{4}{5}$ $\quad \cos s = -\dfrac{3}{5}$ $\quad \tan s = -\dfrac{4}{3}$
21. $\sin s = -\dfrac{5}{13}$ $\quad \cos x = -\dfrac{12}{13}$ $\quad \tan s = \dfrac{5}{12}$

$\quad \csc s = \dfrac{5}{4}$ $\quad \sec s = -\dfrac{5}{3}$ $\quad \cot s = -\dfrac{3}{4}$
$\quad \csc s = -\dfrac{13}{5}$ $\quad \sec s = -\dfrac{13}{12}$ $\quad \cot s = \dfrac{12}{5}$

23. $\sin s = \dfrac{15}{17}$ $\quad \cos s = \dfrac{8}{17}$ $\quad \tan s = \dfrac{15}{8}$

$\quad \csc s = \dfrac{17}{15}$ $\quad \sec s = \dfrac{17}{8}$ $\quad \cot s = \dfrac{8}{15}$

25. $\dfrac{1}{2}$ 27. $-\dfrac{\sqrt{3}}{2}$ 29. $-\dfrac{\sqrt{3}}{3}$ 31. $-\dfrac{2\sqrt{2}}{3}$ 33. $-\dfrac{\sqrt{5}}{2}$

Review Test—Chapter 2

	$\sin\theta$	$\cos\theta$	$\tan\theta$	$\cot\theta$	$\sec\theta$	$\csc\theta$
1.	$\dfrac{5}{13}$	$-\dfrac{12}{13}$	$-\dfrac{5}{12}$	$-\dfrac{12}{5}$	$-\dfrac{13}{12}$	$\dfrac{13}{5}$
2.	$-\dfrac{\sqrt{2}}{2}$	$-\dfrac{\sqrt{2}}{2}$	1	1	$-\sqrt{2}$	$-\sqrt{2}$
3.	given	$-\dfrac{3}{5}$	$\dfrac{4}{3}$	$\dfrac{3}{4}$	$-\dfrac{5}{3}$	$-\dfrac{5}{4}$
4.	$-\dfrac{2\sqrt{2}}{3}$	given	$-2\sqrt{2}$	$-\dfrac{\sqrt{2}}{4}$	3	$-\dfrac{3\sqrt{2}}{4}$
5.	$-\dfrac{\sqrt{3}}{2}$	$-\dfrac{1}{2}$	$\sqrt{3}$	$\dfrac{\sqrt{3}}{3}$	-2	$-\dfrac{2\sqrt{3}}{3}$
6.	$-\dfrac{\sqrt{2}}{2}$	$\dfrac{\sqrt{2}}{2}$	-1	-1	$\sqrt{2}$	$-\sqrt{2}$
7.	$\dfrac{\sqrt{3}}{2}$	$-\dfrac{1}{2}$	$-\sqrt{3}$	$-\dfrac{\sqrt{3}}{3}$	-2	$\dfrac{2\sqrt{3}}{3}$
8.	$-\dfrac{\sqrt{2}}{2}$	$\dfrac{\sqrt{2}}{2}$	-1	-1	$\sqrt{2}$	$-\sqrt{2}$

9. 0.9775 10. 0.8351 11. 0.1974
12. $\alpha = 37°$, $\quad \beta = 53°$, $\quad c = 7.7$
13. $\beta = 53.6°$, $\quad b = .636$, $\quad c = .790$
14. $\alpha = 43.9°$, $\quad \beta = 46.1°$, $\quad b = 7.10$
15. $\alpha = 47.31°$, $\quad b = 5.210$, $\quad c = 7.684$
16. $\alpha = 43.69°$, $\quad \beta = 46.31°$, $\quad a = 88.69$
17. $\beta = 57.4°$, $\quad a = 6.06$, $\quad c = 11.2$
18. 78.2 ft 19. 24.5° 20. 23°
21. 75 ft 22. 2367 ft 23. 980 ft
24. From B = 26.7 km, from A = 22.1 km
25. $\sin\dfrac{5}{4}\pi = -\dfrac{\sqrt{2}}{2}, \cos\dfrac{2}{3}\pi = -\dfrac{1}{2}, \tan\dfrac{11}{6}\pi = -\dfrac{\sqrt{3}}{3}$
26. $\dfrac{5}{4}\pi, -\dfrac{3}{4}\pi$ 27. $\dfrac{5}{3}\pi, -\dfrac{1}{3}\pi$

28. $\sin s = -\dfrac{4}{5}$ $\quad \cos s = -\dfrac{3}{5}$ $\quad \tan s = \dfrac{4}{3}$
29. $\sin s = \dfrac{12}{13}$ $\quad \cos s = -\dfrac{5}{13}$ $\quad \tan s = -\dfrac{12}{5}$

$\quad \csc s = -\dfrac{5}{4}$ $\quad \sec s = -\dfrac{5}{3}$ $\quad \cot s = \dfrac{3}{4}$
$\quad \csc s = \dfrac{13}{12}$ $\quad \sec s = -\dfrac{13}{5}$ $\quad \cot s = -\dfrac{5}{12}$

30. $-\dfrac{1}{2}$ 31. $-\dfrac{\sqrt{3}}{3}$ 32. $-\dfrac{9}{4}$ 33. $-\dfrac{1}{4}$ 34. $\dfrac{2\sqrt{2}}{3}$

Chapter 3

Problem Set 3.1

1. $\dfrac{\sqrt{3}}{2}$ **3.** $-\dfrac{1}{2}$ **5.** $-\dfrac{\sqrt{3}}{2}$ **7.** -1 **9.** $-\dfrac{1}{2}$ **11.** $-\dfrac{\sqrt{3}}{2}$

13.

x	0	$\frac{1}{6}\pi$	$\frac{1}{4}\pi$	$\frac{1}{3}\pi$	$\frac{1}{2}\pi$	$\frac{2}{3}\pi$	$\frac{3}{4}\pi$	$\frac{5}{6}\pi$	π	$\frac{7}{6}\pi$	$\frac{5}{4}\pi$	$\frac{4}{3}\pi$	$\frac{3}{2}\pi$	$\frac{5}{3}\pi$	$\frac{7}{4}\pi$	$\frac{11}{6}\pi$	2π
$\sin x$	0	.50	.71	.87	1	.87	.71	.50	0	$-.50$	$-.71$	$-.87$	-1	$-.87$	$-.71$	$-.50$	0

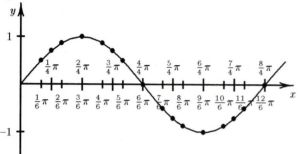

15. $y = \sin x$

$$
\begin{aligned}
X_{\min} &= 0 \\
X_{\max} &= 6.283 \\
X_{\text{scl}} &= 1.571 \\
Y_{\min} &= -1.2 \\
Y_{\max} &= 1.2 \\
Y_{\text{scl}} &= 1
\end{aligned}
$$

17. $y = 2\sin x$

$$
\begin{aligned}
X_{\min} &= -6.283 \\
X_{\max} &= 6.283 \\
X_{\text{scl}} &= 1.571 \\
Y_{\min} &= -2.2 \\
Y_{\max} &= 2.2 \\
Y_{\text{scl}} &= 1
\end{aligned}
$$

19. $y = \sin 2x$

$$
\begin{aligned}
X_{\min} &= -6.283 \\
X_{\max} &= 3.142 \\
X_{\text{scl}} &= .786 \\
Y_{\min} &= -1.2 \\
Y_{\max} &= 1.2 \\
Y_{\text{scl}} &= 1
\end{aligned}
$$

21. $y = -\sin 2x$

$$
\begin{aligned}
X_{\min} &= -6.283 \\
X_{\max} &= 3.142 \\
X_{\text{scl}} &= .786 \\
Y_{\min} &= -1.2 \\
Y_{\max} &= 1.2 \\
Y_{\text{scl}} &= 1
\end{aligned}
$$

23. $\dfrac{9}{2}\pi$ **25.** 10 periods   **27.** 6π **29.** 4π **31.** $y_2 = 2\cos x$

33. period $= \dfrac{1}{256}$ sec, frequency $= 256$ cps, compare $=$ slightly lower pitch

Problem Set 3.2

1. $y = \sin\left(x - \dfrac{1}{4}\pi\right)$

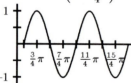

3. $y = \sin\left(x + \dfrac{1}{4}\pi\right)$

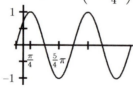

5. $y = -2\cos\left(x - \dfrac{1}{3}\pi\right)$

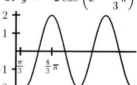

7. $y = \frac{1}{2}\sin\left(\frac{1}{2}x - \frac{1}{6}\pi\right)$

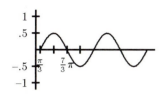

9. $y = \cos\left(x + \frac{1}{2}\pi\right)$

$X_{\min} \ = \ -3.142$
$X_{\max} \ = \ 9.425$
$X_{\text{scl}} \ = \ 1.571$
$Y_{\min} \ = \ -1.2$
$Y_{\max} \ = \ 1.2$
$Y_{\text{scl}} \ = \ 1$

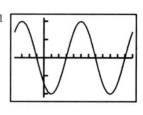

11. $y = \cos\left(x - \frac{1}{2}\pi\right)$

$X_{\min} \ = \ -1.047$
$X_{\max} \ = \ 11.519$
$X_{\text{scl}} \ = \ 1.571$
$Y_{\min} \ = \ -1.2$
$Y_{\max} \ = \ 1.2$
$Y_{\text{scl}} \ = \ 1$

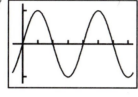

13. $y = -2\sin\left(2x + \frac{1}{4}\pi\right)$

$X_{\min} \ = \ -1.571$
$X_{\max} \ = \ 4.712$
$X_{\text{scl}} \ = \ .393$
$Y_{\min} \ = \ -2.2$
$Y_{\max} \ = \ 2.2$
$Y_{\text{scl}} \ = \ 1$

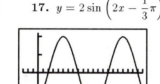

15. $y = -2\sin\left(2x - \frac{1}{4}\pi\right)$

17. $y = 2\sin\left(2x - \frac{1}{3}\pi\right)$

19. $y = -3\cos\left(\frac{1}{2}x + \frac{1}{4}\pi\right)$

first maximum $\left(-\frac{1}{8}\pi, 2\right)$

first minimum $\left(-\frac{1}{12}\pi, -2\right)$

last maximum $\left(\frac{11}{2}\pi, 3\right)$

21. 2 periods

23. $\frac{2}{3}\pi$

25. $y_2 = \frac{1}{2}\cos\left(x - \frac{1}{4}\pi\right)$

27. $y_2 = -\frac{3}{2}\sin 4x$

29. $y_2 = \frac{1}{3}\sin\left(3x + \frac{1}{4}\pi\right)$

31. $y = \cos x - 2$

33. $y = 3\cos x + 2$

35. $y = \frac{1}{2}\sin 2x - 1$

$X_{\min} \ = \ 0$
$X_{\max} \ = \ 6.283$
$X_{\text{scl}} \ = \ .785$
$Y_{\min} \ = \ -1.7$
$Y_{\max} \ = \ .5$
$Y_{\text{scl}} \ = \ .5$

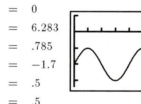

37. $y = -2\sin 3x - 3$

$X_{\min} \ = \ 0$
$X_{\max} \ = \ 4.189$
$X_{\text{scl}} \ = \ .524$
$Y_{\min} \ = \ -5.2$
$Y_{\max} \ = \ 1$
$Y_{\text{scl}} \ = \ 1$

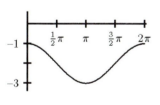

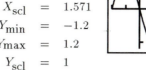
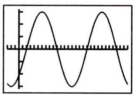
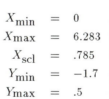
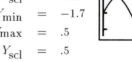

39. $y = |\sin x|$

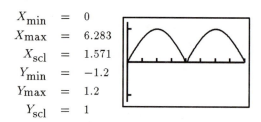

$$
\begin{aligned}
X_{\min} &= 0 \\
X_{\max} &= 6.283 \\
X_{\text{scl}} &= 1.571 \\
Y_{\min} &= -1.2 \\
Y_{\max} &= 1.2 \\
Y_{\text{scl}} &= 1
\end{aligned}
$$

Problem Set 3.3

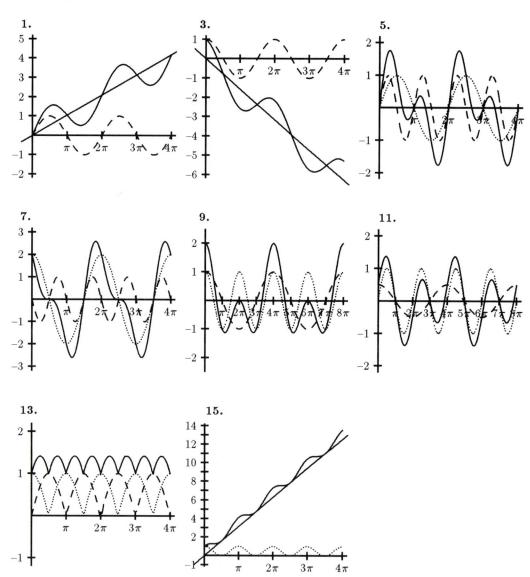

1.

3.

5.

7.

9.

11.

13.

15.

Problem Set 3.4

1. π
3. π
5. 4π
7. $\dfrac{1}{4}\pi$

9. 4π
11. $\dfrac{1}{4}\pi$
13. left $-\dfrac{1}{2}\pi$
    right $\dfrac{1}{2}\pi$
15. left 0
    right π

17. left 0
    right 2π
19. left $-\pi$
    right π
21. left $-\dfrac{5}{6}\pi$
    right $\dfrac{1}{6}\pi$
23. left $-\dfrac{1}{4}\pi$
    right $\dfrac{3}{4}\pi$

---

25. $y = \tan 2x$

$$
\begin{aligned}
X_{\min} &= 0 \\
X_{\max} &= 4.712 \\
X_{\mathrm{scl}} &= .785 \\
Y_{\min} &= -5 \\
Y_{\max} &= 5 \\
Y_{\mathrm{scl}} &= 1 \\
x &= 0
\end{aligned}
$$

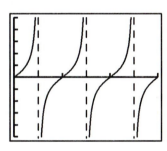

27. $y = \cot \dfrac{1}{2}x$

$$
\begin{aligned}
X_{\min} &= -12.566 \\
X_{\max} &= 6.283 \\
X_{\mathrm{scl}} &= 3.142 \\
Y_{\min} &= -5 \\
Y_{\max} &= 5 \\
Y_{\mathrm{scl}} &= 1 \\
x &= -3\pi
\end{aligned}
$$

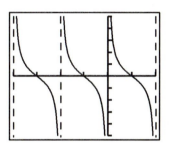

---

29. $y = \tan\left(x - \dfrac{1}{3}\pi\right)$

$$
\begin{aligned}
X_{\min} &= -.5236 \\
X_{\max} &= 8.901 \\
X_{\mathrm{scl}} &= .5236 \\
Y_{\min} &= -5 \\
Y_{\max} &= 5 \\
Y_{\mathrm{scl}} &= 1 \\
x &= \dfrac{1}{3}\pi
\end{aligned}
$$

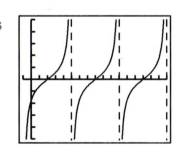

31. $y = \cot\left(x - \dfrac{1}{6}\pi\right)$

$$
\begin{aligned}
X_{\min} &= -.5236 \\
X_{\max} &= 8.901 \\
X_{\mathrm{scl}} &= .5236 \\
Y_{\min} &= -5 \\
Y_{\max} &= 5 \\
Y_{\mathrm{scl}} &= 1 \\
x &= \dfrac{2}{3}\pi
\end{aligned}
$$

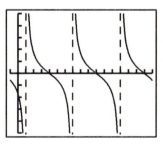

---

33. $y = \tan\left(\dfrac{1}{2}x + \dfrac{1}{6}\pi\right)$

$$
\begin{aligned}
X_{\min} &= -1.571 \\
X_{\max} &= 23.562 \\
X_{\text{scl}} &= 1.047 \\
Y_{\min} &= -5 \\
Y_{\max} &= 5 \\
Y_{\text{scl}} &= 1 \\
x &= -\dfrac{1}{3}\pi
\end{aligned}
$$

35. 5 periods

37. $\dfrac{1}{2}\pi$

39. $y_2 = \tan\left(\dfrac{1}{3}x + \dfrac{1}{36}\pi\right)$

Problem Set 3.5

1. $\dfrac{2\sqrt{3}}{3}$ **3.** $-\dfrac{2\sqrt{3}}{3}$ **5.** 2 **7.** $\sqrt{2}$ **9.** $-\dfrac{2\sqrt{3}}{3}$ **11.** $\dfrac{2\sqrt{3}}{3}$

13. $y = 2\sec\dfrac{1}{2}x$

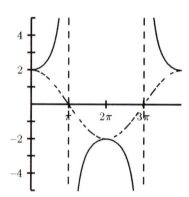

15. $y = 2\sec x$

$$
\begin{aligned}
X_{\min} &= -3.141 \\
X_{\max} &= 9.424 \\
X_{\text{scl}} &= 1.571 \\
Y_{\min} &= -6 \\
Y_{\max} &= 6 \\
Y_{\text{scl}} &= 1
\end{aligned}
$$

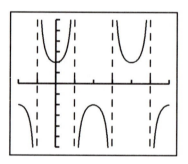

17. $y = \csc 3x$

$$
\begin{aligned}
X_{\min} &= -2.618 \\
X_{\max} &= 1.571 \\
X_{\text{scl}} &= .5236 \\
Y_{\min} &= -6 \\
Y_{\max} &= 6 \\
Y_{\text{scl}} &= 1
\end{aligned}
$$

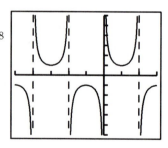

19. $y = \sec\left(x + \dfrac{1}{4}\pi\right)$

$$
\begin{aligned}
X_{\min} &= -1.57 \\
X_{\max} &= 10.996 \\
X_{\text{scl}} &= 0.785 \\
Y_{\min} &= -6 \\
Y_{\max} &= 6 \\
Y_{\text{scl}} &= 1
\end{aligned}
$$

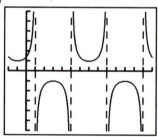

21. $y = \csc\left(x + \dfrac{1}{6}\pi\right)$

$$X_{\min} = -9.425$$
$$X_{\max} = 3.141$$
$$X_{\text{scl}} = 0.524$$
$$Y_{\min} = -6$$
$$Y_{\max} = 6$$
$$Y_{\text{scl}} = 1$$

23. $y = \dfrac{1}{2}\sec\left(2x + \dfrac{1}{3}\pi\right)$

$$X_{\min} = -0.523$$
$$X_{\max} = 5.759$$
$$X_{\text{scl}} = 0.262$$
$$Y_{\min} = -3$$
$$Y_{\max} = 3$$
$$Y_{\text{scl}} = 1$$

25. $y = \dfrac{1}{2}\csc\left(2x + \dfrac{1}{3}\pi\right)$

$$X_{\min} = -5.498$$
$$X_{\max} = 0.785$$
$$X_{\text{scl}} = 0.262$$
$$Y_{\min} = -3$$
$$Y_{\max} = 3$$
$$Y_{\text{scl}} = 1$$

27. $y_2 = 2\sec\left(\dfrac{1}{2}x - \dfrac{7}{24}\pi\right)$

Problem Set 3.6

1. $y = \sin 2x$
$y = \cos\left(2x - \dfrac{1}{2}\pi\right)$

3. $y = -\dfrac{3}{2}\sin\left(x - \dfrac{1}{4}\pi\right)$
$y = \dfrac{3}{2}\cos\left(x + \dfrac{1}{4}\pi\right)$

5. $y = 2\sin\left(2x + \dfrac{1}{3}\pi\right)$
$y = 2\cos\left(2x - \dfrac{1}{6}\pi\right)$

7. $y = 2\tan x$
$y = -2\cot\left(x + \dfrac{1}{2}\pi\right)$

Review Test—Chapter 3

1. $-\dfrac{\sqrt{2}}{2}$ **2.** $-\dfrac{1}{2}$ **3.** 2 **4.** $\dfrac{2\sqrt{3}}{3}$ **5.** $\dfrac{1}{3}\pi$ **6.** 3π **7.** $\dfrac{2}{3}\pi$ **8.** 8π

9. $y = 2\cos\left(2x + \dfrac{1}{3}\pi\right)$

10. $y = \dfrac{1}{2}\sin\left(3x - \dfrac{1}{2}\pi\right)$

11. $y = \tan\left(2x + \dfrac{1}{3}\pi\right)$

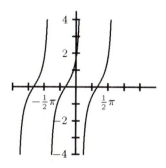

12. $y = \cot\left(\dfrac{1}{2}x - \dfrac{1}{2}\pi\right)$

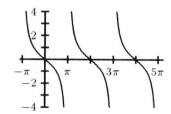

13. $y = 2\cos x$

$$X_{\min} = -3.142$$
$$X_{\max} = 15.708$$
$$X_{scl} = 1.571$$
$$Y_{\min} = -2.2$$
$$Y_{\max} = 2.2$$
$$Y_{scl} = 1$$

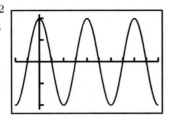

14. $y = -\dfrac{1}{2}\sin x$

$$X_{\min} = -3.142$$
$$X_{\max} = 9.425$$
$$X_{scl} = 1.571$$
$$Y_{\min} = -0.7$$
$$Y_{\max} = 0.7$$
$$Y_{scl} = 0.5$$

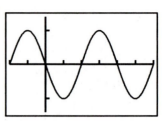

15. $y = 3\sin\dfrac{1}{3}x$

$$X_{\min} = -28.274$$
$$X_{\max} = 9.425$$
$$X_{scl} = 4.712$$
$$Y_{\min} = -3.2$$
$$Y_{\max} = 3.2$$
$$Y_{scl} = 1$$

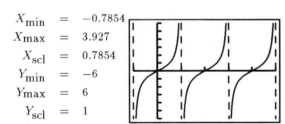

16. $y = 2\cos\dfrac{1}{2}x$

$$X_{\min} = -1.571$$
$$X_{\max} = 23.562$$
$$X_{scl} = 3.142$$
$$Y_{\min} = -2.2$$
$$Y_{\max} = 2.2$$
$$Y_{scl} = 1$$

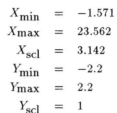

17. $y = \tan 2x$

$$X_{\min} = -0.7854$$
$$X_{\max} = 3.927$$
$$X_{scl} = 0.7854$$
$$Y_{\min} = -6$$
$$Y_{\max} = 6$$
$$Y_{scl} = 1$$

18. $y = \sec 2x$

$$X_{\min} = -3.142$$
$$X_{\max} = 6.283$$
$$X_{scl} = 0.785$$
$$Y_{\min} = -6$$
$$Y_{\max} = 6$$
$$Y_{scl} = 1$$

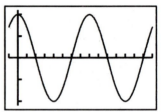

19. a) 3
    b) 6 cm
    c) $\dfrac{1}{3}$

20. $y_2 = -\dfrac{2}{3}\cos x$

21. $y_2 = \dfrac{3}{2}\sin\dfrac{2}{3}x$

22. $y_2 = \tan x$

23. $y_2 = 2\sec\left(\dfrac{1}{2}x - \dfrac{1}{4}\pi\right)$

24. $y_2 = -\dfrac{1}{2}\cos\left(2x + \dfrac{1}{12}\pi\right)$

25. $y = 3\sin\left(2x + \dfrac{1}{4}\pi\right)$
$y = 3\cos\left(2x - \dfrac{1}{4}\pi\right)$

26. $y = 2\tan\left(2x - \dfrac{1}{3}\pi\right)$
$y = -2\cot\left(2x + \dfrac{1}{6}\pi\right)$

27. $y = -\sin 2x + 3$

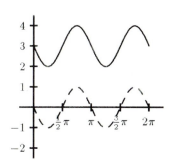

28. $y = 2\sin x + \cos 2x$

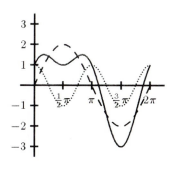

29. $y = \cos x + \dfrac{1}{3}\cos 3x$

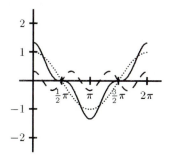

Chapter 4

Problem Set 4.1

1.	is a function	**3.**	not a function	**5.**	is a function
7.	is a function	**9.**	is a function		

11. a) Domain: $x \in$ reals
Range: $y \geq 0$
b) not one-to-one

13. a) Domain: $x \in$ reals
Range: $y \geq 1$
b) not one-to-one

15. a) Domain: $x \geq 0$
Range: $y \in$ reals
b) not one-to-one

17. a) Domain:
$x \leq -2$ or $x \geq 2$
Range: $y \in$ reals
b) not one-to-one

19. a) Domain: $x \in$ reals
Range: $y \geq 2$
b) not one-to-one

21. a) Domain:
$x \neq (2k - 1)\dfrac{1}{2}\pi$
Range: $y \in$ reals
b) not one-to-one

23. a) Domain: $x \geq 0$
Range: $y \in$ reals
b) not one-to-one

25. a) Domain: $x \in$ reals
Range: $y \geq 0$
b) not one-to-one

27. a) Domain: $x \in$ reals
Range: $y \geq 2$
b) not one-to-one

29. a) Domain: $x \in$ reals
Range: $-1 \leq y \leq 1$
b) not one-to-one

31. a) $y = \dfrac{8 - 3x}{2}$
b) is a function
c) is a function

33. a) $y = \sqrt[3]{x + 3}$
b) is a function
c) is a function

35. a) $y = \sqrt[3]{4 - x}$
b) is a function
c) is a function

37. a) $y = 2$
b) is not a function
c) is a function

39. a) $x = 2$
b) is a function
c) is not a function

41. a) $y = \pm 5\sqrt{1 - x^2}$
b) is not a function
c) is not a function

43. $y = \dfrac{x - 3}{2}$

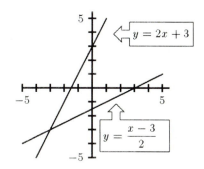

45. $y = \pm\sqrt{x - 2}$

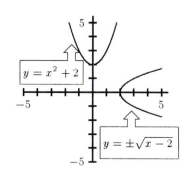

47. $y = \pm\sqrt{3 - x}$

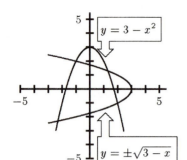

49. $y = 3 - x^2$

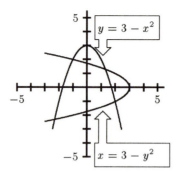

Problem Set 4.2A

1. $\cos x = -\dfrac{\sqrt{3}}{2}$ **3.** $\tan x = 0.001$ **5.** $\sec\theta = 5$ **7.** $\sin\theta = -0.005$

9. $\sin^{-1}\left(\dfrac{\sqrt{3}}{2}\right) = x$ **11.** $\tan^{-1}(1.2) = x$ **13.** $\dfrac{1}{6}\pi$ **15.** $\dfrac{1}{3}\pi$

17. $-\dfrac{1}{3}\pi$ **19.** $\dfrac{1}{6}\pi$ **21.** $-\dfrac{1}{6}\pi$ **23.** $\dfrac{1}{3}\pi$

25. Calculator does not plot $x = -6, -4, -2$ because the domain of $y = \sin^{-1}$ is $-1 \le x \le 1$.

Problem Set 4.2B

1. $\dfrac{1}{2}$ **3.** $-\dfrac{1}{2}$ **5.** $\dfrac{5}{13}$ **7.** $\dfrac{3}{5}$ **9.** $\dfrac{5}{4}$ **11.** $\dfrac{13}{5}$ **13.** $0 \le x \le \pi$

Problem Set 4.3

1. .3067   **3.** .3397   **5.** 1.4466   **7.** 46°   **9.** 71°

11. −41°   **13.** 75°   **15.** 1.01   **17.** .69   **19.** −.82

21. .57   **23.** .8780   **25.** .3497   **27.** 35°   **29.** 37°

Review Test—Chapter 4

1. a) $x \in$ reals
   $y \ge -2$
   b) Yes

2. a) $-5 \le x \le 5$
   $-3 \le y \le 3$
   b) No

3. a) $x \ge 3$
   $y \in$ reals
   b) No

4. a) $-\dfrac{1}{2}\pi < x < \dfrac{1}{2}\pi$
   $y \in$ reals
   b) Yes

5. a) Domain: $x \in$ reals
   Range: $y \ge -1$
   b) No

6. a) Domain: $x \ge 2$
   Range: $y \in$ reals
   b) No

7. a) Domain: $x \in$ reals
   Range: $y \in$ reals
   b) Yes

8. a) $y = \dfrac{x + 6}{3}$
   b) Yes
   c) Yes

9. a) $y = \pm\sqrt{4x^2 + 1}$
   b) No
   c) No

10. a) $y = \sqrt[3]{x - 4}$
   b) Yes
   c) Yes

11. a) $y = \pm\sqrt{x + 3}$
   b) Yes
   c) No

12. $-\dfrac{1}{6}\pi$ **13.** π **14.** $-\dfrac{1}{3}\pi$ **15.** $\dfrac{1}{6}\pi$ **16.** $\dfrac{1}{4}\pi$ **17.** $\dfrac{1}{3}\pi$

18. −60°   **19.** 116°   **20.** 68°   **21.** −1.305   **22.** .360   **23.** −.977

24. −3.6372   **25.** 33°   **26.** 1.8278   **27.** $-\dfrac{3}{4}$ **28.** $\dfrac{5}{3}$ **29.** $\dfrac{4}{5}$

Chapter 5

Problem Set 5.1

1.	J	**3.**	D	**5.**	A	**7.**	E	**9.**	G, M, K, T
11.	D	**13.**	D	**15.**	D	**17.**	F	**19.**	B, L, Q, S

21.–70. Reductions left to the student.

Problem Set 5.2

1. $\sin u$ **3.** $\dfrac{1}{\sin^2 \theta}$ **5.** $\dfrac{\sin \alpha}{\cos \alpha}$ **7.** $\cos \theta$ **9.** $\dfrac{1}{\sin u \cos u}$

11. $\dfrac{\cos \theta}{\sin^2 \theta}$ **13.** $\sin^2 x$ **15.** $\cos^2 x$ **17.** $\dfrac{1}{\cos x}$ **19.** $\dfrac{1}{\cos^2 x}$

21. Identity **23.** Not an identity **25.** Identity **27.** Not an identity **29.** Identity

31. Identity **33.** Not an identity **35.** Identity **37.** Identity **39.** Identity

41.–72. Proofs left to student. Suggested proofs for even-numbered problems in solutions manual.

Problem Set 5.3

1.–32. Proofs left to student. Suggested proofs for even-numbered problems in solutions manual.

33. Identity **35.** Identity **37.** Not an identity **39.** Identity **41.** Not an identity

Review Test—Chapter 5

1.	C	**2.**	G	**3.**	F	**4.**	E	**5.**	D

6.–21. Proofs left to student. Suggested proofs for even-numbered problems in solutions manual.

22. Identity **23.** Not an identity **24.** Not an identity **25.** Identity

Chapter 6

Problem Set 6.1

1. $\dfrac{\sqrt{6} - \sqrt{2}}{4}$ **3.** $\dfrac{\sqrt{2} + \sqrt{6}}{4}$ **5.** $\dfrac{\sqrt{6} + \sqrt{2}}{4}$ **7.** $\dfrac{\sqrt{6} + \sqrt{2}}{4}$

9.–14. Proofs left to student. Suggested proofs for even-numbered problems in solutions manual.

15. $-\sin x$ **17.** $\cos x$ **19.** $\cos x$ **21.** $-\cos x$ **23.** $\cos 90°$

25. $\sin 15°$ **27.** $\dfrac{7}{25}$ **29.** 0 **31.** $\dfrac{56}{65}$ **33.** $\dfrac{56}{65}$

35.–46. Proofs left to student. Suggested proofs for even-numbered problems in solutions manual.

Problem Set 6.2

1.–4. Proofs left to student. Suggested proofs for even-numbered problems in solutions manual.

5. $\dfrac{\sqrt{3}}{2}$ **7.** $\sqrt{5}$ **9.** undefined

Problem Set 6.3

1. $-2 - \sqrt{3}$ **3.** $2 + \sqrt{3}$ **5.** $2 - \sqrt{3}$ **7.** $\sqrt{3} - 2$

9. $2 - \sqrt{3}$ **11.** $\dfrac{1 + \tan \theta}{1 - \tan \theta}$ **13.** $\tan \theta$

15.–26. Proofs left to student. Suggested proofs for even-numbered problems in solutions manual.

Problem Set 6.4

1. K **3.** A **5.** E, P **7.** F, T **9.** C **11.** N
13. H **15.** I **17.** N **19.** M **21.** $\cos x$ **23.** $-\sin x$
25.–34. Proofs left to student. Suggested proofs for even-numbered problems in solutions manual.

Review Test—Chapter 6

1. $2 - \sqrt{3}$ **2.** $\dfrac{\sqrt{2} - \sqrt{6}}{4}$ **3.** $\dfrac{\sqrt{6} + \sqrt{2}}{4}$ **4.** $\dfrac{\sqrt{3}}{2}$ **5.** $-\cos x$

6. $\sin x$ **7.** $\cos x$ **8.** $\dfrac{1 + \tan x}{1 - \tan x}$ **9.** $\dfrac{\cos x - \sqrt{3}\sin x}{2}$ **10.** $\dfrac{\sqrt{3}\sin x - \cos x}{2}$

11. $\dfrac{6\sqrt{2} - 4}{15}$ **12.** $\dfrac{4 + 6\sqrt{2}}{15}$ **13.** $\dfrac{61}{45}$ **14.** $\dfrac{\sqrt{3}}{2}$ **15.** $\dfrac{2}{\sqrt{2}} = \sqrt{2}$

16.–22. Proofs left to student. Suggested proofs for even-numbered problems in solutions manual.

Chapters 1–6 Cumulative Review

1. $200°$ **2.** $\dfrac{7}{6}\pi$ **3.** $420°$ **4.** 3π **5.** 80.42 inches **6.** $88\dfrac{\text{rad}}{\text{sec}}$

	$\sin\theta$	$\cos\theta$	$\tan\theta$	$\csc\theta$	$\sec\theta$	$\cot\theta$
7.	$\dfrac{3\sqrt{13}}{13}$	$-\dfrac{2\sqrt{13}}{13}$	$-\dfrac{3}{2}$	$\dfrac{\sqrt{13}}{3}$	$-\dfrac{\sqrt{13}}{2}$	$-\dfrac{2}{3}$
8.	$\dfrac{\sqrt{3}}{2}$	$-\dfrac{1}{2}$	$-\sqrt{3}$	$\dfrac{2\sqrt{3}}{3}$	-2	$-\dfrac{\sqrt{3}}{3}$
9.	$-\dfrac{1}{2}$	$-\dfrac{\sqrt{3}}{2}$	$\dfrac{\sqrt{3}}{3}$	-2	$-\dfrac{2\sqrt{3}}{3}$	$\sqrt{3}$

10. 1.1 **11.** 1.4 **12.** $\alpha = 42°, \beta = 48°, b = 3.8$
13. $\beta = 51.6°, a = 0.391, c = 0.629$ **14.** 219 feet   **15.** 14.9 miles, N35.2°E

16. $\sin s = -\dfrac{3}{5}$ $\cos s = \dfrac{4}{5}$ $\tan s = -\dfrac{3}{4}$ **17.** $\sqrt{3}$

$\csc s = -\dfrac{5}{3}$ $\sec s = \dfrac{5}{4}$ $\cot s = -\dfrac{4}{3}$

18. $y = 3\sin\dfrac{1}{2}x$

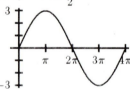

19. $y = \dfrac{5}{2}\cos\left(2x + \dfrac{1}{3}\pi\right)$

20. $\dfrac{\pi}{4}$ **21.** 3π

22. $y = \cot\left(\dfrac{1}{2}x\right)$

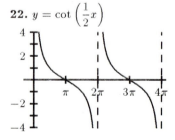

23. $y = \dfrac{1}{2}\tan\left(x - \dfrac{\pi}{4}\right)$

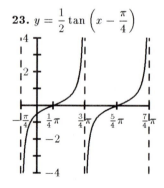

24. $y = 2 \sin 3x$

$$X_{\min} = -3.141$$
$$X_{\max} = 1.047$$
$$X_{\text{scl}} = 0.524$$
$$Y_{\min} = -2$$
$$Y_{\max} = 2$$
$$Y_{\text{scl}} = 1$$

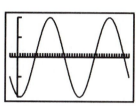

25. $y = \frac{1}{2} \cos \frac{1}{3} x$

$$X_{\min} = -9.425$$
$$X_{\max} = 28.274$$
$$X_{\text{scl}} = 4.712$$
$$Y_{\min} = -.5$$
$$Y_{\max} = .5$$
$$Y_{\text{scl}} = .5$$

26. $y = \frac{3}{4} \sin (3x + \pi)$

$$X_{\min} = -1.047$$
$$X_{\max} = 5.236$$
$$X_{\text{scl}} = 0.524$$
$$Y_{\min} = -1$$
$$Y_{\max} = 1$$
$$Y_{\text{scl}} = .25$$

27. $y = -2 \cos \left(\frac{1}{2} x - \frac{1}{12} \pi \right)$

$$X_{\min} = -1.571$$
$$X_{\max} = 23.562$$
$$X_{\text{scl}} = 0.524$$
$$Y_{\min} = -2$$
$$Y_{\max} = 2$$
$$Y_{\text{scl}} = 1$$

28. $y = 2 \sec \left(2x + \frac{1}{3} \pi \right)$

$$X_{\min} = -5.498$$
$$X_{\max} = 0.785$$
$$X_{\text{scl}} = 0.262$$
$$Y_{\min} = -4$$
$$Y_{\max} = 4$$
$$Y_{\text{scl}} = 1$$

29. $y = \cot \left(\frac{1}{2} x - \frac{1}{12} \pi \right)$

$$X_{\min} = -10.996$$
$$X_{\max} = 7.854$$
$$X_{\text{scl}} = 0.524$$
$$Y_{\min} = -4$$
$$Y_{\max} = 4$$
$$Y_{\text{scl}} = 1$$

30. $y = -2 \sin x + 2$

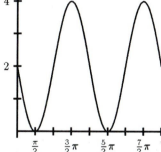

31. $y = \frac{1}{2} \cos x + \sin 2x$

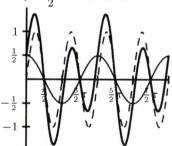

32. $y = \frac{3}{2} \cos \left(3x - \frac{7}{4} \pi \right)$

33. $y = 3 \cos \left(\frac{1}{2} x - \frac{1}{6} \pi \right)$ or

$y = 3 \sin \left(\frac{1}{2} x + \frac{1}{3} \pi \right)$

34. Domain: $x \in$ reals
Range: $y \geq 4$
Not one-to-one

35. Domain: $x \geq -1$
Range: $y \in$ reals
Not one-to-one

36. a) $y = \frac{1}{2} x - \frac{5}{2}$
b) Yes
c) Yes

37. a) $y = \pm \sqrt{3 - 2x^2}$
b) Not a function
c) Not a function

38. $30°$ or $\frac{\pi}{6}$

39. $-45°$ or $-\frac{\pi}{4}$

40. a) $68.35°$
b) 1.193

41. a) $-70.74°$
b) -1.235

42. $53.1°$

43. 1.074

44. $\frac{12}{5}$

45. $\frac{5}{3}$

46.– 47. Reductions left to the student.

48.– 50. Proofs left to student. Suggested proofs in solutions manual.

51. Identity   **52.** Not an identity   **53.** $\dfrac{1+\sqrt{3}}{2\sqrt{2}}$ **54.** $\sin\theta$

55. $\dfrac{1+\tan\theta}{1-\tan\theta}$ **56.** $\dfrac{1}{2}$ **57.** $-\dfrac{4+6\sqrt{2}}{15}$

58.– 59. Proofs left to student. Suggested proofs in solutions manual.

Chapter 7

Problem Set 7.1

1. 0   **3.** $\sqrt{3}$ **5.** 1   **7.** $-\dfrac{1}{2}$ **9.** $\cos 30°$ **11.** $\tan\dfrac{1}{3}\pi$

13. $2\sin A$ **15.** $-\dfrac{24}{25}$ **17.** $-\dfrac{24}{7}$ **19.** $-\dfrac{119}{169}$ **21.** Identity   **23.** Not an Identity

25.–38.   Proofs left to student. Suggested proofs for even-numbered problems in solutions manual.

Problem Set 7.2

1. M   **3.** J   **5.** P   **7.** O   **9.** E

11. K   **13.** C   **15.** C   **17.** E   **19.** O

21. $\dfrac{\sqrt{\sqrt{3}+2}}{2}$ **23.** $\dfrac{\sqrt{3}}{3}$ **25.** $\dfrac{\sqrt{2+\sqrt{3}}}{2}$ **27.** $2-\sqrt{3}$ **29.** $\sin 15°$

31. $\cos 67\dfrac{1}{2}°$ **33.** $\tan 105°$ **35.** $\dfrac{2\sqrt{5}}{5}$ **37.** 2   **39.** $-\dfrac{\sqrt{26}}{26}$

41.–54.   Proofs left to student. Suggested proofs for even-numbered problems in solutions manual.

Problem Set 7.3

1. Identity   **3.** Not an identity   **5.** Identity   **7.** Identity

9.–32.   Proofs left to student. Suggested proofs for even-numbered problems in solutions manual.

Review Test—Chapter 7

1. $-\dfrac{\sqrt{3}}{2}$ **2.** -1 **3.** $-\sqrt{3}$ **4.** $\dfrac{\sqrt{2-\sqrt{3}}}{2}$

5. $\dfrac{\sqrt{2-\sqrt{3}}}{2}$ **6.** $\sqrt{2}-1$ **7.** $-\dfrac{24}{25}$ **8.** $-\dfrac{7}{25}$

9. $\dfrac{24}{7}$ **10.** $\dfrac{2\sqrt{5}}{5}$ **11.** $\dfrac{\sqrt{5}}{5}$ **12.** 2

13. Not an identity   **14.** Not an identity   **15.** Identity   **16.** Identity

17.–26.   Proofs left to student. Suggested proofs for even-numbered problems in solutions manual.

Chapter 8

Problem Set 8.1

1. $30°, 150°$ **3.** $60°, 120°, 240°, 300°$ **5.** $\frac{1}{3}\pi, \frac{2}{3}\pi$

7. $\frac{1}{4}\pi, \frac{3}{4}\pi, \frac{5}{4}\pi, \frac{7}{4}\pi$ **9.** $54.74°, 305.26°$ **11.** $35.26°, 215.26°$

13. $1.9513, 5.0929$ **15.** $.6847, 2.4569, 3.8263, 5.5985$

17. a) $150° + k \cdot 360°, 210° + k \cdot 360°$
    b) $150°, 210°$

19. a) $30° + k \cdot 180°, 150° + k \cdot 180°$
    b) $30°, 150°, 210°, 330°$

21. a) $30° + k \cdot 180°, 150° + k \cdot 180°$
    b) $30°, 150°, 210°, 330°$

23. a) $\frac{1}{3}\pi + 2k\pi, \frac{2}{3}\pi + 2k\pi$
    b) $\frac{1}{3}\pi, \frac{2}{3}\pi$

25. a) $\frac{1}{6}\pi + k\pi, \frac{5}{6}\pi + k\pi$
    b) $\frac{1}{6}\pi, \frac{5}{6}\pi, \frac{7}{6}\pi, \frac{11}{6}\pi$

27. a) $\frac{5}{6}\pi + k\pi$
    b) $\frac{5}{6}\pi, \frac{11}{6}\pi$

29. a) $40.89° + k \cdot 180°$
    b) $40.89°, 220.89°$

31. a) $24.09° + k \cdot 180°, 155.91° + k \cdot 180°$
    b) $24.09°, 155.91°, 204.09°, 335.91°$

33. a) $2.4189 + 2k\pi, 3.8643 + 2k\pi$
    b) $2.4189, 3.8643$

35. a) $.9235 + k\pi, 2.2181 + k\pi$
    b) $.9235, 2.2181, 4.0651, 5.3597$

Problem Set 8.2

1. $0°, 30°, 150°, 180°$ **3.** $90°, 120°, 270°, 300°$ **5.** $30°, 150°, 270°$

7. $\frac{1}{3}\pi, \frac{5}{6}\pi, \frac{4}{3}\pi, \frac{11}{6}\pi$ **9.** $0, \frac{1}{3}\pi, \frac{2}{3}\pi, \pi, \frac{4}{3}\pi, \frac{5}{3}\pi$ **11.** $\frac{1}{6}\pi, \frac{1}{2}\pi, \frac{3}{2}\pi, \frac{11}{6}\pi$

13. $19.47°, 160.53°$ **15.** $109.47°, 250.53°$ **17.** $1.1593, 5.1239$

19. $.8481, 2.2935, 3.8713, 5.5535$ **21.** a) $210.00° + k \cdot 360°, 330.00° + k \cdot 360°$
    b) $210.00°, 330.00°$

23. a) $60.00° + k \cdot 360°, 199.47° + k \cdot 360°, 300.00° + k \cdot 360°, 340.53° + k \cdot 360°$
    b) $60.00°, 199.47°, 300.00°, 340.53°$

25. a) $30.00° + k \cdot 360°, 150.00° + k \cdot 360°, k \cdot 360°$
    b) $0.00°, 30.00°, 150.00°$

27. a) $138.59° + k \cdot 360°, 221.41° + k \cdot 360°$
    b) $138.59°, 221.41°$

29. a) $1.3181 + 2k\pi, 2.7211 + k\pi, 4.9651 + 2k\pi$
    b) $1.3181, 2.7211, 4.9651, 5.8627$

31. a) $1.2310 + 2k\pi, 3.3943 + 2k\pi, 5.0522 + 2k\pi, 6.0305 + 2k\pi$
    b) $1.2310, 3.3943, 5.0522, 6.0305$

33. a) $2.2728 + 2k\pi, 4.0104 + 2k\pi$
    b) $2.2728, 4.0104$

35. a) $2.3270 + 2k\pi, 3.9562 + 2k\pi$
    b) $2.3270, 3.9562$

Problem Set 8.3

1. a) $30° + k \cdot 180°, 60° + k \cdot 180°$
   b) $30°, 60°, 210°, 240°$

3. a) $10° + k \cdot 60°$
   b) $10°, 70°, 130°, 190°, 250°, 310°$

5. a) $15° + k \cdot 120°, 105° + k \cdot 120°$
   b) $15°, 105°, 135°, 225°, 255°, 345°$

7. a) $10° + k \cdot 60°$
   b) $10°, 70°, 130°, 190°, 250°, 310°$

9. a) $\frac{1}{6}\pi + k\pi, \frac{5}{6}\pi + k\pi, k\pi$
   b) $\frac{1}{6}\pi, \frac{5}{6}\pi, \pi, \frac{7}{6}\pi, \frac{11}{6}\pi, 0$

11. a) $\frac{1}{12}\pi + \frac{1}{3}k\pi, \frac{1}{4}\pi + \frac{1}{2}k\pi$
    b) $\frac{1}{12}\pi, \frac{1}{4}\pi, \frac{5}{12}\pi, \frac{3}{4}\pi, \frac{13}{12}\pi, \frac{5}{4}\pi, \frac{17}{12}\pi, \frac{7}{4}\pi$

13. a) $\frac{1}{12}\pi + k\pi, \frac{5}{12}\pi + k\pi, \frac{1}{2}\pi + k\pi$
    b) $\frac{1}{12}\pi, \frac{5}{12}\pi, \frac{1}{2}\pi, \frac{13}{12}\pi, \frac{17}{12}\pi, \frac{3}{2}\pi$

15. a) $\frac{1}{18}\pi + \frac{2}{3}k\pi, \frac{5}{18}k\pi + \frac{2}{3}k\pi, \frac{3}{8}\pi + \frac{1}{2}k\pi$
    b) $\frac{1}{18}\pi, \frac{5}{18}\pi, \frac{3}{8}\pi, \frac{13}{18}\pi, \frac{7}{8}\pi, \frac{17}{18}\pi, \frac{11}{8}\pi, \frac{25}{18}\pi, \frac{29}{18}\pi, \frac{15}{8}\pi$

17. a) $.8040 + \frac{1}{3}k\pi$
    b) $.8040, 1.8512, 2.8984, 3.9456, 4.9928, 6.0400$

19. a) $.3146 + \frac{1}{2}k\pi, 1.2562 + \frac{1}{2}k\pi$
    b) $.3146, 1.2562, 1.8854, 2.8270, 3.4562, 4.3978, 5.0270, 5.9686$

21. a) $.3524 + \frac{1}{4}k\pi, .6948 + \frac{1}{4}k\pi$
    b) $.3524, .6948, 1.1378, 1.4802, 1.9232, 2.2656, 2.7086, 3.0510, 3.4940, 3.8364, 4.2794, 4.6218,$
       $5.0648, 5.4072, 5.8502, 6.1926$

23. a) $.3614 + \frac{1}{2}k\pi$
    b) $.3614, 1.9322, 3.5030, 5.0738$

25. a) $.6042 + \frac{1}{2}k\pi, 1.4902 + \frac{1}{2}k\pi$
    b) $.6042, 1.4902, 2.1750, 3.0610, 3.7458, 4.6318, 5.3166, 6.2026$

27. a) $.2432 + \frac{1}{3}k\pi, .8040 + \frac{1}{3}k\pi$
    b) $.2432, .8040, 1.2904, 1.8512, 2.3376, 2.8984, 3.3848, 3.9456, 4.4320, 4.9928, 5.4792, 6.0400$

Problem Set 8.4

1. a) $\frac{1}{6}\pi + k\pi, \frac{3}{4}\pi + k\pi$
   b) $\frac{1}{6}\pi, \frac{3}{4}\pi, \frac{7}{6}\pi, \frac{7}{4}\pi$

3. a) $k\pi$
   b) $0, \pi$

5. a) $\frac{1}{3}\pi + 4k\pi, \frac{5}{3}\pi + 4k\pi, 3\pi + 4k\pi$
   b) $\frac{1}{3}\pi, \frac{5}{3}\pi$

7. a) $\frac{1}{6}\pi + 2k\pi, \frac{1}{2}\pi + k\pi, \frac{5}{6}\pi + 2k\pi$
   b) $\frac{1}{6}\pi, \frac{1}{2}\pi, \frac{5}{6}\pi, \frac{3}{2}\pi$

9. a) $2k\pi, \frac{1}{3}\pi + 2k\pi, \frac{5}{3}\pi + 2k\pi$
   b) $0, \frac{1}{3}\pi, \frac{5}{3}\pi$

11. a) $\frac{7}{6}\pi + 2k\pi, \frac{3}{2}\pi + 2k\pi, \frac{11}{6}\pi + 2k\pi$
    b) $\frac{7}{6}\pi, \frac{3}{2}\pi, \frac{11}{6}\pi$

13. a) $\frac{1}{6}\pi + 2k\pi, \frac{1}{2}\pi + 2k\pi, \frac{5}{6}\pi + 2k\pi$
    b) $\frac{1}{6}\pi, \frac{1}{2}\pi, \frac{5}{6}\pi$

15. a) $.2846 + 2k\pi, 2.8570 + 2k\pi$
    b) $.2486, 2.8570$

17. a) $2.8198 + k\pi, \frac{1}{2}k\pi$
    b) $0, 1.5708, 2.8198, 3.1416, 4.7124, 5.9614$

19. a) $.6591 + k\pi, 1.5708 + k\pi, 2.4826 + k\pi$
    b) $.6591, 1.5708, 2.4826, 3.8001, 4.7124, 5.6242$

21. a) $k\pi, 1.1503 + k\pi, 1.9913 + k\pi$
b) $0, 1.1503, 1.9913, 3.1416, 4.2919, 5.1329$

23. a) $.7854 + k\pi$
b) $.7854, 3.9270$

25. a) $1.4118 + 2k\pi, 4.8714 + 2k\pi$
b) $1.4118, 4.8714$

27. $x = 1.0472, 2.0944, 4.1888, 5.236$

$$
\begin{aligned}
X_{\min} &= 0 \\
X_{\max} &= 6.238 \\
X_{scl} &= 1.047 \\
Y_{\min} &= -4 \\
Y_{\max} &= 2 \\
Y_{scl} &= 1
\end{aligned}
$$

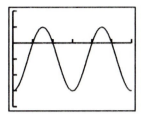

29. $x = .3218, 2.0346, 3.4634, 5.1762$

$$
\begin{aligned}
X_{\min} &= 0 \\
X_{\max} &= 6.238 \\
X_{scl} &= .3218 \\
Y_{\min} &= -5 \\
Y_{\max} &= 10 \\
Y_{scl} &= 1
\end{aligned}
$$

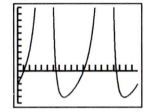

31. $x = 1.1071, 3.1416, 4.2487$

$$
\begin{aligned}
X_{\min} &= 0 \\
X_{\max} &= 6.238 \\
X_{scl} &= 1.571 \\
Y_{\min} &= -7 \\
Y_{\max} &= 7 \\
Y_{scl} &= 1
\end{aligned}
$$

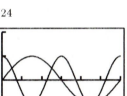

33. $x = .5236$

$$
\begin{aligned}
X_{\min} &= 0 \\
X_{\max} &= 6.238 \\
X_{scl} &= 1.571 \\
Y_{\min} &= -5 \\
Y_{\max} &= 5 \\
Y_{scl} &= 1
\end{aligned}
$$

35. $x = .5236, 2.6180, 4.7124$

$$
\begin{aligned}
X_{\min} &= 0 \\
X_{\max} &= 6.238 \\
X_{scl} &= .785 \\
Y_{\min} &= -2 \\
Y_{\max} &= 2 \\
Y_{scl} &= 1
\end{aligned}
$$

Problem Set 8.5

1. $41.3°$ or $48.7°$ **3.** $19.2°$ or $70.8°$ **5.** $45°$ **7.** $12°$

9. $60°$ **11.** $90°$ **13.** $64.8°$ **15.** $61.3°$

17. 92% **19.** $0 \le \theta \le 75.5°$ **21.** $34°$ **23.** $49.1°$

25. $14.3°$ **27.** 1.27 miles; $29.9°$ **29.** $38.68°$ or $141.32°$

Chapter 8—Review Test

1. a) $60° + k \cdot 360°, 120° + k \cdot 360°$
b) $60°, 120°$

2. a) $30° + k \cdot 180°, 150° + k \cdot 180°$
b) $30°, 150°, 210°, 330°$

3. a) $30° + k \cdot 360°, 150° + k \cdot 360°,$
$270° + k \cdot 360°$
b) $30°, 150°, 270°$

4. a) $k \cdot 180°, 240° + k \cdot 360°,$
$300° + k \cdot 360°$
b) $0°, 180°, 240°, 300°$

5. a) $10° + k \cdot 120°, 50° + k \cdot 120°$
b) $10°, 50°, 130°, 170°, 250°, 290°$

6. a) $90° + k \cdot 180°, 150° + k \cdot 180°$
b) $90°, 150°, 270°, 330°$

7. a) $k \cdot 360°, 60° + k \cdot 360°, 300° + k \cdot 360°$
b) $0°, 60°, 300°$

8. a) $90° + k \cdot 180°$
b) $90°, 270°$

9. a) $\frac{4}{3}\pi + 2k\pi, \frac{5}{3}\pi + 2k\pi$

b) $\frac{4}{3}\pi, \frac{5}{3}\pi$

10. a) $\frac{1}{4}\pi + \frac{1}{2}k\pi$

b) $\frac{1}{4}\pi, \frac{3}{4}\pi, \frac{5}{4}\pi, \frac{7}{4}\pi$

11. a) $\frac{1}{6}\pi + k\pi, \frac{1}{3}\pi + k\pi$

b) $\frac{1}{6}\pi, \frac{1}{3}\pi, \frac{7}{6}\pi, \frac{4}{3}\pi$

12. a) $\frac{1}{3}\pi + k\pi, \frac{2}{3}\pi + k\pi$

b) $\frac{1}{3}\pi, \frac{2}{3}\pi, \frac{4}{3}\pi, \frac{5}{3}\pi$

13. a) $\frac{1}{12}\pi + \frac{1}{2}k\pi$

b) $\frac{1}{12}\pi, \frac{7}{12}\pi, \frac{13}{12}\pi, \frac{19}{12}\pi$

14. a) $\frac{1}{12}\pi + \frac{1}{2}k\pi, \frac{1}{4}\pi + \frac{1}{2}k\pi$

b) $\frac{1}{12}\pi, \frac{1}{4}\pi, \frac{7}{12}\pi, \frac{3}{4}\pi, \frac{13}{12}\pi, \frac{5}{4}\pi, \frac{19}{12}\pi, \frac{7}{4}\pi$

15. a) $\frac{1}{4}\pi + k\pi$

b) $\frac{1}{4}\pi, \frac{5}{4}\pi$

16. a) $\frac{1}{3}\pi + 2k\pi, \frac{1}{2}\pi + k\pi, \frac{5}{3}\pi + 2k\pi$

b) $\frac{1}{3}\pi, \frac{1}{2}\pi, \frac{5}{3}\pi, \frac{3}{2}\pi$

17. a) $19.47° + k \cdot 360°, 160.53° + k \cdot 360°$
b) $19.47°, 160.53°$

18. a) $50.77° + k \cdot 180°, 129.23° + k \cdot 180°$
b) $50.77°, 129.23°$

19. a) $30.00° + k \cdot 360°, 150.00° + k \cdot 360°$
b) $30.00°, 150.00°$

20. a) $18.00° + k \cdot 360°, 162.00° + k \cdot 360°,$
$234.00° + k \cdot 360°, 306.00° + k \cdot 360°$
b) $18.00°, 162.00°, 234.00°, 306.00°$

21. a) $13.80° + k \cdot 120°, 106.20° + k \cdot 120°$
b) $13.80°, 106.20°, 133.80°,$
$226.20°, 253.80°, 346.20°$

22. a) $36.09° + k \cdot 180°, 77.91° + k \cdot 180°$
b) $36.09°, 77.91°, 216.09°, 257.91°$

23. a) $26.57° + k \cdot 180°, 153.43° + k \cdot 180°$
b) $26.57°, 153.43°, 206.57°, 333.43°$

24. a) $49.11° + k \cdot 180°, 130.89° + k \cdot 180°$
b) $49.11°, 130.89°, 229.11°, 310.89°$

25. a) $3.7571 + 2k\pi, 5.6677 + 2k\pi$
b) $3.7571, 5.6677$

26. a) $.4636 + k\pi, 2.6779 + k\pi$
b) $.4636, 2.6779, 3.6052, 5.8195$

27. a) $3.8713 + 2k\pi, 4.7124 + 2k\pi, 5.5535 + 2k\pi$
b) $3.8713, 4.7124, 5.5535$

28. a) $.4914 + k\pi, 2.0622 + k\pi$
b) $.4914, 2.0622, 3.6330, 5.2038$

29. a) $.8156 + \frac{1}{3}k\pi$
b) $.8156, 1.8628, 2.9100, 3.9572,$
$5.0044, 6.0516$

30. a) $.0200 + \frac{1}{3}k\pi, .3291 + \frac{1}{3}k\pi$
b) $.0200, .3291, 1.0672, 1.3763, 2.1144, 2.4235,$
$3.1616, 3.4707, 4.2088, 4.5179, 5.2560, 5.5651$

31. a) $.3844 + 2k\pi, 1.5708 + k\pi, 2.7572 + 2k\pi$
b) $.3844, 1.5708, 2.7572, 4.7124$

32. a) $3.8078 + 2k\pi, 5.6170 + 2k\pi$
b) $3.8078, 5.6170$

33. undefined

$X_{\min}$ = 0
$X_{\max}$ = 6.283
X_{scl} = 0.785
$Y_{\min}$ = −1
$Y_{\max}$ = 6
Y_{scl} = 1

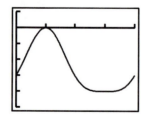

34. $x = .9046, 5.3787$

$X_{\min}$ = 0
$X_{\max}$ = 6.238
X_{scl} = 1.571
$Y_{\min}$ = −2
$Y_{\max}$ = 2
Y_{scl} = 1

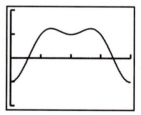

35. $x = 1.5708$

$X_{\min}$ = 0
$X_{\max}$ = 6.238
X_{scl} = 1.571
$Y_{\min}$ = −5
$Y_{\max}$ = 1
Y_{scl} = 1

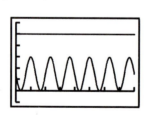

36. $x = .5752, 2.5656, 4.3072, 5.1173$

$X_{\min}$ = 0
$X_{\max}$ = 6.238
X_{scl} = .785
$Y_{\min}$ = −5
$Y_{\max}$ = 5
Y_{scl} = 1

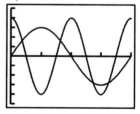

37. $25.8°$ **38.** $25.2°$ or $64.8°$ **39.** $63.4°$ or $116.6°$

Selected Answers

Chapter 9

Problem Set 9.1

1. $\alpha = 35.2°$, $a = 5.22$, $c = 5.28$ **3.** $\gamma = 25.5°$, $a = 21.6$, $c = 9.41$

5. $\gamma = 90.09°$, $b = 1.395$, $c = 3.341$ **7.** $\beta = 51.8°$, $a = 505$, $b = 447$

9. $\alpha = 37.5°$, $b = .493$, $c = .337$ **11.** $\alpha = 68.1°$, $b = 179$, $c = 396$

13. $\beta = 50.9°$, $a = 58.4$, $c = 26.7$ **15.** $\gamma = 115.05°$, $a = 207.8$, $b = 246.9$

17. $\beta = 155.7°$, $b = 60.6$, $c = 39.1$ **19.** $\gamma = 34.9°$, $b = 11.4$, $c = 6.68$

Problem Set 9.2

1. $\alpha = 79.7°$, $\beta = 24.3°$, $a = 107$ **3.** $\alpha = 133°$, $\gamma = 27°$, $a = 64$
                                             $\alpha' = 7°$, $\gamma' = 153°$, $a' = 11$

5. $\beta = 49.8°$, $\gamma = 85.0°$, $c = 10.8$ **7.** $\alpha = 118.3°$, $\beta = 33.5°$, $a = 66.3$
   $\beta' = 130.2°$, $\gamma' = 4.6°$, $c' = .868$ $\alpha' = 5.3°$, $\beta' = 146.5°$, $a' = 6.96$

9. no solution   **11.** $\beta = 27.6°$, $\gamma = 106.0°$, $c = 23.6$

13. $\beta = 73.0°$, $\gamma = 48.7°$, $b = .881$ **15.** $\alpha = 71.0°$, $\beta = 45.6°$, $b = 6.95$
                                             $\alpha' = 109.0°$, $\beta' = 7.6°$, $b' = 1.29$

17. no solution

Problem Set 9.3

1. 128 feet   **3.** $9.2°$ **5.** 527 meters   **7.** 98 feet   **9.** 187 feet

Problem Set 9.4

1. 25   **3.** 14.4   **5.** 1.023

7. $c = 70.6$, $\alpha = 54.1°$, $\beta = 77.0°$ **9.** $c = 0.0408$, $\alpha = 19.55°$, $\beta = 61.98°$

11. $b = 1.20$, $\alpha = 42.5°$, $\gamma = 27.1°$ **13.** $\alpha = 34.6°$

15. $\gamma = 58.1°$ **17.** $\alpha = 41.1°$, $\beta = 55.6°$, $\gamma = 83.3°$

19. $\alpha = 76.0°$, $\beta = 42.2°$, $\gamma = 61.8°$ **21.** $\alpha = 41.2°$, $\beta = 33.1°$, $\gamma = 105.7°$

23. $\alpha = 87.06°$, $\beta = 34.60°$, $\gamma = 58.34°$

Problem Set 9.5A

1. 1350 feet   **3.** 46 miles   **5.** 1st road N32°E; 2nd road S50°E

7. $44.05°, 52.62°, 83.33°$

Problem Set 9.5B

1. $\alpha = 68.38°$, $a = 30.91$, $c = 25.18$ **3.** $\alpha = 15.00°$, $\beta = 146.58°$, $b = 828.5$

5. $\beta = 51.25°$, $a = 791.1$, $b = 626.5$ **7.** No triangle

9. $\beta = 63.95°$, $\gamma = 63.16°$, $b = 13.55$ **11.** $\beta = 60.51°$, $a = 61.52$, $c = 206.2$
   $\beta' = 10.27°$, $\gamma' = 116.84°$, $b' = 2.690$

13. 44 miles   **15.** 87° or 33°

17. N21°W or S87°E   **19.** 0.76 miles per minute
   or 45 miles per hour

Problem Set 9.6

1. 121 **3.** 5560 **5.** 112 **7.** 90 **9.** 2.067

Review Test—Chapter 9

1. $\alpha = 74°$, $a = 49.4$, $c = 50.7$ **2.** $\alpha = 56.9°$, $a = 301$, $c = 334$

3. $\alpha = 25.5°$, $b = 45.2$, $c = 18.9$ **4.** no solution

5. $\beta = 27.1°$, $\gamma = 144.5°$, $c = 259.6$ **6.** 3.66
$\beta' = 152.9°$, $\gamma' = 18.7°$, $c' = 143.3$

7. 60.1 **8.** 0.881

9. $c = 6.74$, $\alpha = 52.3°$, $\beta = 80.5°$ **10.** $a = 705$, $\beta = 18.8°$, $\gamma = 28.5°$

11. $\alpha = 12.9°$, $\beta = 121.9°$, $\gamma = 45.3°$ **12.** 183°

13. 174.6 feet **14.** From A : 7.91 miles,
from B : 6.85 miles

15. 3800 ft **16.** 33.9

17. 10.68

Chapter 10

Problem Set 10.1

1. 18 miles N26°W **3.** 2100 lbs **5.** 74°
15° with the 900-lb force
11° with the 1250-lb force

7. 270 miles **9.** 58 miles, 242° **11.** 292 lbs, 31°

Problem Set 10.2

1. Northerly 43 mph, **3.** Ground speed 364 mph,
easterly 133 mph course 233°

5. Speed of the wind 22 knots, **7.** Horizontal 80 mph,
direction from 96° vertical 60 mph

9. Speed of the wind 74 mph, **11.** S87°E
direction from 238°

13. Airspeed 179 mph,
course 121°

Problem Set 10.3

1. $\langle 1, 4 \rangle$ **3.** $\langle -9, 16 \rangle$ **5.** $\sqrt{29}$ **7.** $\sqrt{17}$ **9.** $\sqrt{73}$

11. $\sqrt{5}$ **13.** $v = \left\langle \dfrac{2}{\sqrt{13}}, -\dfrac{3}{\sqrt{13}} \right\rangle$ **15.** $v = \left\langle -\dfrac{1}{\sqrt{26}}, -\dfrac{5}{\sqrt{26}} \right\rangle$ **17.** $4i - 2j$ **19.** $3i + j$

Problem Set 10.4

1. $\langle 79.9, 60.18 \rangle$ **3.** $\langle -120.4, 159.7 \rangle$ **5.** $\langle -2, 10 \rangle$ **7.** $\langle 14, 6 \rangle$ **9.** 5

11. $\sqrt{104}$ **13.** 9 **15.** $\theta = 76.68°$ **17.** 1083 ft-lbs

19. Horizontal 32 lbs, **21.** Weight 242 lbs, **23.** Force 52 lbs,
vertical 24 lbs 32 lbs against each other, work 1300 ft-lbs
work 60.3 ft-lbs

25. Work 41,366 ft-lbs

Review Test—Chapter 10

1. $\langle 5,2 \rangle, \sqrt{29}$

2. $\langle 4,2 \rangle, 2\sqrt{5}$

3. $\left\langle \dfrac{3}{5}, -\dfrac{4}{5} \right\rangle$

4. $\left\langle -\dfrac{5}{\sqrt{34}}, -\dfrac{3}{\sqrt{34}} \right\rangle$

5. $-3i + j$

6. $2i + 7j$

7. $\langle 53,37 \rangle$

8. $\langle 71,-71 \rangle$

9. $\langle 34,-16 \rangle$

10. $2\sqrt{5}$

11. 29

12. $37°$

13. 4

14. 812 ft-lbs

15. Resultant: 76 kg, angle: $37°$

16. 66 lb, $72°$

17. $92°$

18. 120 lbs, N38°E 600 ft-lbs

19. Northerly: 31 mph, westerly: 177 mph

20. Horizontal: 50 lb, vertical: 34 lb, work 150,000 ft-lbs

21. Sliding: 190 lb, perpendicular: 408 lb, W : 1140 ft-lbs

22. S3°W

23. 120,000 ft-lbs

24. 29,000 ft-lbs

25. N71°E

26. 253 mph, 318°

Chapter 11

Problem Set 11.1

1. $3i$

3. $-6i$

5. $-3\sqrt{2}i$

7. $5\sqrt{3}i$

9. i

11. i

13. $-i$

15. -1

17. $-2 + 8i$

19. $10 - \sqrt{2}i$

21. $20 + 2\sqrt{2}i$

23. $24 + 21i$

25. $35 + 35i$

27. $11 - \sqrt{5}i$

29. $20 + 32i$

31. $3 - 8i$

33. $-\dfrac{1}{3} + \dfrac{\sqrt{3}}{6}i$

35. $\dfrac{24}{13} + \dfrac{36}{13}i$

37. $\dfrac{1}{2} + 2i$

39. $-\dfrac{4\sqrt{2}}{3} + \dfrac{8}{3}i$

41. $-\dfrac{3\sqrt{3}}{19} + \dfrac{12}{19}i$

43. $\dfrac{5}{6} + \dfrac{\sqrt{2}}{24}i$

45. $\dfrac{1}{5} - \dfrac{3\sqrt{6}}{5}i$

Problem Set 11.2

1. $\sqrt{5}$

3. $\sqrt{15}$

5. $\sqrt{2}(\cos 315° + i\sin 315°)$

7. $2(\cos 270° + i\sin 270°)$

9. $13(\cos 67.4° + i\sin 67.4°)$

11. $2(\cos 45° + i\sin 45°)$

13. $12(\cos 100° + i\sin 100°)$

15. $45\sqrt{2}(\cos 10° + i\sin 10°)$

17. $2(\cos 40° + i\sin 40°)$

19. $\cos 260° + i\sin 260°$

21. $24\underline{/160°}$

23. $5\sqrt{2}\underline{/50°}$

25. $2\sqrt{5}\underline{/45°}$

27. $\dfrac{2\sqrt{2}}{3}\underline{/24°}$

29. $\dfrac{\sqrt{6}}{3}\underline{/330°}$

31. $-12 + 0i$

33. $-8.356 - 1.473i$

35. $-1.5 + 0i$

37. $2\underline{/270°}$

39. $8\sqrt{6}\underline{/105°}$

41. $\dfrac{2\sqrt{2}}{5}\underline{/195°}$

43. $\dfrac{\sqrt{77}}{7}\underline{/156.1°}$

Problem Set 11.3

1. $8\underline{/90°}$
3. $32\underline{/40°}$
5. $4\underline{/240°}$
7. $324\underline{/180°}$

9. $81\underline{/0°}$
11. $-8.46 - 3.08i$
13. $8 - 13.86i$
15. $\underline{/90°}, \underline{/210°}, \underline{/330°}$

17. $2\underline{/67.5°}, 2\underline{/157.5°}, 2\underline{/247.5°}, 2\underline{/337.5°}$

19. $\sqrt[8]{12}\underline{/37.5°}, \sqrt[8]{12}\underline{/127.5°}, \sqrt[8]{12}\underline{/217.5°}, \sqrt[8]{12}\underline{/307.5°}$

21. $3i, -2.6 - 1.5i, 2.6 - 1.5i$ 23. $2i, -2i$

25. $1.27 + 0.79i, -0.79 + 1.27i, -1.27 - 0.79i, 0.79 - 1.27i$

27. $\sqrt{2} + \sqrt{2}i, -\sqrt{2} + \sqrt{2}i, -\sqrt{2} - \sqrt{2}i, \sqrt{2} - \sqrt{2}i$
     or $1.41 + 1.41i, -1.41 + 1.41i, -1.41 - 1.41i, 1.41 - 1.41i$

29. $\sqrt{3} + i, -\sqrt{3} + i, -2i$
     or $1.73 + i, -1.73 + i, -2i$

31. $-0.32 + 1.33i, -0.99 - 0.94i, 1.31 - 0.39i$

Review Test—Chapter 11

1. $15 - \sqrt{3}i$
2. $22 - \sqrt{2}i$
3. $22 - 4\sqrt{3}i$

4. $49 + 11\sqrt{6}i$
5. $-\dfrac{6}{13} + \dfrac{17}{13}i$
6. $\dfrac{6}{7} - \dfrac{19\sqrt{3}}{21}i$

7. $\sqrt{151}$
8. $\sqrt{41}$
9. $45(\cos 190° + i \sin 190°)$

10. $18\sqrt{2}(\cos 40° + i \sin 40°)$
11. $1(\cos 70° + i \sin 70°)$
12. $\dfrac{5\sqrt{2}}{4}(\cos 240° + i \sin 240°)$

13. $12\sqrt{2}\underline{/200°}$
14. $100\sqrt{2}\underline{/80°}$
15. $\dfrac{4}{3}\underline{/70°}$

16. $\dfrac{\sqrt{3}}{2}\underline{/300°}$
17. $15 - 15\sqrt{3}i$
18. $2\sqrt{2} + 2\sqrt{6}i$

19. $8\sqrt{3}(\cos 180° + i \sin 180°)$
20. $65(\cos 210.5° + i \sin 210.5°)$
21. $\dfrac{3\sqrt{2}}{4}(\cos 165° + i \sin 165°)$

22. $\sqrt{2}(\cos 197° + i \sin 197°)$
23. $256\underline{/240°}$
24. $81\underline{/0°}$

25. $2\sqrt{2}\underline{/150°}, 2\sqrt{2}\underline{/330°}$
26. $2\sqrt[10]{2}\underline{/27°}, 2\sqrt[10]{2}\underline{/99°}, 2\sqrt[10]{2}\underline{/171°}, 2\sqrt[10]{2}\underline{/243°}, 2\sqrt[10]{2}\underline{/315°}$

27. $41.57i$ 28. $.52 + 1.63i, -1.67 - .36i, 1.15 - 1.26i$

29. $0 + i, 0 - i, .87 + .50i, -.87 + .50i, -.87 - .50i, .87 - .50i$

30. $.11 + 1.31i, -1.31 + .11i, -.11 - 1.31i, 1.31 - .11i$

Chapter 12

Problem Set 12.1

1. 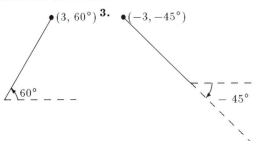 •(3, 60°) **3.** •(−3, −45°)

5. 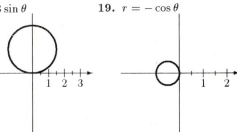 (−3, −150°)•

7. 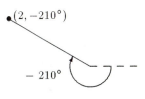 •(2, −210°)

9. $(1, -330°)$, $(-1, -150°)$, $(-1, 210°)$ **11.** $(5, 30°)$, $(5, -330°)$, $(-5, -150°)$

13. $r = \sin\theta$ 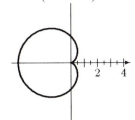   **15.** $r = -\sin\theta$ 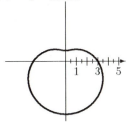   **17.** $r = 3\sin\theta$ 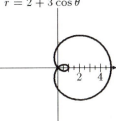   **19.** $r = -\cos\theta$

21. $r = 2(1 - \cos\theta)$ **23.** $r = 3 - 2\sin\theta$ **25.** $r = 2 + 3\cos\theta$

27. polar axis   **29.** polar axis, line $\theta = \dfrac{1}{2}\pi$, origin (pole)

31. line $\theta = \dfrac{1}{2}\pi$ **33.** polar axis   **35.** polar axis

Problem Set 12.2

1.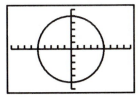

maximum $x = 5$
maximum $y = 5$
minimum $x = -5$
minimum $y = -5$
  Symmetrical to: x-axis
  y-axis
  origin

3.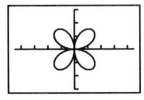

maximum $x = 1.54$
maximum $y = 1.54$
minimum $x = -1.54$
minimum $y = -1.54$
  Symmetrical to: x-axis
  y-axis
  origin

5.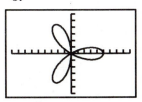

maximum $x = 5$
maximum $y = 4.38$
minimum $x = -2.8$
minimum $y = -4.38$
  Symmetrical to: x-axis

7.

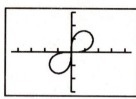

maximum $x = 1.61$
maximum $y = 1.61$
minimum $x = -1.61$
minimum $y = -1.61$
  Symmetrical to: origin

9.

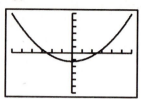

maximum $x = +\infty$
maximum $y = +\infty$
minimum $x = -\infty$
minimum $y = -1.25$
  Symmetrical to: y-axis

Problem Set 12.3

1. $2r \cos \theta + r \sin \theta = 6$ **3.** $r = -6 \sin \theta$ **5.** $r = \dfrac{4}{\sin \theta}$ **7.** $-3r \cos \theta + 2r \sin \theta = 6$

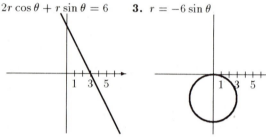

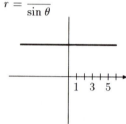

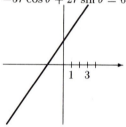

9.   slope $= \dfrac{5}{3}$, y-intercept $= -2$ **11.**   slope $= \dfrac{3}{4}$, y-intercept $= -\dfrac{1}{2}$

13. $r = 6$ **15.** $r = -2 \sin \theta$ **17.** $r = \dfrac{3}{\sin \theta}$ **19.** $r = \dfrac{3}{1 + \cos \theta}$

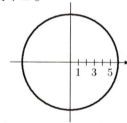

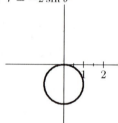

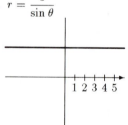

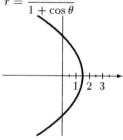

21. $r = 2\theta$ **23.** $r^2 = 9 \sin 2\theta$

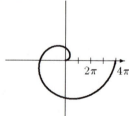

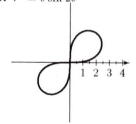

Review Test—Chapter 12

1. $(2, 240°), (-2, 60°), (-2, -300°)$

2. $(-4, -150°), (4, 30°), (4, -330°)$

3. $r = 4\cos\theta$

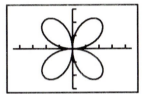

4. $r = -3\sin\theta$

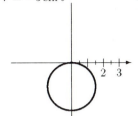

5. $r = 4 - 3\cos\theta$

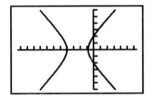

6. $r = 3 + 4\sin\theta$

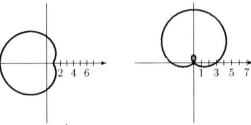

7. polar axis

8. line $\theta = \dfrac{1}{2}\pi$, origin (pole)

9. $r = 3\sin 2\theta$

maximum $x \approx 2.12$
maximum $y \approx 2.12$
minimum $x \approx -2.12$
minimum $y \approx -2.12$

10. $r = \dfrac{4}{2 - 3\cos\theta}$

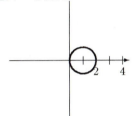

maximum $x = +\infty$
maximum $y = +\infty$
minimum $x = -\infty$
minimum $y = -\infty$

11. $3r\cos\theta - 2r\sin\theta = 6$

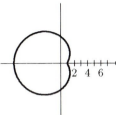

12. $r = 2\cos\theta$

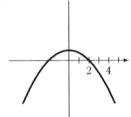

13. slope $= \dfrac{4}{5}$, y-intercept $= -\dfrac{6}{5}$

14. slope $= \dfrac{1}{2}$, y-intercept $= -\dfrac{3}{2}$

15.

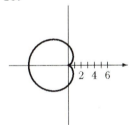

16.

Chapters 1–12 Cumulative Review

1. $20°$ **2.** $\dfrac{1}{4}\pi$ **3.** $45.837°$ **4.** 3.857 radians

5. 2.3 ft **6.** 282.74 ft **7.** $v = 102.67\dfrac{\text{ft}}{\text{sec}}$

$\omega = 82.14\dfrac{\text{rad}}{\text{sec}}$

8. $\sin\theta = \dfrac{2\sqrt{2}}{3}$ $\csc\theta = \dfrac{3\sqrt{2}}{4}$ **9.** $\sin\theta = -\dfrac{\sqrt{3}}{2}$ $\csc\theta = -\dfrac{2\sqrt{3}}{3}$

$\cos\theta = -\dfrac{1}{3}$ $\sec\theta = -3$ $\cos\theta = \dfrac{1}{2}$ $\sec\theta = 2$

$\tan\theta = -2\sqrt{2}$ $\cot\theta = -\dfrac{\sqrt{2}}{4}$ $\tan\theta = -\sqrt{3}$ $\cot\theta = -\dfrac{\sqrt{3}}{3}$

10. $\sin\theta = -\dfrac{1}{2}$ $\csc\theta = -2$ **11.** $b \approx 9.61$

$\cos\theta = -\dfrac{\sqrt{3}}{2}$ $\sec\theta = -\dfrac{2\sqrt{3}}{3}$ $\beta \approx 50.495°$

$\tan\theta = \dfrac{\sqrt{3}}{3}$ $\cot\theta = \sqrt{3}$ $\alpha \approx 39.5°$

12. 437.32 ft **13.** 792 paces **14.** $x = \dfrac{1}{3}\pi$ **15.** $1\dfrac{1}{2}$ periods

S77.7°E $y = \dfrac{1}{2}$ **16.** 7 periods

17. Starting point: $= \dfrac{1}{8}\pi$

Quarter point: $= \dfrac{3}{8}\pi$

Midpoint: $= \dfrac{5}{8}\pi$

Three-quarter point: $= \dfrac{7}{8}\pi$

End point: $= \dfrac{9}{8}\pi$

18. Starting point: $= -\dfrac{2}{3}\pi$

Quarter point: $= -\dfrac{1}{6}\pi$

Midpoint: $= \dfrac{1}{3}\pi$

Three-quarter point: $= \dfrac{5}{6}\pi$

End point: $= \dfrac{4}{3}\pi$

19. $y_2 = \dfrac{3}{2}\sin 6x$ **20.** $y_2 = \cos\dfrac{1}{2}x$

21. $y = 2\sin\left(x + \dfrac{1}{4}\pi\right)$ or **22.** $y = \dfrac{1}{2}\tan\left(x - \dfrac{2}{3}\pi\right)$ or

$y = 2\cos\left(x - \dfrac{1}{4}\pi\right)$ $y = -\dfrac{1}{2}\cot\left(x - \dfrac{1}{6}\pi\right)$

23. a) Domain: $x \in$ reals **24.** a) Domain: $x \in$ reals

Range: $y \geq -\dfrac{3}{2}$ Range: $y \in$ reals

b) Not one-to-one b) One-to-one

25. a) $y = \dfrac{1}{2}x + 4$ **26.** a) $y = \pm\sqrt{x - 2}$

b) Function b) Function

c) Function c) Not a function

27. a) $-\dfrac{1}{2}\pi \le y \le \dfrac{1}{2}\pi$

    b) $0 \le y \le \pi$

    c) $-\dfrac{1}{2}\pi < y < \dfrac{1}{2}\pi$

28. $-60°$ or $-\dfrac{1}{3}\pi$

29. $45°$ or $\dfrac{1}{4}\pi$

30. $120°$ or $\dfrac{2}{3}\pi$

31. $-\dfrac{4}{3}$

32. $\dfrac{13}{5}$

33.– 34. Reductions left to the students.

35. Not an identity   **36.** Identity   **37.** Identity

38. $\dfrac{\sqrt{6} - \sqrt{2}}{4}$

39. $\dfrac{\sqrt{3}\cos\theta + \sin\theta}{2}$

40. $\dfrac{\tan\theta + 1}{1 - \tan\theta}$ or $\dfrac{1 + \tan\theta}{1 - \tan\theta}$

41. $\dfrac{6\sqrt{2} - 4}{15}$

42. Proof left to the students

43. $\sqrt{3}$

44. $\dfrac{1}{2}\sqrt{2 + \sqrt{3}}$

45.– 46. Proofs left to the students.

47. a) $90° + 180°k$
         $30° + 360°k$
         $330° + 360°k$

    b) $30°, 90°, 270°, 330°$

48. a) $k\pi$

    b) $0, \pi$

49. a) $30° + 360°k$
         $150° + 360°k$
         $270° + 360°k$

    b) $30°, 150°, 270°$

50. a) $\dfrac{5}{18}\pi + \dfrac{1}{3}\pi k$

    b) $\dfrac{5}{18}\pi, \dfrac{11}{18}\pi, \dfrac{17}{18}\pi, \dfrac{23}{18}\pi, \dfrac{29}{18}\pi, \dfrac{35}{18}\pi$

51. a) $54.74° + 360°k$
         $305.26° + 360°k$

    b) $54.74°, 305.26°$

52. a) $0.8861 + k\pi$
         $2.2555 + k\pi$

    b) $0.8861, 2.2555, 4.0277, 5.3971$

53. a) $0.3218 + k\pi$
         $2.0344 + k\pi$

    b) $0.3218, 2.0344, 3.4633, 5.1760$

54. a) $23.51° + 120°k$
         $60° + 120°k$
         $96.49° + 120°k$

    b) $23.51°, 60°, 96.49°, 143.51°,$
         $180°, 216.49°, 263.51°, 300°, 336.49°$

55. $\gamma = 82.8°$
     $b \approx 60.4$
     $c \approx 64.7$

56. $\alpha = 78.2°$
     $b \approx 8.56$
     $c \approx 10.9$

57. $c \approx 58.4$ $c' \approx 11$
     $\beta \approx 38.4°$ $\beta' \approx 141.6°$
     $\gamma \approx 113.2°$ $\gamma' \approx 10°$

58. $h \approx 3221.5$ ft

59. $b \approx 6.49$

60. $a \approx 1.294$
     $\beta \approx 38.5°$
     $\gamma \approx 73.6°$

61. $\alpha \approx 28.3°$
     $\beta \approx 132.5°$
     $\gamma \approx 19.2°$

62. $x \approx 13.9$ miles

63. $\alpha \approx 57.1°$
     $b \approx 70.6$ lbs

64. parallel $F \approx 39.5$ lbs
     perpendicular $F \approx 185.8$ lbs
     work ≈ 790.0 ft-lbs

65. 172.06 mph
     $210.57°$

66. $\sqrt{34}$

67. $\left\langle \dfrac{2}{\sqrt{13}}, -\dfrac{3}{\sqrt{13}} \right\rangle$

68. $16, 43.2°$

69. $22 + i\sqrt{6}$

70. $-\dfrac{10}{11} + \dfrac{25\sqrt{2}}{22}i$

71. $r = \sqrt{55}$

72. $45(\cos 80° + i\sin 80°)$

73. $\sqrt{5}(\cos 236° + i\sin 236°)$

74. $-3\sqrt{2} + 3\sqrt{6}\,i$

75. $\dfrac{3}{2} - \dfrac{\sqrt{3}}{2}i$

76. $42(\cos 211.6° + i\sin 211.6°)$

77. $\sqrt{10}(\cos 108.4° + i\sin 108.4°)$

78. $144\underline{/120°}$ **79.** $2\underline{/52.5°}$, $2\underline{/142.5°}$, $2\underline{/232.5°}$, $2\underline{/322.5°}$

80.

r	-2	-1.3	-0.5	0.5	3	5.5	6.5	7.3	8	7.3	6.5	5.5	3	0.5	-0.5	-1.3	-2
θ	0	$\frac{1}{6}\pi$	$\frac{1}{4}\pi$	$\frac{1}{3}\pi$	$\frac{1}{2}\pi$	$\frac{2}{3}\pi$	$\frac{3}{4}\pi$	$\frac{5}{6}\pi$	π	$\frac{7}{6}\pi$	$\frac{5}{4}\pi$	$\frac{4}{3}\pi$	$\frac{3}{2}\pi$	$\frac{5}{3}\pi$	$\frac{7}{4}\pi$	$\frac{11}{6}\pi$	2π

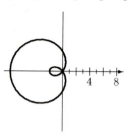

81. $\left(-\dfrac{5\sqrt{2}}{2}, -\dfrac{5\sqrt{2}}{2}\right)$ **82.** $(5, 143°)$

83. Symmetric to the line: $\theta = \dfrac{1}{2}\pi$ **84.** Symmetric to: polar axis and origin

85. $x = \dfrac{4\cos\theta}{3 - 2\sin\theta}$

$y = \dfrac{4\sin\theta}{3 - 2\sin\theta}$

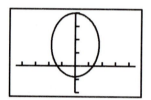

maximum $x \approx 1.8$
maximum $y = 4$
minimum $x \approx -1.8$
minimum $y \approx -0.8$

86. $r = 2\sin\theta$

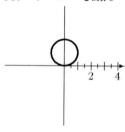

87. $m = \dfrac{1}{2}$

$y\text{-intercept} = -\dfrac{3}{2}$

Index

Index

Complex numbers
    addition of, 328
    algebraic operations with, 326
    complex number system, 327
    conjugate of, 329
    DeMoivre's Theorem, 339
    equality of, 328
    finding roots, 340
    graphical representation of, 331
    imaginary unit, 326
    modulus of, 331
    polar form of, 331, 336
    product of, 328, 335
    pure imaginary numbers, 326
    quotient of, 337
    rectangular form of, 332
    trigonometric form of, 331
Conditional equation
    definition of, 152
Conversion factors
    Reminder, 14
Corresponding angles, 52
$\cos^{-1} x$
    domain and range, 139
Cosecant
    definition of, 30
Cosine
    definition of, 30
Cosine curve
    amplitude, 81
    effect of B, 83
    generic box, 79, 83
    phase shift, 90
Cosine x
    inverse of, 137
Cotangent
    definition of, 30
Cotangent curve
    generic box, 110
Coterminal angles, 26

Degrees, 4
DeMoivre's Theorem, 339, 340
Dependent variable, 126
Diameter of the earth, 18
Distance of the moon from the earth, 19
Domain, 126
Dot product, 312, 314, 315
    algebraic, 315
    geometric, 314

Earth
    diameter of, 18

Equilateral triangle, 37
Eratosthenes, 16

Frequency, 88
Function
    definition of, 128
    one-to-one, 129
    vertical line test, 128
Fundamental identities, 152

Geometry Reminder
    angles, 3, 8
    circles, 4
    parallel lines, 52–53
    parallelograms, 299
    similar triangles, 31
    symmetry of a circle, 65
    triangles, 36
Grads, 5
Graphs
    addition of ordinates, 102
    cotangent function, 109
    of generic sine and cosine curves, 76
    secant and cosecant functions, 114
    tangent and cotangent functions, 105
    tangent function, 108
    $y = \cos^{-1} x$, 138
    $y = \sin^{-1} x$, 136
    $y = \tan^{-1} x$, 140

Half-line, 2
Heron's formula, 291
Hypotenuse, 36

Identities
    definition of, 152
    double-angle, 204
    fundamental, 156, 194, 217–218
    half-angle, 209
    opposite angle, 161
    product, 154, 156
    product-to-sum, 215
    proving, 163, 169, 216
    Pythagorean, 155, 156
    ratio, 154, 156
    reciprocal, 153, 156
    sum and difference formula, 180, 181, 183, 189
    sum-to-product, 215
Imaginary numbers, 326
Inclined planes, 312, 318
Independent variable, 126
Initial side, 3, 30
Instantaneous linear velocity, 21

$y = \sin(\theta)$

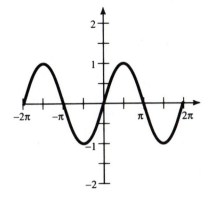

$y = \cos(\theta)$

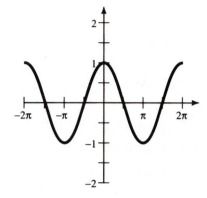

$y = \tan(\theta)$

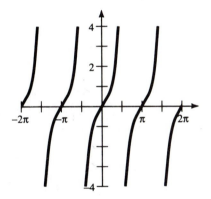

$y = \cot(\theta)$

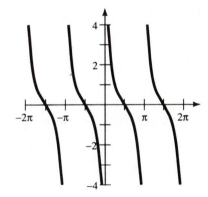

$y = \sec(\theta)$

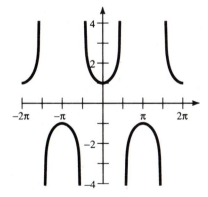

$y = \csc(\theta)$

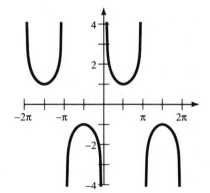

■ Reciprocal Identities

$$\sin \theta = \frac{1}{\csc \theta} \qquad\qquad \csc \theta = \frac{1}{\sin \theta}$$

$$\cos \theta = \frac{1}{\sec \theta} \qquad\qquad \sec \theta = \frac{1}{\cos \theta}$$

$$\tan \theta = \frac{1}{\cot \theta} \qquad\qquad \cot \theta = \frac{1}{\tan \theta}$$

■ Ratio Identities

$$\tan \theta = \frac{\sin \theta}{\cos \theta} \qquad\qquad \cot \theta = \frac{\cos \theta}{\sin \theta}$$

■ Pythagorean Identities

$$\sin^2\theta + \cos^2\theta = 1$$

$$1 + \tan^2\theta = \sec^2\theta$$

$$\cot^2\theta + 1 = \csc^2\theta$$

■ Negative Angle Identities

$$\cos(-\theta) = \cos \theta$$

$$\sin(-\theta) = -\sin \theta$$

$$\tan(-\theta) = -\tan \theta$$

■ Sum and Difference Identities

$$\cos(\alpha - \beta) = \cos \alpha \cos \beta + \sin \alpha \sin \beta$$

$$\cos(\alpha + \beta) = \cos \alpha \cos \beta - \sin \alpha \sin \beta$$

$$\sin(\alpha + \beta) = \sin \alpha \cos \beta + \cos \alpha \sin \beta$$

$$\sin(\alpha - \beta) = \sin \alpha \cos \beta - \cos \alpha \sin \beta$$

$$\tan(\alpha + \beta) = \frac{\tan \alpha + \tan \beta}{1 - \tan \alpha \tan \beta}$$

$$\tan(\alpha - \beta) = \frac{\tan \alpha - \tan \beta}{1 + \tan \alpha \tan \beta}$$